TWO-DIMENSIONAL MATERIALS FOR PHOTODETECTOR

Edited by **Pramoda Kumar Nayak**

Two-dimensional Materials for Photodetector
http://dx.doi.org/10.5772/intechopen.68252
Edited by Pramoda Kumar Nayak

Contributors

Qijin Chi, Yingying Tang, Xianyi Cao, Ashok K Sood, John Zeller, Inturu Omkaram, Sanghyun Lee, Kent Price, Konstantin Zhuravlev, Dmitry Esaev, Igor Marchishin, Natalia Valisheva, Aleksander Toropov, Dmitry Dmitriev, Dmitry Protasov, Dmitry Gulyaev, Sergey Dvoretsky, Victor Ovsyuk, Anatoly Savchenko, Vladimir Fateev, Michael Demyanenko, Yasir Noori, Christopher Woodhead, Robert Young, Zhi Yang, Jinjuan Dou, Minqiang Wang, Jack Jia-Sheng Huang, Yu-Heng Jan, Emin Chou, HsiangSzu Chang, Jin-Wei Shi, Gorkem Memisoglu, Burhan Gulbahar, Chih Wei Luo, Phuoc Huu Le

Published by InTech

Janeza Trdine 9, 51000 Rijeka, Croatia

Publishing Process Manager Dajana Pemac
Technical Editor SPi Global
Cover InTech Design team

First published April, 2018
Printed in Croatia
Legal deposit, Croatia: National and University Library in Zagreb

Additional hard copies can be obtained from orders@intechopen.com

Two-dimensional Materials for Photodetector, Edited by Pramoda Kumar Nayak
p. cm.
Print ISBN 978-953-51-3951-5
Online ISBN 978-953-51-3952-2

For EU product safety concerns:
IN TECH d.o.o., Prolaz Marije Krucifikse Kozulić 3, 51000 Rijeka, Croatia,
info@intechopen.com or visit our website at intechopen.com.

Meet the editor

Pramoda Kumar Nayak is currently working as an adjunct professor (Ramanujan Fellow) at the Department of Physics, Indian Institute of Technology Madras, India. He received his doctoral degree in Physics from the Indian Institute of Technology Guwahati, India. After the completion of his PhD degree, he worked as a postdoctoral researcher and visiting scientist in several institutions around the globe including the Institute for Plasma Research, India; National Cheng Kung University, Taiwan; National Tsing Hua University, Taiwan; and Ulsan National Institute of Science and Technology, Republic of Korea. His research interest is focused on two-dimensional materials, van der Waals heterostructures, topological insulators, and quantum dots. He has published 45 international journals, 5 book chapters, 1 book, and 2 edited volumes so far in his research career.

Contents

Preface

Two-dimensional (2D) materials including graphene, hexagonal boron nitride, transition metal dichalcogenides (TMDs), group III–VI and IV–VI layered semiconductors, topological insulators, and so on are the thinnest forms of materials to ever occur in nature and have the potential to dramatically alter and revolutionize our material world. In the monolayer form, all these materials exhibit unexpected physical and chemical properties due to quantum confinement effect. These materials can control or interact with light converting the photons into electrical signals for their attractive applications in photonics, electronics, and optoelectronics. Some of the unique properties of these materials including wide photoresponse wavelength, passivated surfaces, strong interaction with incident light, and high mobility have created tremendous interest in photodetector application.

Excitement surrounds the promise of replacing conventional bulk photodetectors with devices based on 2D materials, allowing better integration, flexibility, and potentially improved performance. These materials are supposed to be the heart of a multitude of next-generation photodetector technologies. Very recently, artificial heterostructures made up of 2D materials are also used to tailor the performance of photodetectors. Keeping the above aspects in mind, the present book entitled *Two-Dimensional Materials for Photodetector* has been edited to provide state-of-the-art knowledge in photodetector technology based on these exciting 2D materials.

This book consists of 10 chapters contributed by a team of experts in this exciting field. Chapter 1 starts with the working principle, device structure, and performance of the three kinds of 2D photodetectors including graphene, TMDs, and halide perovskite. In Chapter 2, the authors review recent advances in photodetector technology based on TMDs only, which provide advantages of using 2D materials as photodetector in terms of mobility enhancement, external quantum efficiency, photo responsivity, and detectivity.

Modern nanoscale III–IV semiconductor photodetectors are important building blocks for high-speed optical communications. In Chapter 3, the authors review the state-of-the-art 2.5G, 10G, and 25G avalanche photodiodes, which are available in commercial applications, and discuss the key device parameters including avalanche breakdown voltage, dark current, temperature dependence, bandwidth, and sensitivity of these detectors. Recent studies have shown that 2D metal halide perovskites hold a promising potential for the development of new-generation photodetectors. In this context, the authors present the summary and highlights of the latest researches on 2D metal halide perovskite-based photodetectors in Chapter 4 with emphasis on synthesis methods, structural characterization, optoelectronic properties, theoretical analysis, and simulations.

Infrared (IR) photodetectors based on strained layer superlattices (SLS) have drawn much interest from academic and commercial sectors in recent decades. In Chapter 5, the authors review recent development of 2D SLS detector arrays based on InAs/Ga(In)Sb heterojunction for IR applications. In Chapter 6, the authors demonstrate the technology of manufacturing focal plane array based on the AlGaAs/GaAs quantum-well infrared photodetector. In Chapter 7, the authors discuss on the development, characterization, modeling, and simulation of type-II superlattice-structured materials including InAs/GaSb and InAs/GaInSb and device systems for photodetector application.

Graphene has numerous advantages for realizing high-performance nano-electromechanical systems such as nanoscale sensors including strain sensors, optical modulators, or energy harvesters. In this context, graphene-based acousto-optic sensors with vibrating resonance energy transfer and their applications have been presented in Chapter 8. The low inherent light absorption of 2D materials is an outstanding issue to be solved for them to be used for efficient photodetection. In Chapter 9, the authors introduce different approaches to increase light absorption and collection using engineered structures such as patterning the substrate with photonic crystal and using 2D-compatible solid immersion lenses. The optoelectronic properties of topological insulators (TIs) are important subjects for the development of optoelectronic devices. Therefore, the issues associated with electron-phonon interaction, carrier lifetime, and carrier dynamics are very important for optimizing device performance. Chapter 10 provides a brief introduction of the TIs, time-resolved pump-probe spectroscopy, and some ultrafast dynamic properties of Bi-based topological insulators.

I hope that this book will be useful to a large number of researchers in a variety of disciplines including materials science, electrical engineering, nanophotonics, and condensed matter physics.

I am very pleased to get the opportunity to have served as an editor of this book. I would also like to acknowledge the help given by IntechOpen publisher, in particular Ms. Dajana Pemac, publishing process manager, for her assistance, patience, and support throughout the whole process of this book project.

Dr. Pramoda Kumar Nayak
Department of Physics
Indian Institute of Technology Madras
Chennai, India

Chapter 1

Graphene, Transition Metal Dichalcogenides, and Perovskite Photodetectors

Zhi Yang, Jinjuan Dou and Minqiang Wang

Additional information is available at the end of the chapter

http://dx.doi.org/10.5772/intechopen.74021

Abstract

Recent years have witnessed a tremendous progress in 2D materials photodetector. Their unique properties including wide photoresponse wavelength, passivated surfaces, strong interaction with incident light, and high mobility enable dramatical superiority in photodetector application. The photophysics, device structure, working mechanism, and performance of three kinds of 2D materials photodetectors including graphene, transition metal dichalcogenides (TMDs), and halide perovskite are discussed in detail to give a profound understanding for the readers. In addition, we highlight the challenges and opportunities faced in studying 2D materials photodetector, and various strategies are summarized to help on the development of photodetection research and find ways to approach major problems.

Keywords: graphene, transition metal dichalcogenides, perovskite, hybrid structure, photodetectors, responsivity, carrier transfer

1. Introduction

Two-dimensional (2D) materials are a class of materials derived from layered van der Waals solids. The in-plane atoms are held together by tight covalent or ionic bonds along 2D directions to form atomic layers, while the atomic layers are bonded together by weak van der Waals interactions along the out-of-plane direction [1–3]. Nicolosi et al. summarized different types of layered materials, which can be grouped into diverse families. The simplest are the atomically thin, hexagonal sheets of graphene and hexagonal boron nitride (h-BN). Other layered materials also include transition metal dichalcogenides (TMDs), black phosphorus (BP), metal halides (such as PbI_2 and $MgBr_2$), layered metal oxides (such as MnO_2 and MoO_3), layered double hydroxides (LDHs) [such as $Mg_6Al_2(OH)_{16}$], layered III-VIs (such as InSe and GaS), layered V-VIs (such as Bi_2Te_3 and Sb_2Se_3), and halide perovskites [4]. Graphene is composed of a single layer of carbon atoms bonded together in a hexagonal honeycomb lattice, which was firstly obtained by exfoliating graphite using an adhesive tape in 2004 [5]. Then graphene becomes the most famous 2D layered material on the basis of its appealing electronic, optical, mechanical, and thermal properties, and thus Geim and Novosolov were awarded the Nobel Prize in Physics in 2010.

2D materials have many distinctive properties including mechanical stiffness, strength and elasticity and high electrical and thermal conductivity. In this chapter, the optical and electrical properties of 2D materials are cared for photodetector application. Compared with traditional bulk materials, 2D materials have several superiorities as photodetectors. Firstly, 2D materials are sensitive over a wide range of the electromagnetic spectrum including ultraviolet (UV), visible, infrared (IR), and terahertz (THz) (see **Figure 1a**) [6]. The bandgap and photoresponse range of different 2D material photodetectors are summarized in **Figure 1b**. Graphene photodetector can sense over an ultrawide spectrum from UV to IR and even THz due to its zero bandgap and linear dispersion near the Dirac point [7, 8]. TMD photodetectors have a photoresponse over a spectrum from visible to near infrared [9]. The bandgap of BP can be increased from 0.3 eV of bulk to around 2 eV of monolayer, corresponding to near infrared and visible [10]. Halide perovskite photodetectors have a photoresponse over a spectrum from UV to near infrared [11]. Secondly, the surfaces of 2D materials are free of dangling bonds, facilitating the construction of vertical heterojunctions without the restriction of lattice matching and avoiding surface leakage current due to their naturally passivated surfaces, and noise current can be decreased to fairly small order [12]. Thirdly, 2D semiconductors with ultrathin thickness can still strongly interact with incident light resulting from the singularities effect near the conduction and valence band edges [13]. Therefore, 2D semiconductor films are ideal materials for photodetector application.

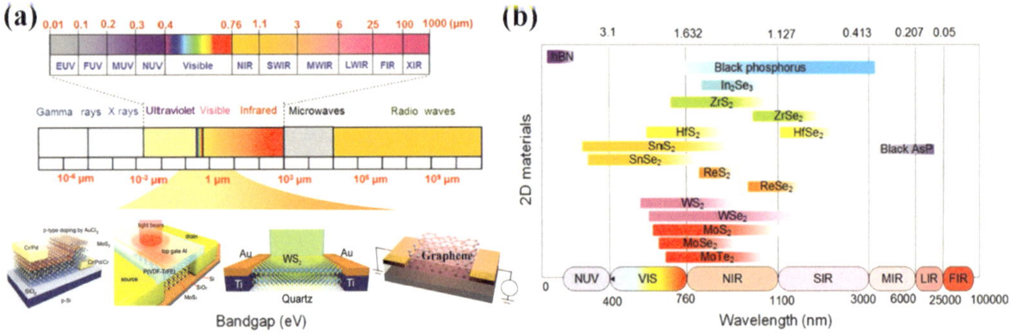

Figure 1. (a) The electromagnetic spectrum of 2D materials photodetector and (b) the gap value and detection range of different 2D materials [6].

2. Graphene photodetector

Graphene is an appealing material for optoelectronic applications. The concentrations of electrons and holes can be as high as 10^{13} cm^{-2}, and their mobilities μ can exceed 15,000 cm^2 V^{-1} s^{-1} even under ambient conditions [14]. Due to its zero bandgap and linear dispersion near the Dirac point, graphene exhibits an extremely broadband absorption from UV to THz. Despite of the atomic layer thickness, graphene absorbs 2.3% of incident photons without significant wavelength dependence [6].

2.1. The physical mechanisms

The physical mechanisms of graphene photodetector including the photoconductive effect, the photovoltaic effect, the photo-thermoelectric effect, the bolometric effect, the photogating effect, and the plasma-wave-assisted mechanism (also called Dyakonov-Shur mechanism) have been reported (**Figure 2**) [15]. Based on the sensing mechanism, photodetectors can be divided into two categories: photon detectors and thermal detectors. Photodetectors that operate via the photovoltaic effect are usually called photodiodes. A photodiode photodetector can function at two modes under illumination: the photovoltaic mode at zero bias and the photoconductive mode under reverse bias [3, 16]. The charge-separation mechanism is shown in **Figure 3**. Photovoltaic effect is based on the separation of photogenerated electron-hole pairs by a built-in electric field at junctions, and electrons and holes are collected at opposite electrodes, which generates a considerable photocurrent (short-circuit current). A photodiode working in photovoltaic mode has the lowest dark current, and an improved detectivity can be obtained. In photoconductive mode, the external electric field with a moderate biasing voltage has the same direction as the built-in one, which increases the separation efficiency of the electron-hole pairs, as well as the response speed due to reduced carrier transit time and lowered diode capacitance. The photo-thermoelectric effect is also called Seebeck effect. Pronounced thermoelectric photocurrent can be generated in graphene by hot carriers, and the temperature differences between carriers are able to produce a significant photoresponse [17]. The bolometric effect is associated with the resistance change produced by heating associated with the incident photons. Instead of generating photocurrent under zero bias, it only modifies the conductance of graphene under high external bias [18–19]. A d.c. voltage is generated

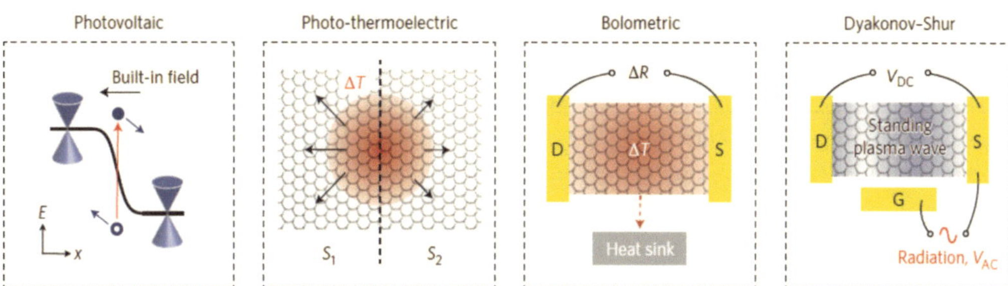

Figure 2. Schematic of the four kinds of graphene photodetection mechanisms [15].

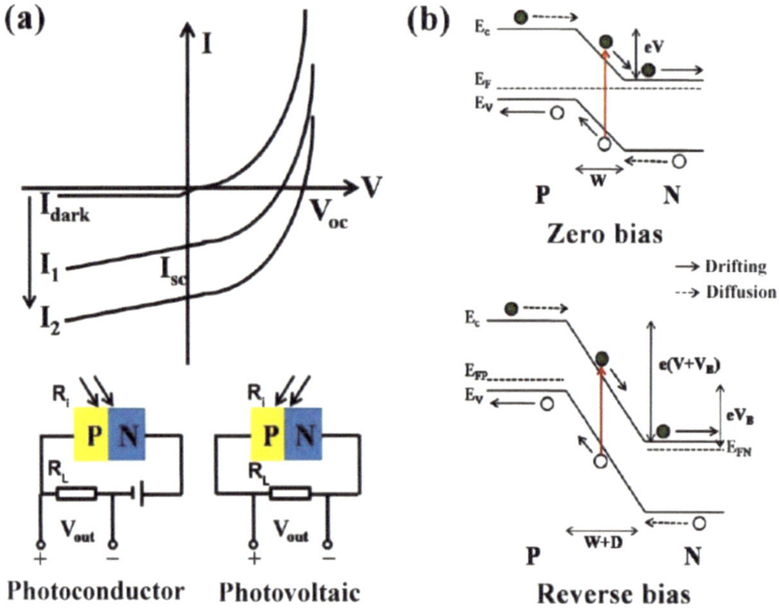

Figure 3. (a) The current-voltage curve of photodiode, and (b) the energy-band diagram of photovoltaic mode and photoconductor mode.

in a field-effect transistor in response to a THz radiation oscillation field through plasma-wave rectification, which was first proposed by Dyakonov and Shur [20]. The graphene photodetector based on plasma-wave-assisted mechanism can be used to integrate with silicon photonics.

2.2. The structure and performance

Here, all graphene photodetectors are based on the photoconductor and photovoltaic effects. The external bias voltage is not desired for graphene photodetectors because a large dark current can be generated as graphene has high conductivity. The built-in electric fields can be introduced by constructing a homogeneous p-n junction, a metal-graphene Schottky junction, or a tunneling barrier junction. Surface chemical modification methods can be adopted for the controllable n- or p-type doping in graphene. Liu et al. fabricated a graphene p-n junction photodiode photodetector through a selected-area chemical doping process, showing its IR detection ability [21]. A metal-graphene-metal (MGM) photodetector can be achieved by taking advantage of the work-function difference between graphene and a contacting metal. In 2009, Xia et al. reported the first graphene photodetector with a 40 GHz bandwidth and a 0.5 mA W^{-1} photoresponsivity (**Figure 4**(**a**, **b**)) [22]. Then they fabricated a MGM photodetector with asymmetric electrodes (Ti and Pd) to extend the operation region, and a photoresponsivity of 6.1 mA W^{-1}, a bandwidth of 16 GHz, and a data rate of 10 Gbit s^{-1} were obtained (**Figure 4** (**c**, **d**)) [23]. In this device, photocurrent is generated by local illumination of one of the metal/graphene interfaces of a back-gated graphene FET. The field arises from charge transfer from the respective contact metal to graphene, and it can be adjusted by choosing proper metal. A graphene double-layer tunneling barrier junction can be used to achieve ultra-broadband and high-sensitivity photodetection. The device is composed of two graphene layers sandwiching a

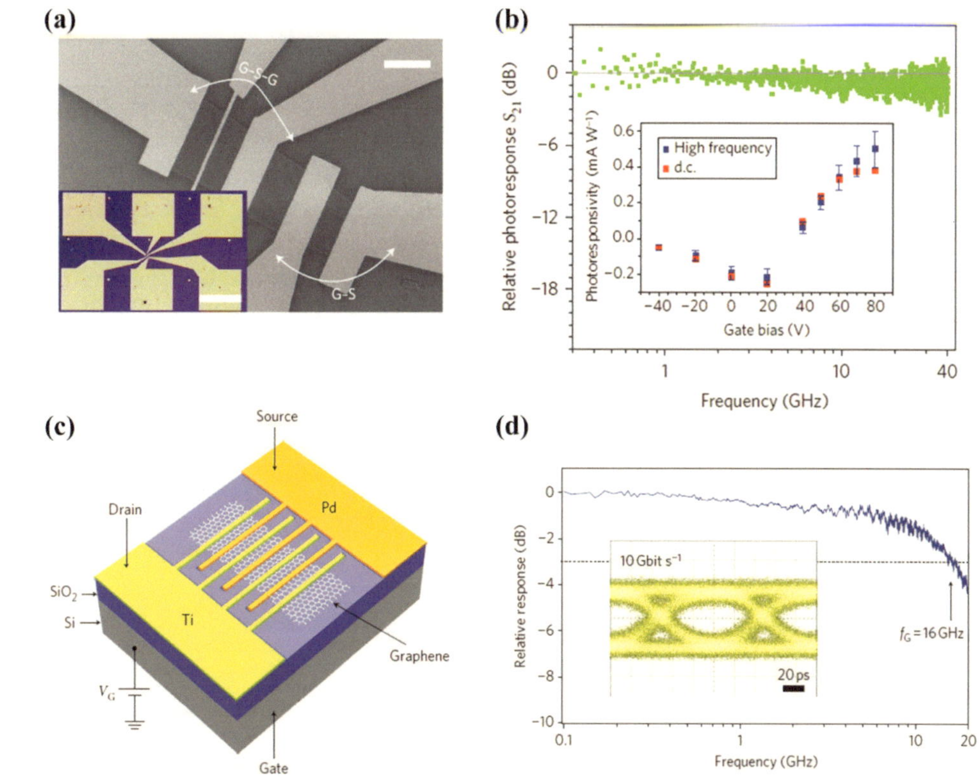

Figure 4. The metal-graphene Schottky junction photodetectors: (a, b) symmetric electrodes and (c, d) asymmetric electrodes [22–23].

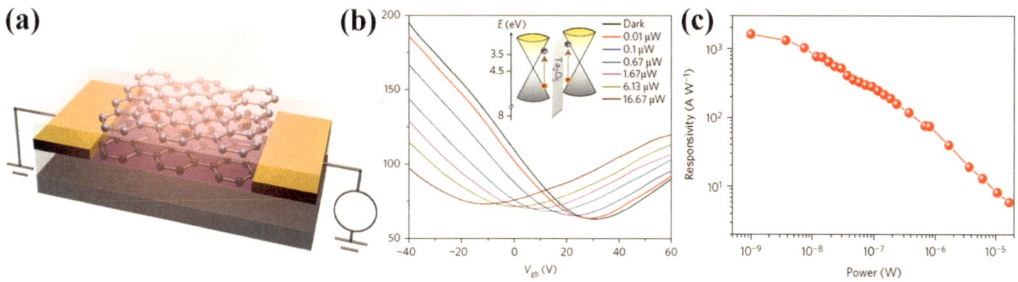

Figure 5. Graphene double-layer tunneling barrier junction photodetectors: (a) schematic of device structure, (b) I-V_g characteristics of the photodetector under different laser powers, and (c) photoresponsivity versus illumination power [24].

5-nm-thick Ta_2O_5 tunnel barrier (**Figure 5**) [24]. The top and bottom graphene layer exhibit shifted Fermi levels about 0.12 eV due to the different extents of interaction with substrates. Under illumination, photogenerated hot carriers in the top graphene tunnel into the bottom layer, leading to a charge build-up on the gate. The devices demonstrated room-temperature photodetection from the visible to the mid-infrared range, with mid-infrared responsivity higher than 1 A/W. Under low excitation power, the device shows a remarkable responsivity of greater than 1000 A/W at 1 V, suggesting the built-in amplification mechanism.

3. TMD photodetector

2D transition metal dichalcogenides (TMDs) include MX_2 (M = Mo, W; X = S, Se, Te) which is formed by covalently bonded X-M-X 2D hexagonal trilayers, and neighboring layers bond with each other via weakly van der Waals forces [25]. Bulk TMDs possess an indirect band gap of 0.9–1.4 eV. Only monolayer TMDs have direct band gap, and the indirect band gap of TMDs is thickness dependent resulting from strong interlayer coupling and quantum confinement effect, contributing to efficient light absorption and emission [26]. The absorption spectrum of TMDs is well matched to the solar spectrum, and the absorption coefficient is typically on the order of 10^4–10^6 cm^{-1} so that more than 95% of the sunlight is absorbed for TMD films with sub-micrometer thickness. The mobility of TMDs increases with the number of layers, typically ranging from 0.2 to 34,000 $cm^2 V^{-1} s^{-1}$ [27]. The charge-carrier density depends on the doping levels and recombination centers, and the typical value is 10^{12} cm^{-2} [28]. Moreover, monolayer TMDs have long-lived excitons because of high binding energies, such as 320 meV for excitons in WS_2 and 20 meV for trions in MoS_2, which are several times higher than those in bulk [29, 30].

3.1. MoS_2 photodetector

Monolayer MoS_2 has a direct bandgap of 1.8 eV, large carrier mobility above 200 $cm^2 V^{-1} s^{-1}$, strong photoluminescence, good chemical stability and mechanical flexibility, and so on [3]. MoS_2-based photodetectors are always in the form of a FET configuration. In 2012, Yin et al. developed a mechanically exfoliated monolayer MoS_2 phototransistor, which has a photoresponsivity of 7.5 mA/W for 750 nm light and a response time of 50 ms [31]. As shown in **Figure 6a**, the cut-off wavelength of photodetector is at ≈670 nm, which agrees well with the bandgap of the monolayer MoS_2. After that, several factors have been investigated to explore how they affect the photoresponse performance. First, the charge trapping in MoS_2 or at the MoS_2/substrate interface plays a key role in the photoresponse process. Lopez-Sanchez et al. obtained a high responsivity up to 880 A/W from a monolayer MoS_2 phototransistor (**Figure 6b**), but the response time can be several seconds, which is influenced by the surroundings of MoS_2 [32]. This is a specific example to demonstrate the effect of trap states on responsivity and response time. Second, the responsivity and response speed are highly dependent on the layer number of MoS_2. Khan et al. reported an increased responsivity and response speed with increasing numbers of MoS_2 layers. They also demonstrated that responsivity is significantly enhanced in N_2 gas environment compared with that in atmospheric environment, indicating that the charge-carrier density can be affected by the oxygen (**Figure 6c, d**) [33]. Third, the preparation methods including mechanically exfoliated, chemical vapor deposition (CVD), liquid exfoliation, and solution synthesis can have a significant effect on photoresponse (**Figure 6e, f**). Similar performance has been reported for mechanically exfoliated and CVD methods, while inferior performance is obtained for solution-processed MoS_2. Besides, electrode contact and surface modification are also reported to affect the performance of MoS_2 photodetector, and several strategies are employed to enhance the photoresponse performance [32, 34, 35]. The photodetector with low Schottky barrier exhibits a very high photo gain, but response speed is slow. On the contrary, a very short response time can be obtained for the photodetector with high Schottky-contact barrier, but the gain decreases by several orders of magnitude [36].

http://dx.doi.org/10.5772/intechopen.74021

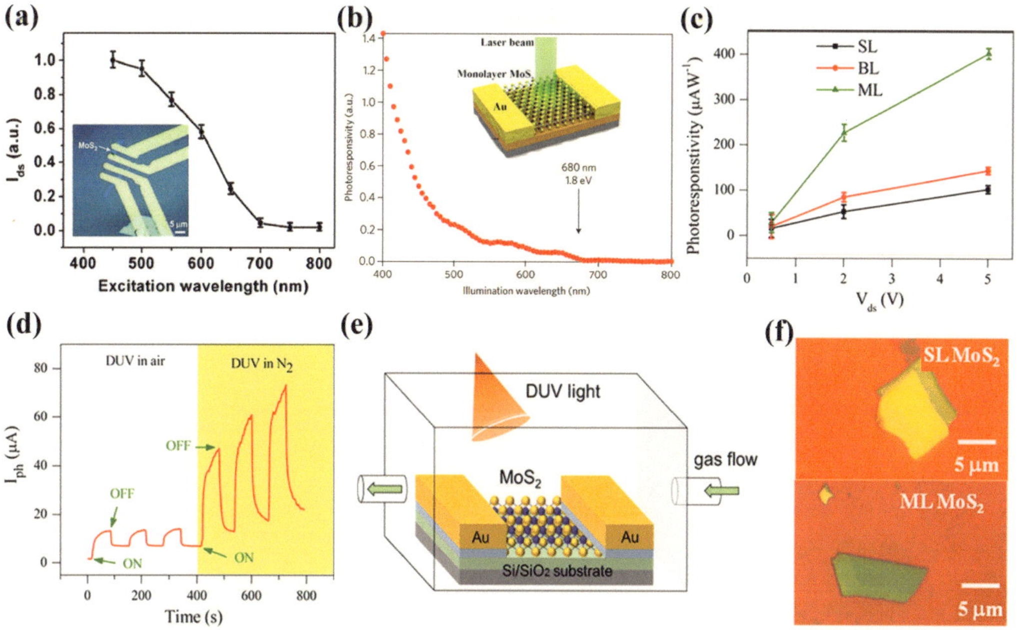

Figure 6. MoS_2 photodetectors based on FET configuration. (a) The dependence of I_{ds} of a single-layer MoS_2 phototransistor on excitation wavelength. Inset shows an optical image of single-layer MoS_2 photodetector device [31] and (b) responsivity of a monolayer MoS_2 phototransistor as a function of incident power. Inset shows the structure and photocurrent spatial map of the MoS_2 photodetector [32]; (c) photoresponsivity of single-layer, bilayer, and multilayer MoS_2 photodetectors as a function of V_{ds}; and (d) time-dependent photoresponse in atmospheric and N_2 environments. (e) Schematic of MoS_2 photodetector prepared by CVD method. (f) Optical images of single layer and multilayer of MoS_2 [33].

3.2. WS_2 photodetector

Monolayer WS_2 has a direct bandgap of 1.98 eV instead of an indirect bandgap of 1.4 eV in the bulk, but its carrier mobility of 0.2 $cm^2\ V^{-1}\ s^{-1}$ is lower than bulk and it also has strong photoluminescence due to high exciton binding energy [26]. WS_2 is also reported to have much smaller effective electron mass, providing better transistor performance than Si as well as MoS_2 [37]. Besides, WS_2 as a potential electronic material is inert, nontoxic, and environmentally friendly. The photoresponse performance of WS_2 photodetector is also affected by the same factors as MoS_2 including the charge trapping induced by substrate and atmosphere environment, the layer number, the preparation methods, electrode contact, and surface modification. Perea-López et al. have studied the photoresponse of CVD-grown few-layer WS_2 photodetector (**Figure 7a**). The responsivity is highly dependent on the excitation wavelength, and the highest photoresponse wavelength is 647 nm. The device shows a maximum photocurrent of 0.6 μA, a responsivity of 92 μA W^{-1}, and a fast response speed of ≈5 ms. The low responsivity is attributed to the high resistances ranging between 10^9 and 10^{12} Ω resulting from the low mobility [38]. Lee et al. reported a mechanically exfoliated WS_2 photodetector with a photoresponsivity of 0.27 A/W and an on/off ratio of 10^2–10^3 [37]. Huo et al. reported that the responsivity of mechanically exfoliated WS_2 photodetector can increase from 5.7 A W^{-1} in vacuum to a maximum of 884 A W^{-1} in an NH_3 atmosphere (**Figure 7b**). The high sensitivity on the surrounding environment can be ascribed to the electron transfer to absorbed gas molecules, demonstrating the role of charge trapping [39]. As shown in **Figure 7c**, Yao et al. prepared

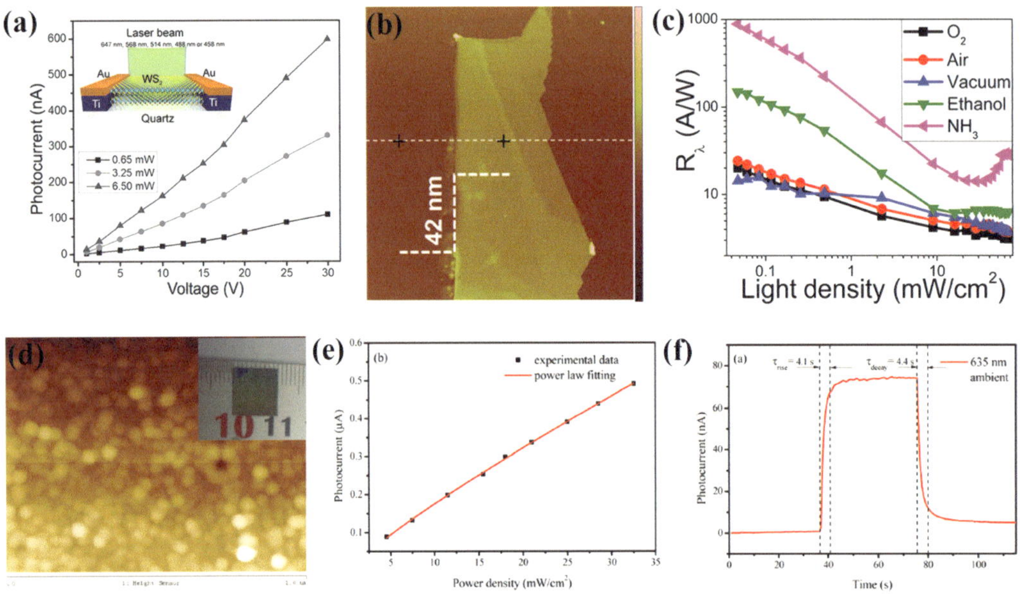

Figure 7. The photoresponse performance of WS_2 photodetectors. (a) IV curve acquired with different laser power irradiation. Inset shows schematic of a few-layered WS_2 photodetector [38], (b) AFM image of the multilayer WS_2 nanoflakes, (c) the relationship between the light intensity and photoresponsivity under various gas atmospheres [39], (d) AFM image and photographs of a large-area multilayered WS_2 film on a 1 × 1 cm^2 Si substrate, (e) power-dependent photocurrent, and (f) temporal photoresponse of the WS_2 photoresistor [40].

multilayer WS_2 films with the size of cm^2 by pulsed-laser deposition method. Benefiting from the large optical absorbance and high carrier mobility (31 $cm^2\ V^{-1}\ s^{-1}$), the responsivity of the device approaches a high value of 0.51 A W^{-1} in an ambient environment, and it can be further enhanced to 0.7 A W^{-1} in a vacuum environment [40].

4. Perovskite photodetector

Recent years have witnessed the rapid development of the organic-inorganic hybrid halide perovskites. Their exceptional properties including an adjustable spectral absorption range, high carrier mobility, long diffusion lengths, and the affordability of fabrication render them one of the most exceptional optoelectronic materials for photoelectrical device applications especially in photodetectors [41]. Organic-inorganic hybrid halide perovskites have the same chemical formula ABX_3. The A cation can be organic species including $CH_3CH_2NH_3^+$ (EA), $HC(NH_2)_2^+$ (FA), and CH_3NH_3 (MA) or inorganic species such as Cs, Rb, K, and Na, and their ionic radiuses decrease gradually. The A cation is thought no direct contribution toward electronic properties, but its size can alter the degree of distortion that in turn affects electronic properties [42]. The B metal cation can be group IV metal such as Pb^{2+}, Sn^{2+}, and Ge^{2+} in a divalent oxidation state, and the Pb^{2+} has superior stability than Sn^{2+} and Ge^{2+}. The halide anion can be the most effectively varied component. In general, with increasing ionic radius, a narrower bandgap can be obtained, resulting in that absorption spectra shift to longer wavelengths.

Perovskites have a large absorption coefficient, so strong optical absorbance across the entire visible spectrum can be achieved for a thin thickness less than 500 nm, facilitating the collection of charge carriers [43]. Moreover, $MAPbI_3$ has been reported with the carrier diffusion lengths up to 100 nm for both electrons and holes, and exceeding 1 μm has been reported in the $MAPbI_{3-x}Cl_x$, resulting from long carrier lifetime and large carrier mobility [44, 45]. So far, perovskite photodetectors have been widely reported from different aspects including composition engineering such as the organic-inorganic hybrid $MAPbX_3$ (X = Cl, Br, I) and all-inorganic $CsPbX_3$ (X = Cl, Br, I) perovskites, morphology such as compact polycrystalline films and nanocrystal films (quantum dots, nanocubes, nanowires, nanoplatelets, and nanosheets), device structure such as photoconductor and photodiode, and so on. Compared with organic-inorganic perovskite, all-inorganic perovskite exhibits better air stability.

4.1. 2D perovskite photodetector

2D halide perovskites with different morphologies including microdisks, nanoplatelets, and nanosheets (NSs) have been synthesized via both CVD and colloidal chemistry methods. Many features including high photoluminescence (PL) quantum yield, quantum confinement effect, increased exciton binding energy, and long electron diffusion length have been demonstrated. In 2015, Jang et al. synthesized $MAPbBr_3$ nanoplatelets with the length of 70 nm and the thickness of 15 nm by colloidal chemistry method, and mixed halide perovskite ($MAPbBr_{3-x}Cl_x$ and $MAPbBr_{3-x}I_x$) nanoplatelets were obtained by anion-exchange reaction in solution. The full-range band gap tuning over a wide range (1.6–3 eV) was demonstrated in **Figure 8a** [46]. By comparing photoresponse performance in **Figure 8b** and **c**, $MAPbBrI_2$ has a larger photocurrent than $MAPbBr_3$ under the same irradiation condition, which may result from its stronger light absorption. Next, Liu et al. synthesized triangle 2D $MAPbI_3$ NSs with thinner thickness of several nanometers and larger lateral sizes of several micrometers based on two-step CVD method. A photoconductor photodetector was fabricated in the form of a FET configuration (**Figure 8d** and **e**). High responsivities of 22 12 A W^{-1} at 1 V bias voltage were obtained under irradiations of 405 nm laser with 0.07 mW cm^{-2}. The response or recovery time was shorter than 20 or 40 ms upon 405 nm irradiation (**Figure 8f** and **g**) [47]. The excellent photoresponse properties make this 2D perovskite promise for photodetection applications.

2D all-inorganic perovskites are also reported to fabricate high-performance photodetectors. In particular, $CsPbBr_3$ NSs has been widely studied on synthesis, photophysics, electrical property, and device application due to its high air stability. 2D $CsPbBr_3$ perovskite is formed by incorporating Cs^+ cations into a 2D sheet of $PbBr_4^{2-}$ octahedra of which six Br atoms surround each Pb atom and four in-plane Br atoms are being shared by two octahedrons [11]. Bulk $CsPbBr_3$ has been reported to possess a thermodynamically orthorhombic structure at room temperature, and it can transform to tetragonal and cubic phases at 88 and 130°C, respectively [48]. However, the cubic phase could be preserved at room temperature owing to a large number of surfaces. Both cubic and orthorhombic phases are reported for $CsPbBr_3$ NSs by different groups [49, 50]. In terms of well-controlled synthesis, Akkerman et al. synthesized few-layer $CsPbBr_3$ NSs with monolayer-level thickness control [51]. Shamsi et al. synthesized few-layer $CsPbBr_3$ NSs with lateral dimensions up to a micrometer size. [52] For photophysics, Li et al. measured PL decay time of 2 ns and exciton binding energy of 120 meV for $CsPbBr_3$

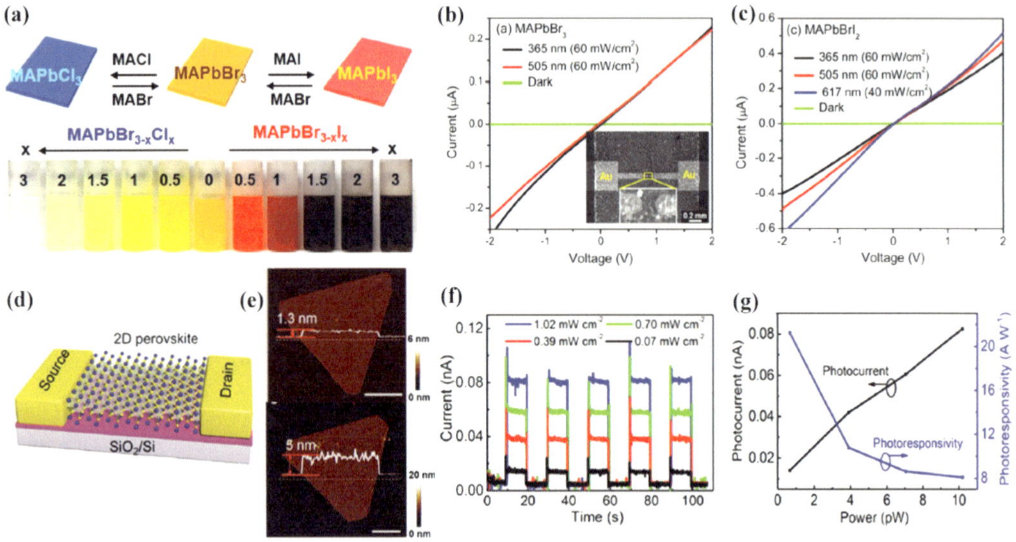

Figure 8. (a) Schematic for $MAPbX_3$ NSs synthesized by anion-exchange reaction. I–V characteristics of (b) $MAPbBr_3$ and (c) $MAPbBrI_2$ NSs film under different irradiation and dark conditions, (d) schematic of the 2D $MAPbI_3$ NSs photodetector [46], (e) AFM images of $MAPbI_3$ NSs with different thicknesses, (f) time-dependent photocurrent measurement, and (g) photocurrent and photoresponsivity versus optical illumination power under the 405 nm irradiation [47].

nanoplates with a thickness of 3.4 nm and the lateral dimensions of 20.2 nm [53]. Our group reported a larger exciton binding energy of 231.6 meV for $CsPbBr_3$ NSs with larger lateral dimensions of 700 nm and a thickness of 1.8 nm based on the absorption spectrum fitting with a quantum-well absorption model (**Figure 9a–c**). At the same time, high carrier mobility of 77.9 $cm^2 V^{-1} s^{-1}$ was obtained by the Hall effect and transient photoconductivity measurements [54]. There are some reports on 2D $CsPbBr_3$ NSs photodetector application. In 2016, Song et al. fabricated flexible $CsPbBr_3$ NS film visible-light photodetectors with a light on/off ratio of $>10^3$, a R of 0.64 A W^{-1}, and an EQE of 54% under 442 nm light at a power of 0.35 mW cm^{-2}. The response and recovery time were measured to be 0.019 or 0.025 ms, respectively. A high stability and outstanding flexibility (>10,000 cycles) were also obtained [55]. This work demonstrates that all-inorganic perovskite $CsPbBr_3$ NSs have a great potential in photodetector application. Next, Lv et al. reported a similar result on $CsPbBr_3$ NSs photodetectors [56]. Then, our group reported that the responsivity could be improved by the annealing, which was demonstrated to increase coupling energy but decrease exciton binding energy resulting from ligand rearrangement. Tunneling instead of resonant energy transfer as a dominant way of exciton dissociation contributed to a high charge transfer rate. This exciton dissociation mechanism was proposed based on microscopic KPFM results (**Figure 9d**). As shown in **Figure 9e** and **f**, the response and recovery time were short, and the largest responsivity of 0.53 A W^{-1} could be obtained under the 525 nm irradiation with the intensity of 0.009 mW cm^{-2} [54].

4.2. Perovskite/graphene hybrid photodetector

Despite that the perovskites with different compositions, morphologies, and structures have been reported as photodetectors with high performance, their responsivity can be further

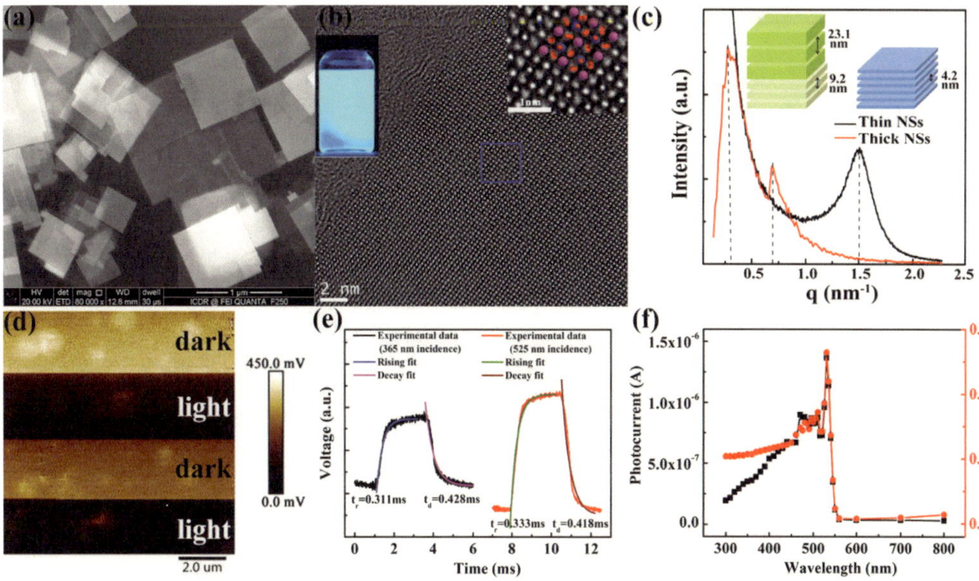

Figure 9. (a) Scanning electron microscope (SEM) image, (b) transmission electron microscopy (TEM) image, and (c) small-angle X-ray scattering (SAXS) patterns of $CsPbBr_3$ NSs, (d) Kelvin probe force microscopy (KPFM) surface potential mapping, (e) transient photoresponse curves, and (f) the wavelength-dependent photocurrent and responsivity curves of $CsPbBr_3$ NS film photodetector [54].

improved by constructing hybrid structure. A promising strategy is based on hybrid phototransistors, where light is efficiently absorbed by perovskite layer while photogenerated carriers can transfer into conducting layer such as graphene, and high gain can be obtained from one carrier that recirculate many times within the lifetimes of the opposite charges [3]. For this hybrid photodetector, graphene is firstly transferred on the FET configuration substrate, and then perovskites including polycrystalline film or nanocrystal can be deposited on the graphene. In 2015, Lee et al. fabricated the $MAPbI_3$/graphene hybrid photodetector, and a dramatic PL quenching was observed for hybrid film resulting from effective charge-carrier transfer through π-π interaction between perovskite and sp^2 hybridized graphene (**Figure 10a, b**). The device exhibited a broad spectral photoresponsivity across the visible range. The photoresponsivity and effective quantum efficiency (EQE) were 180 A/W and 5×10^4% at a relatively high illumination power of 1 μW, respectively (**Figure 10c**). The device also exhibited good on-off switching, and response and recovery times were 87 and 540 ms (**Figure 10d**) [57]. The success of this novel perovskite/graphene hybrid photodetector is a powerful evidence to demonstrate efficient hybrid structure. Instead of compact polycrystalline perovskite film, Wang et al. reported a high-performance perovskite/graphene hybrid photodetector consisting of graphene covered with a layer of dispersive $MAPbBr_2I$ islands. A photoconductive gain of 10^9, a responsivity of 6.0×10^5 A W^{-1}, and a broadband detection across UV-visible are obtained. The photoresponse mechanism was discussed deeply. Under illumination, electron-hole pairs were generated in perovskite and then separated at the heterojunction interface. The holes were transferred to graphene sheets, and electrons were trapped in perovskite with a quite long lifetime, inducing a photogating effect [58]. Then, Qian et al. fabricated a nitrogen-doped graphene quantum dots (NGQDs)-$MAPbI_3$/reduced graphene oxide (rGO) transistor, exhibiting a high responsivity of 1.92×10^4 A/W and

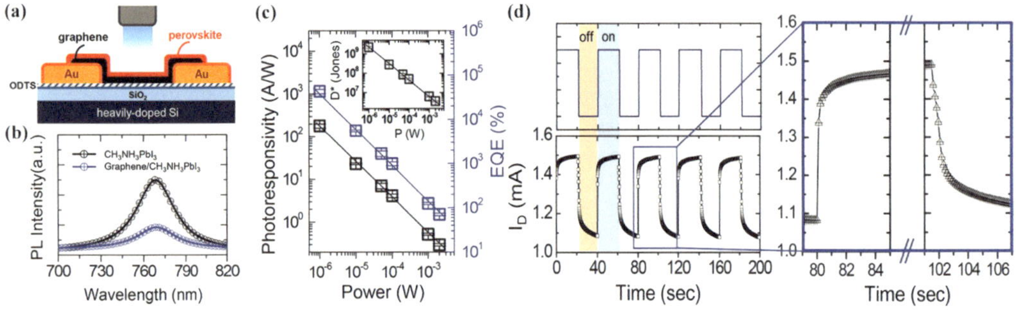

Figure 10. Graphene/MAPbI$_3$ hybrid photodetector. (a) Schematic diagram of the device, (b) PL spectra of the pristine MAPbI$_3$ and graphene/MAPbI$_3$ hybrid films upon excitation at 532 nm, and (c) responsivity and EQE vs. illumination power at 520 nm wavelength. The inset shows the photodetectivity vs. illumination power, (d) Photo-switching characteristics of the photodetector under alternating dark and light illumination (500 μW, 520 nm). The gate and drain voltages were 0 and 0.1 V, respectively [57].

a short response and recovery time of 10 ms. The outstanding performance was attributed to that NGQDs offer an effective path for electron transfer from perovskite to the rGO [59]. Despite high responsivity, perovskite/graphene hybrid photodetectors suffer from a large dark current about mA order because of ultrahigh conductivity of bottom graphene layer, resulting in high noise level.

Besides the lateral photoconductor structure, a vertical graphene/perovskite/graphene (GPG) photodetector has also been fabricated, and higher responsivity is achieved due to its shorter carrier extraction length. Duan's group fabricated a series of single 2D material heterojunction. Wang et al. reported a graphene/$CsPbBr_3$ microplate/graphene sandwiched vertical photodetector by growing $CsPbBr_3$ microplates on an exfoliated graphene flake, followed by the dry transfer of a second layer of graphene (**Figure 11a**). As shown in **Figure 11b**, the photocurrent rapidly increased with the light power, and the photocurrent versus bias curves were nonlinear, indicating the presence of a contact barrier. The responsivity decreases with the increasing light power density, and the maximum responsivity exceeds 10^5 A/W (**Figure 11c**) [60]. Besides, Cheng et al. fabricated a graphene/$MAPbI_3$/graphene photodetector, delivering a high responsivity of 950 A W^{-1} and a high photoconductive gain of 2200 at 1 V bias under a 1.52 nW irradiation of 532 nm laser (**Figure 11d** and **e**). It also exhibited good and reproducible photo-switching behavior with response and recovery time of 22 and 37 ms. The photocurrent mapping showed uniformly contribution from the overlapping area, indicating that the majority of photoinduced carriers travel in vertical direction (**Figure 11f**) [61].

4.3. Perovskite/TMDs hybrid photodetector

Compared with perovskite/graphene hybrid photodetector, perovskite/TMD hybrid photodetector has a significant lower dark current. TMD layer serves as both light absorbers and carrier transport layer, while perovskite as sensitized layer can enhance light absorbance and extend the detection range. The heterojunction formed between perovskite and TMD layer

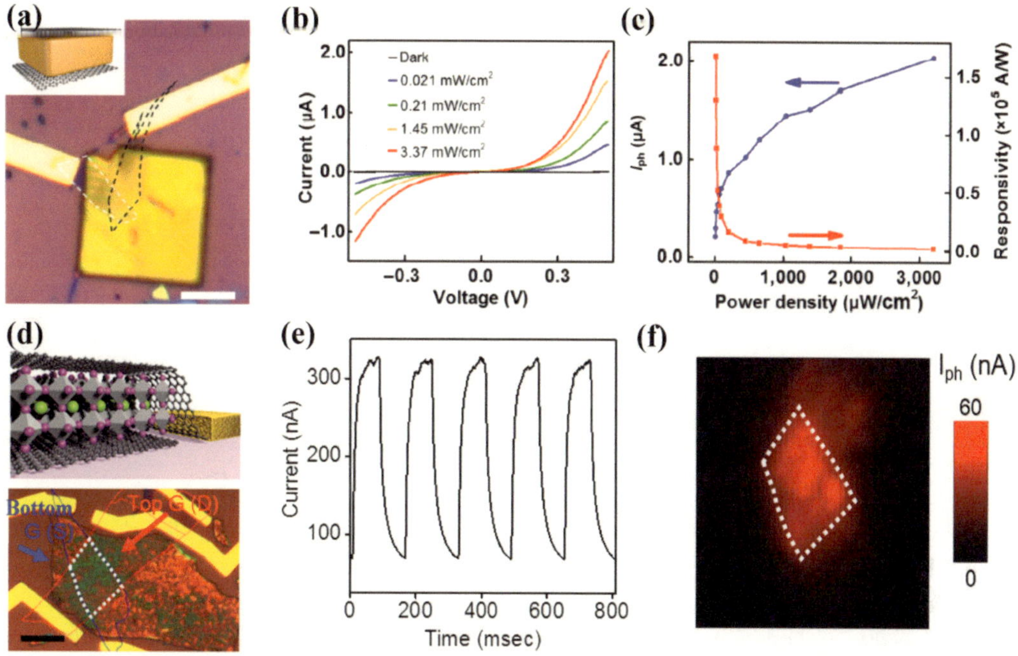

Figure 11. Graphene/$CsPbBr_3$ microplate/graphene photodetector. (a) Optical microscope image of a vertical heterostructure. The inset shows a device schematic, (b) photocurrent versus bias in the dark and under different illumination light power densities, (c) power density dependence of the photocurrent and responsivity [60], graphene/$MAPbI_3$ nanoflakes/graphene photodetector, (d) device schematic, (e) photo-switching characteristics of the photodetector under alternating dark and light illumination, and (f) the corresponding photocurrent mapping [61].

can promote the carrier transfer, contributing to a high responsivity and fast response speed. In 2016, Ma et al. fabricated a hybrid perovskite $MAPbI_3$ and TMDs material WS_2 planar photoconductor detector for the first time. Both $MAPbI_3$ and WS_2 layers contacted with Au electrodes, which could transport photocarriers as parallel channels (**Figure 12a**). From the band structures of hybrid bilayer in **Figure 12b**, photogenerated charges were separated at the interface due to the Fermi level difference, which reduced the charge recombination and enhanced the carrier separation. The significant PL quenching for both $MAPbI_3$ and WS_2 was observed, indicating efficient charge transfer at the interface and exciton dissociation (**Figure 12d**). Furthermore, they proposed the photoresponse mechanism. In dark, the charge transfer at the interface lowers (raises) the Fermi level of WS_2 (perovskite), leading to the formation of depletion regions, suppressing the dark current. Under the light illumination, excitons are generated in the perovskite film, and electrons are transferred to WS_2 film, which raises (lowers) the Fermi level of WS_2 (perovskite) and reduces the Schottky barriers at contacts, resulting in the high photocurrent (**Figure 12c**). The device exhibited a high on/off ratios of $\approx 10^5$ and a high responsivity of ≈ 17 A W^{-1}. Owing to the high mobility of the WS_2 monolayer and the efficient interfacial charge separation, the response speed of the hybrid photodetector was enhanced by four orders of magnitude compared to the reference perovskite single layer (**Figure 12e** and **f**) [62].

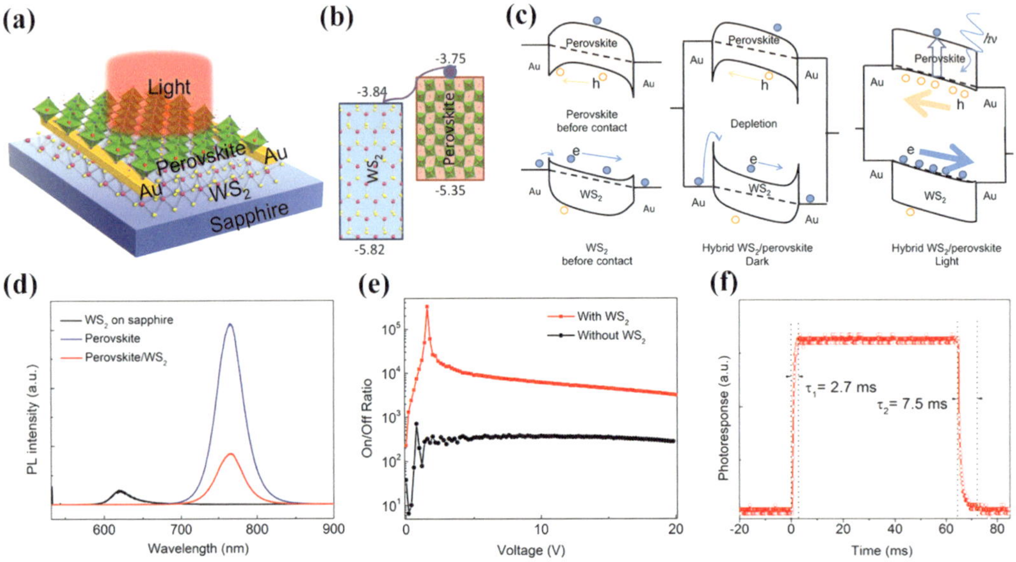

Figure 12. $WS_2/MAPbI_3$ hybrid photodetector. (a) Schematic device structure; (b) band structures of hybrid bilayer; (c) working mechanism of hybrid photodetector; (d) PL spectra of WS_2, perovskite, and hybrid bilayer upon excitation at 532 nm; (e) bias dependence of the on/off ratios; and (f) time-dependent photoresponse [62].

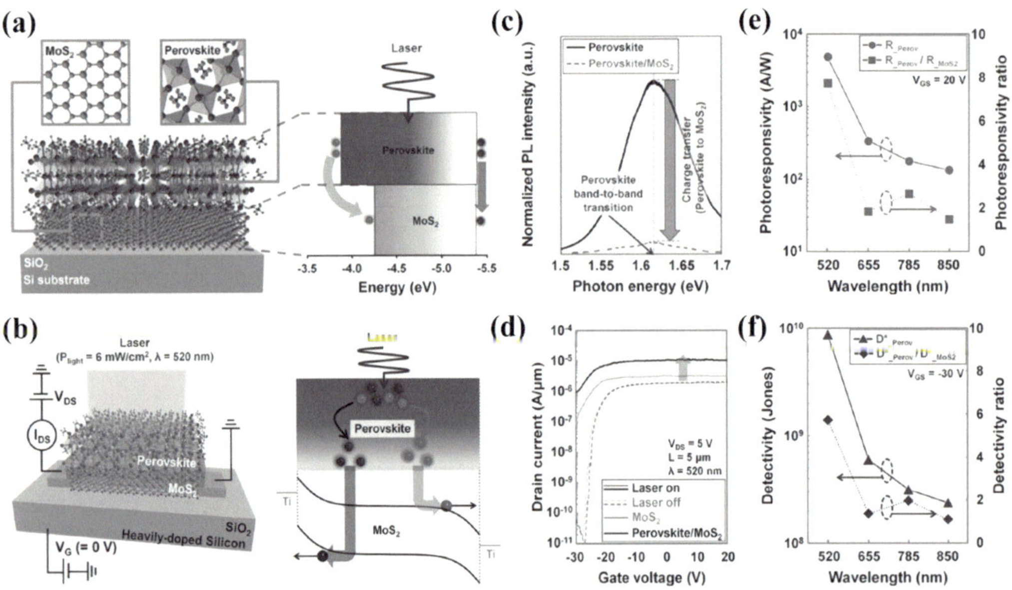

Figure 13. $MoS_2/MAPbI_3$ hybrid photodetector. (a) Schematic device structure and energy-band diagram, (b) schematic device structure and energy-band diagram of photodetector under 5 V bias voltage and laser exposure, (c) PL spectra of perovskite and perovskite/MoS_2 hybrid bilayer, (d) IV characteristics of MoS_2 and perovskite/MoS_2 photodetectors under both dark and illuminated conditions, (e) photoresponsivity as a function of the incidence wavelength, and (f) photo-switching characteristics [63].

At the same time, Kang et al. developed an ultrahigh-performance MoS_2/$MAPbI_3$ hybrid photodetector by using mechanically exfoliated 25-nm-thick MoS_2 nanoflakes [63]. Similar with the WS_2/$MAPbI_3$ hybrid bilayer, the photogenerated electrons in perovskite are thought to reach the MoS_2 layer through diffusion (**Figure 13a**). The significant PL quenching for hybrid bilayer was also observed, indicating the charge transfer at the interface (**Figure 13c**). Under the bias voltage and laser illumination, electron-hole pairs generated in perovskite were separated based on the ambipolar charge transport property of the perovskite and diffuse into the MoS_2 region, followed by the carrier collection step caused by the electric field in the Ti-MoS_2 junctions (**Figure 13b**). The photocurrent of hybrid photodetector has increased by one order than MoS_2 photodetector when V_G was −30 V(**Figure 13d**). The photoresponsivity of the hybrid photodetector and their ratios as a function of wavelength is shown in **Figure 13e**. In the case of the 520 nm laser, the photoresponsivity was highly enhanced by factors of 7.7 (from 636 to 4.9×10^3 A W^{-1}). The response and recovery time were also decreased for hybrid photodetector, but they are quite long compared with WS_2/$MAPbI_3$ hybrid photodetector (**Figure 13f**). Then, Lu et al. reported a WSe_2/$MAPbI_3$ hybrid photodetector with a photoresponsivity of 110 A W^{-1}, a high EQE of 2.5×10^4%, and a high D* of 2.2×10^{11} Jones [64]. In all, the outstanding photoresponse performance of perovskite/TMDs hybrid photodetector demonstrates this strategy as an efficient route to improve the performance of 2D material photodetector.

5. Summary and outlook

The distinctive optical and electrical properties enable 2D materials as an ideal and powerful candidate in photodetector application. The superiorities of 2D materials include a wide range of photoresponse wavelength (graphene is from UV to THz, TMDs is from visible to near infrared, halide perovskite is from UV to near infrared), facilitating the construction of vertical heterojunctions due to their naturally passivated surfaces, and the strong interaction with incident light resulting from the singularities effect near the conduction and valence band edges. This chapter has introduced the photophysics, device structure, working mechanism, and performance of three kinds of 2D materials photodetectors including graphene, TMDs, and halide perovskite. For graphene photodetector, four kinds of physical mechanisms are discussed, and the graphene p-n junction and metal-graphene Schottky junction are based on photovoltaic and photoconductor effects, respectively. These two effects also work for most of the 2D materials photodetectors. Generally, graphene photodetector has a wide sensitive wavelength and fast response speed, but it also suffers from a low responsivity and a high dark current. For TMD photodetector, there are the same performance influence factors, such as the charge trapping induced by substrate and atmosphere environment, the layer number, the preparation methods, electrode contact, and surface modification. However, the photoresponse performance is also determined by the photophysical parameters including mobility and absorption coefficient. For example, MoS_2 photodetector has a higher responsivity than WS_2 because of its high mobility. For 2D halide photodetector, the high absorption and mobility promote the high photoresponse performance of $MAPbBr_3$ nanoplatelets and $CsPbBr_3$ nanosheets. Furthermore, both the perovskite/graphene and perovskite/TMD hybrid structures are employed to enhance the photoresponse based on reduced the charge recombination and enhanced the carrier separation due to band difference. In summary, the 2D materials show their unique superiority in photodetector application.

In spite of the great success for 2D materials photodetectors, it is admitted that lots of challenges still remain. The dark current of graphene photodetector is too large, resulting in a high noise. The sensitive to environment such as oxygen and water is an unfavorable factor to TMDs and perovskite photodetectors, which is the origin of the device instability. To solve these problems, the novel 2D materials, approaches, and hybrid structure should be developed to meet the requirement for practical applications. First, the responsivity and response speed can be enhanced by exploring novel 2D materials with controllable preparation method. Second, novel approaches for chemical doping and surface treatment should be investigated, which has been proved as a feasible strategy for improving or manipulating device performance. Third, new hybrid photodetectors should be focused to demonstrate the increased responsivity resulting from efficient carriers transfer. Both materials engineering and device structure design are important to achieve high and reliable photoresponse performance for pursuing practical application of 2D materials photodetectors in the future.

Acknowledgements

This work was supported by Natural Science Foundation of China (NSFC grant nos. 61604122, 51572216, and 61774124) and China Postdoctoral Science Foundation (grant 2017M613139).

Author details

Zhi Yang*, Jinjuan Dou and Minqiang Wang

*Address all correspondence to: yangzhi029@xjtu.edu.cn

Electronic Materials Research Laboratory (EMRL), Key Laboratory of Education Ministry, International Center for Dielectric Research, Shaanxi Engineering Research Center of Advanced Energy Materials and Devices, Xi'an Jiaotong University, Xi'an, China

References

[1] Novoselov KS, Jiang D, Schedin F, Booth TJ, Khotkevich VV, Morozov SV, Geim AK. Two-dimensional atomic crystals. In: Proceedings of the National Academy of Sciences. 2005;**102**:10451

[2] Xu M, Liang T, Shi M, Chen H. Graphene-like two-dimensional materials. Chemical Reviews. 2013;**113**:3766

[3] Xie C, Mak C, Tao X, Yan F. Photodetectors based on two-Dimensional layered materials beyond graphene. Advanced Functional Materials. 2017;**27**:1603886

[4] Nicolosi V, Chhowalla M, Kanatzidis MG, Strano MS, Coleman JN. Liquid exfoliation of layered materials. Science. 2013;**340**:1226419

[5] Novoselov KS, Geim AK, Morozov SV, Jiang D, Zhang Y, Dubonos SV, Grigorieva1 IV, Firso AA. Electric field effect in atomically thin carbon films. Science. 2004;**306**:666

[6] Wang J, Fang H, Wang X, Chen X, Liu W, Hu W. Recent progress on localized field enhanced two-dimensional material photodetectors from ultraviolet—visible to infrared. Small. 2017;**13**:1700894

[7] Meric I, Han MY, Young AF, Ozyilmaz B, Kim P, Shepard KL. Current saturation in zero-bandgap, top-gated graphene field-effect transistors. Nature Nanotechnology. 2008;**3**:654

[8] Bonaccorso F, Sun Z, Hasan T, Ferrari AC. Graphene photonics and optoelectronics. Nature Photonics. 2010;**4**:611

[9] Chhowalla M, Shin HS, Eda G, Li LJ, Loh KP, Zhang H. The chemistry of two-dimensional layered transition metal. Nature Chemistry. 2013;**5**:263

[10] Tran V, Soklaski R, Liang Y, Yang L. Layer-controlled band gap and anisotropic excitons in few-layer black phosphorus. Physical Review B. 2014;**89**:235319

[11] Chen S, Shi G. Two-dimensional materials for halide perovskite-based optoelectronic devices. Advanced Materials. 2017;**29**:1605448

[12] Geim AK, Grigorieva IV. Van der Waals heterostructures. Nature. 2013;**499**:419

[13] Britnell L, Ribeiro RM, Eckmann A, Jalil R, Belle BD, Mishchenko A, et al. Strong light-matter interactions in heterostructures of atomically thin films. Science. 2013;**340**:1311

[14] Geim AK, Novoselov KS. The rise of graphene. Nature Materials. 2007;**6**:183

[15] Koppens FHL, Mueller T, Avouris P, Ferrari AC, Vitiello MS, Polini M. Photodetectors based on graphene, other two-dimensional materials and hybrid systems. Nature Nanotechnology. 2014;**9**:780

[16] Konstantatos G, Sargent EH. Nanostructured materials for photon detection. Nature Nanotechnology. 2010;**5**:391

[17] Bistritzer R, MacDonald AH. Electronic cooling in graphene. Physical Review Letters. 2009;**102**:206410

[18] Yan J, Kim MH, Elle JA, Sushkov AB, Jenkins GS, Milchberg HM, et al. Dual-gated bilayer graphene hot-electron bolometer. Nature Nanotechnology. 2012;**7**:472

[19] Wang X, Gan G. The nonlinear electronic transport in multilayer graphene on silicon-on-insulator substrates. Chinese Physics B. 2017;**26**:034203

[20] Dyakonov M, Shur M. Shallow water analogy for a ballistic field effect transistor: New mechanism of plasma wave generation by dc current. Physical Review Letters. 1993;**71**:2465

[21] Liu N, Tian H, Schwartz G, JBH Tok TR, Bao Z. Large-area, transparent, and flexible infrared photodetector fabricated using P-N junctions formed by n-doping chemical vapor deposition grown graphene. Nano Letters. 2014;**14**:3702

[22] Xia F, Mueller T, Lin Y, Avouris P. Ultrafast graphene photodetector. Nature Nanotechnology. 2009;**4**:839

[23] Mueller T, Xia F, Avouris P. Graphene photodetectors for high-speed optical communications. Nature Photonics. 2010;**4**:297

[24] Liu C, Chang Y, Norris TB, Zhong Z. Graphene photodetectors with ultra-broadband and high responsivity at room temperature. Nature Nanotechnology. 2014;**9**:273

[25] Dou Y, Zhang L, Xu X, Sun Z, Liao T, Dou SX. Atomically thin non-layered nanomaterials for energy storage and conversion. Chemical Society Reviews. 2017;**46**:7338-7373

[26] Velicky M, Toth PS. 2D Materials in electrochemistry. Applied Materials Today. 2017;**8**:68

[27] Radisavljevic B, Radenovic A, Brivio J, Giacometti V, Kis A. Single-layer MoS2 transistors. Nature Nanotechnology. 2011;**6**:147

[28] Braga D, Gutiérrez Lezama I, Berger H, Morpurgo AF. Quantitative determination of the band gap of WS2 with ambipolar ionic liquid-gated transistors. Nano Letters. 2012;**12**:5218

[29] Chernikov A, Berkelbach TC, Hill HM, Rigosi A, Li Y, Aslan OB, et al. Exciton binding energy and nonhydrogenic rydberg series in monolayer WS2. Physical Review Letters. 2014;**113**:076802

[30] Mak KF, He K, Lee C, Lee GH, Hone J, Heinz TF, et al. Tightly bound trions in monolayer MoS2. Nature Materials. 2013;**12**:207

[31] Yin Z, HH Li HHL, Jiang L, Shi Y, Sun Y, et al. Single-Layer MoS2 phototransistors. ACS Nano. 2012;**6**:74

[32] Lopez-Sanchez O, Lembke D, Kayci M, Radenovic A, Kis A. Ultrasensitive photodetectors based on monolayer MoS2. Nature Nanotechnology. 2013;**8**:497

[33] Khan MF, Iqbal MW, Iqbal MZ, Shehzad MA, Seo Y, Eom J. Temperature dependent phonon shifts in single-layer WS2. ACS Applied Materials & Interfaces. 2014;**6**:21645

[34] Tsai DS, Liu KK, Lien DH, Tsai ML, Kang CF, Lin CA, Li LJ, He JH. Few-layer MoS2 with high broadband photogain and fast optical switching for use in harsh environments. ACS Nano. 2013;**7**:3905

[35] Lu J, Lu JH, Liu H, Liu B, Chan KX, Lin J, Chen W, Loh KP, Sow CH. Improved photoelectrical properties of MoS2 films after laser micromachining. ACS Nano. 2014;**8**:6334

[36] Zhang W, Chiu MH, Chen CH, Chen W, Li LJ, Wee ATS. Monolayer MoSe2 grown by chemical vapor deposition for fast photodetection. ACS Nano. 2014;**8**:8653

[37] Hwan Lee S, Lee D, Sik Hwang W, Hwang E, Jena D, Jong Yoo W. High-performance photocurrent generation from two-dimensional WS2 field-effect transistors. Applied Physics Letters. 2014;**104**:193113

[38] Perea-López N, Elías AL, Berkdemir A, Castro-Beltran A, Gutiérrez HR, Feng S, et al. Photosensor device based on few-layered WS2 films. Advanced Functional Materials. 2013;**23**:5511

[39] Huo N, Yang S, Wei Z, Li SS, Xia JB, Li J. Photoresponsive and gas sensing field-effect transistors based on multilayer WS2 nanoflakes. Scientific Reports. 2014;**4**:5209

[40] Yao JD, Zheng ZQ, Shao JM, Yang GW. Stable, highly-responsive and broadband photodetection based on large-area multilayered WS2 films grown by pulsed-laser deposition. Nanoscale. 2015;**7**:14974

[41] Chena Q, Marcoa ND, Yang Y, Song T, Chen C, Zhao H, et al. Under the spotlight: The organic–inorganic hybrid halide perovskite for optoelectronic applications. Nano Today. 2015;**10**:355

[42] Eperon GE, Stranks SD, Menelaou C, Johnston MB, Herz LM, Snaith HJ. Formamidinium lead trihalide: A broadly tunable perovskite for efficient planar heterojunction solar cells. Energy & Environmental Science. 2014;**7**:982

[43] Kojima A, Teshima K, Shirai Y, Miyasaka T. Organometal halide perovskites as visible-light sensitizers for photovoltaic cells. Journal of the American Chemical Society. 2009;**131**:6050

[44] Stranks SD, Eperon GE, Grancini G, Menelaou C, Alcocer MJP, Leijtens T, Herz LM, Petrozza A, Snaith HJ. Electron-hole diffusion lengths exceeding 1 micrometer in an organometal trihalide perovskite absorber. Science. 2013;**342**:341

[45] Xing G, Mathews N, Sun S, Lim SS, Lam YM, Grätzel M, Mhaisalkar S, Sum TC. Long-range balanced electron- and hole-transport lengths in organic-inorganic CH3NH3PbI3. Science. 2013;**342**:344

[46] DM Jang KP, DH Kim JP, Shojaei F, Kang HS, et al. Nano Letters. 2015;**15**:1591

[47] Liu J, Xue Y, Wang Z, Xu ZQ, Zheng C, Weber B, et al. Two-dimensional CH3NH3PbI3 perovskite: Synthesis and optoelectronic application. ACS Nano. 2016;**10**:3536

[48] Wang Y, Guan X, Li D, Cheng H, Duan X, Lin Z, Duan X. Chemical vapor deposition growth of single-crystalline cesium lead halide microplatelets and heterostructures for optoelectronic applications. Nano Research. 2017;**10**:1223

[49] Yu Y, Zhang D, Kisielowski C, Dou L, Kornienko N, Bekenstein Y, et al. Atomic resolution imaging of halide perovskites. Nano Letters. 2016;**16**:7530

[50] Dang Z, Shamsi J, Palazon F, Imran M, Akkerman QA, Park S, et al. In situ transmission electron microscopy study of electron beam-induced transformations in colloidal cesium lead halide perovskite nanocrystals. ACS Nano. 2017;**11**:2124

[51] Akkerman QA, Motti SG, Kandada ARS, Mosconi E, D'Innocenzo V, Bertoni G, et al. Solution synthesis approach to colloidal cesium lead halide perovskite nanoplatelets with monolayer-level thickness control. Journal of the American Chemical Society. 2016;**138**:1010

[52] Shamsi J, Dang Z, Bianchini P, Canale C, Di Stasio F, Brescia R, et al. Colloidal synthesis of quantum confined single crystal CsPbBr3 nanosheets with lateral size control up to the micrometer range. Journal of the American Chemical Society. 2016;**138**:7240

[53] Li J, Luo L, Huang H, Ma C, Ye Z, Zeng J, He H. 2D behaviors of excitons in cesium lead halide perovskite nanoplatelets. Journal of Physical Chemistry Letters. 2017;**8**:1161

[54] Yang Z, Wang MQ, Qiu HW, Yao X, XZ Lao SJX, Lin ZH, Sun LY, Shao JY. Engineering the exciton dissociation in quantum-confined 2D CsPbBr3 nanosheet films. Advanced Functional Materials. 2018;**28**:1705908

[55] Song J, Xu L, Li J, Xue J, Dong Y, Li X, Zeng H. Monolayer and few-layer all-inorganic perovskites as a new family of two-dimensional semiconductors for printable optoelectronic devices. Advanced Materials. 2016;**28**:4861

[56] Lv L, Xu Y, Fang H, Luo W, Xu F, Liu L, et al. Generalized colloidal synthesis of high-quality, two-dimensional cesium lead halide perovskite nanosheets and their applications in photodetectors. Nanoscale. 2016;**8**:13589

[57] Lee Y, Kwon J, Hwang E, Ra C-H, Yoo WJ, Ahn J-H, et al. High-performance perovskite–graphene hybrid photodetector. Advanced Materials. 2015;**27**:41

[58] Wang Y, Zhang Y, Lu Y, Xu W, Mu H, Chen C, et al. Hybrid graphene–perovskite phototransistors with ultrahigh responsivity and gain. Advanced Optical Materials. 2015;**3**:1389

[59] Qian L, Sun Y, Wu M, Xie D, Ding L, Shi G. A solution-processed high-performance phototransistor based on a perovskite composite with chemically modified graphenes. Advanced Materials. 2017;**29**:1606175

[60] Wang Y, Guan X, Li D, Cheng H, Duan X, Lin Z, Duan X. Chemical vapor deposition growth of single-crystalline cesium lead halide microplatelets and heterostructures for optoelectronic applications. Nano Research. 2017;**10**:1223

[61] Cheng HC, Wang G, Li D, He Q, Yin A, Liu Y, et al. van der waals heterojunction devices based on organohalide perovskites and two-dimensional materials. Nano Letters. 2016;**16**:367

[62] Ma C, Shi Y, Hu W, Chiu MH, Liu Z, Bera A, et al. Heterostructured WS2/CH3NH3PbI3 photoconductors with suppressed dark current and enhanced photodetectivity. Advanced Materials. 2016;**28**:3683

[63] Kang DH, Pae SR, Shim J, Yoo G, Jeon J, Leem JW, et al. An ultrahigh-performance photodetector based on a perovskite–transition-metal-dichalcogenide hybrid structure. Advanced Materials. 2016;**28**:7799

[64] Lu J, Carvalho A, Liu H, Lim SX, Castro Neto AH, Sow CH. Hybrid bilayer WSe2–CH3NH3PbI3 organolead halide perovskite as a high-performance photodetector. Angewandte Chemie, International Edition. 2016;**55**:11945

Chapter 2

Transition Metal Dichalcogenide Photodetectors

Inturu Omkaram, Young Ki Hong and Sunkook Kim

Additional information is available at the end of the chapter

http://dx.doi.org/10.5772/intechopen.72295

Abstract

Two Dimensional (2D) materials has triggered to have transition metal dichalcogenides (TMDCs) emerging as a new class of materials that can control or interact with light to convert the photons to electrical signals for its attractive applications in photonics, electronics and optoelectronics. 2D materials along with gapless Graphene interact with light over the wavelength region of the different spectral regions having the short wavelength of the UV and extreme UV, Visible, near IR, mid IR and THz due to excellent light absorption, enabling ultrafast and ultrasensitive detection of light in photodetectors. Next generation photodetectors are possible promising candidates for high sensitivity and TMDCs based photodetectors are the heart of the multitude of technologies to understand the principle of photodetection mechanisms and device performances. Phototransistors/photoconductors show wide varied detection performances with responsivities ranging from 10^{-7} A/W - 10^{7} A/W on single or few layer TMDCs having response time between 10^{-5} s to 10^{3} s. The semiconducting TMDCs like MoS_2, $MoSe_2$, WS_2, WSe_2 and ReS_2 are gaining suitable applications in optoelectronic devices and the device design, mechanism and enhancing the performance of photodetectors are introduced and discussed systematically in this chapter. In spite of the growing demands on TMDC based devices the origin of the photoresponse characteristics is attractive and encouraging to understand and provide a path to the subject of investigation and guidelines for the future development of this rapidly growing field.

Keywords: 2D semi-conductors, transition metal dichalcogenides (TMDCs), photodetectors, optoelectronic 2D devices

1. Introduction

Recently, 2D nanomaterials are rapidly expanding as one of the prime goal of materials research from different disciplines such as physics, materials science, chemistry and electrical engineering. 2D materials with new entrants have been widely researched for its unique electronic & optoelectronic components, sensors, biomedical and drug delivery applications [1–5]. It is a fascinating counterpart to gapless graphene that has successful isolation, the ideas and methods have rapidly established from these early studies were prolonged to various layered materials [6–8]. Alike to graphite, the two dimensional crystals established on atomically thin films of layered semiconductors, such as the family of transition metal dichalcogenides (TMDCs) offer an attractive platform for various optoelectronic applications [9–21]. TMDCs are highly attractive due to the existence of an appropriate energy band gap (1.2–1.8 eV) that has numerous unique properties that can be synthesized into thick atomic planes when compared to the bulk counterparts. The semiconducting 2D metal chalcogenides are of peculiar interest with the essential possibilities to explore the band gap discoveries by varying the number of layers making them an exciting application for devices [7]. Investigations on band structure have dramatically changed from bulk layer to single layer samples having the direct band gap semiconducting TMDCs suitable for optoelectronic applications. The quantum size effects play a prominent role due to their nanosize in expressing the distinctive properties of the material that are not noticed in its bulk form having inspiring fascinating applications in the development of transparent and flexible optoelectronic devices [22–24].

In optoelectronics, graphene has been utilized for the realization of photodetectors and optical modulators. 2D materials offer complementary to graphene that lacks the interlayer interactions having potential applications in the nano and optoelectronics to evaluate their electronic band structure and their spaced energy levels [25–27]. Interest in two-dimensional (2D) TMDCs materials can be defined as any material in which the bond strength between atoms within a plane are much stronger than the bonds out of the plane that are exfoliated easily to present interesting electrical-optical behavior. Optoelectronics integrate the physics of light and these electronic devices can control light that converts photons or plasmons to electrical signals. When light is incident on a semiconductor, they create free carriers (electron-hole pairs) on the exciton binding energy in the semiconductor. The photon energy greater than the band gap is gained by the electrons that move the common barrier in between the metal and semiconductor, the energy gained by the electron which is called the work function is given by kinetic energy [28]

$$E_e = \frac{hc}{\lambda} - \phi_m \tag{1}$$

where λ is the incident wavelength, φ_m is termed to be the metal work function and C the light velocity. Since the photoelectric effect is based on the photon energy hν, the wavelength of interest is related to the energy transition ΔE in the device operation with this relationship $\lambda = \frac{hc}{\Delta E} = \frac{1.24}{\Delta E}$ (eV) μm where ΔE is the difference of energy levels transition. The discrete energies ΔE of the semiconductor causes the excitation with the greater photon energy and the bound excitons generate a photocurrent when it is separated by an applied electric field with two major classes of semiconductor photo detectors. The two categories are likely to be the photodiodes and phototransistors that are transparent and flexible thin film electronic devices having its great importance in the transparent displays, UV Detectors, wearable electronics and solar cells [29–31].

There have been partial 2D discussion on the focus of TMDCs in recent years. This chapter examines the properties of 2D materials from a new perspective related to the optoelectronic properties of 2D materials. In a pioneering work of MoS_2 phototransistors the exfoliated single-layer exhibits a photoresponsivity of 7.5 mA/W at 50 V of the gate voltage [32]. Phototransistors founded on CVD MoS_2 multilayer by local bottom gate structures show a supreme responsivity of 342.6 A/W [33] and monolayer/WS_2 multilayers devices show a photoresponsivity value of 880 A/W and a photodetector with high detectivity of 2.2×10^{12}–7.7×10^{11} Jones and a maximum photoresponsivity up to 2570 A/W have been reported [17, 23]. The synthesis route and the structural characterization of TMDCs having fascinating properties of 2D materials provides a discussion of the TMDC-based devices with photo-responsivity, the main emphasis is on describing relevant methods and important outcomes as well as some of the novel applications of 2D photo detectors [34–36]. It is an intent to give a comprehensive overview of the recent experimental results related to 2D photo detectors and its applications. In particular, this chapter covers 2D photo detector materials synthesis and characterization, device (phototransistor) fabrication, mechanism, performance of photo-transistors in optoelectronics. This method attempts to highlight the various synthesis approaches for 2D materials like bottom up synthesis routes including chemical vapor deposition (CVD), hydrothermal and layer-by layer conversation and bottom down approaches like chemical or mechanical exfoliation had their corresponding characterization by which the current understanding of 2D materials has some tools for future contributions. The achievement of 2D materials created explosive interest to enrich the performance of TMDC photo detectors, having the practice of periodic elements challenges novel discoveries for the exciting new physics and ultimately thin devices.

2. Synthesis methods for 2D materials

2D TMDCs can be divided into two groups as

(2.1) top down

(2.2) bottom up.

With top down approaches, there are mechanical and chemical exfoliations. In mechanical exfoliation the approach starts from bulk TMDC materials which naturally comprise of many layers where scotch tape is used to peel nanosheets of bulk TMDCs. It has advantage of producing high purity and clean single crystal flakes but still limited to control the size and thickness as the materials are easy to crack owing to grain margins, material flaws and process prompted stress. For chemical exfoliation, ion assisted exfoliation can be used. Bulk TMDCs powder is submerged in lithium ion containing solution (like n-butyllithium) and the lithium ion to intercalate into the bulk TMDCs. The intercalated material is exposed to water and water will react vigorously to evolve H_2 gas with in the lithium layers thus separating the layers rapidly producing high yield of monolayer TMDCs. It also failed to enlarge the size of the monolayer which will restrict in practical applications.

The label bottom up approaches itself notates to construct from the bottom and the best methods are physical vapor deposition (PVD)/Chemical vapor deposition (CVD). In PVD, the bulk TMDCs is frenzied in a tube furnace until it thermally evaporates. A flow of inert gas would carry the vapor stream to a cooler region and deposit it on to the substrate. Single crystals are developed with grain dimension of several tens of micrometer, additionally layer number can be controlled by varying the amount of precursor used. In CVD two precursors are used (1) transition metal compound (M) (2) chalcogenide (X) precursor, the chalcogenide is placed upstream in the tube furnace at a lower temperature than the transition metal compound at higher temperature region in the tube furnace downstream. The furnace heats up so that the precursors are thermally evaporated, reacts to form MX_2 and is deposited on a substrate which is further downstream as compared to transition metal compound with inert gas as carrier gas. The advantage of CVD growth is that large area uniform polycrystalline thin film (cms) are grown with a possibly to grow single crystal by varying precursors.

2.1. Top down approaches

2.1.1. Mechanical exfoliation

It is a topdown method that is proved to be quite successful that is most traditional and simplest to prepare high quality single layer of graphene and TMDCs materials. Geim and Novoselov [37] used a block of graphite by cleaving it into different layers of graphene and transferring it onto the substrates using scotch tape. The TMDCs samples are prepared by peeling off from their parent bulk crystals with micromechanical cleavage by adhesive tape as shown in **Figure 1(a)**. This process was quick, cost efficient and the tape was repeated—peeled material is applied to a substrate. Geim and Novoselov [37] cleaved the layered and revealed that this method can also be applied to Boron Nitride &TMDCs [12].

Due to some limitations in this method of mechanical cleaving, the morphology control was not systematically carried with flake size, the thicknesses of the nanosheets obtained were small and tens of microns, it's a problematic for massive production outside the laboratories has its own limitations in applications.

2.1.2. Electric chemical exfoliation

It is an effective topdown method that is inspired by the exfoliation methods that was advanced where lesser Li ions being charged are intercalated to different layered TMDCs bulk materials. Zheng et al. [38] developed the bulk TMDCs is positioned on the cathode and a lithium intercalated material were in the electrolyte (water or ethanol) as shown in **Figure 1(b)**. In this set up charge is allowed to pass and the enforcement of the charged ions in the layered material by intercalating the cathode. Lithium ions are induced inside the layers of TMDC materials at change and Hydrogen gas is produced when the intercalated bulk material is reacted with water in sonication to obtain the 2D TMDCs [13]. This separates the TMDC layers during charging process and require low temperature to fully control the procedure by

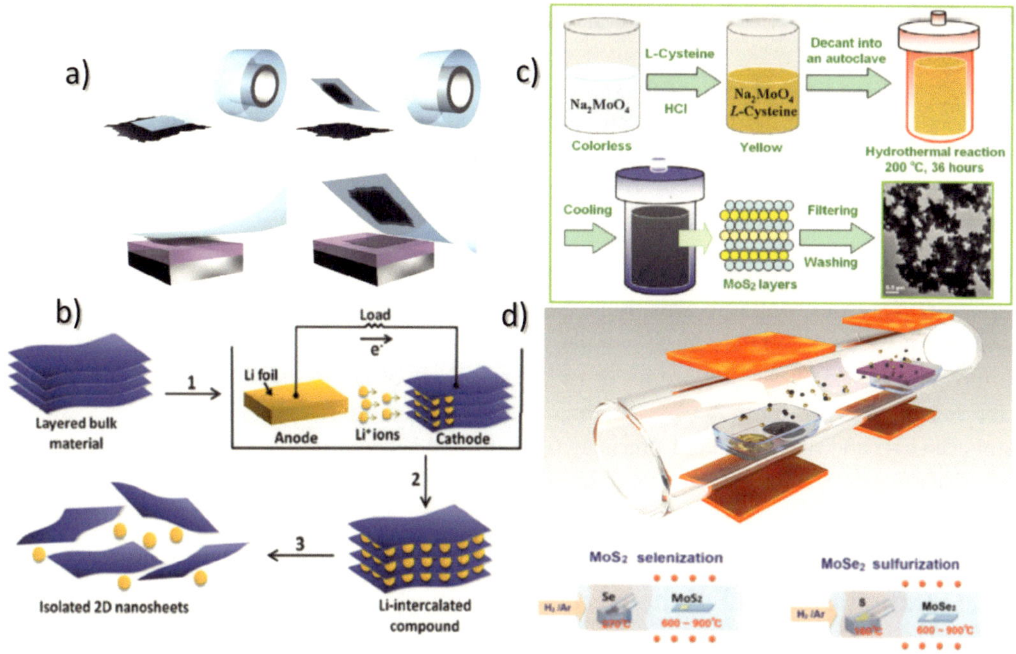

Figure 1. (a) Mechanical exfoliation method. (b) Lithiation and exfoliation process in a lithium ion battery system. (c) Preparation of MoS_2 with the use of a simple hydrothermal method. (d) Synthesis of the hexagonal $MoSe_2$ multilayer nanoparticles with MoO_3 and Se sources and the experimental set-up of the selenization and sulfurization process. Adapted from Novoselov and Castro Neto [12], Zhang et al. [13], Guo et al. [39], Jung et al. [99], Shi et al. [14].

referring to a discharging curve. The disadvantage by these intercalation methods is due to the impurities lead to the modifications in the electrical properties.

2.2. Bottom up approaches

2.2.1. Hydro-thermal method

It is an owing bottom-up technique for synthesis of two-dimensional TMDCs. It is normally supported in a pressurized autoclave with the reactions in aqueous solutions to procedure TMDC layered single crystal [39] as depicted in **Figure 1(c)**. The temperature and pressure on the solution in an autoclave can be elevated to the boiling point of water and touched to the pressure of a vapor saturation. MX_2 TMDC layered material, for instance, is produced by adding Mo, W and S, Se into autoclave and heating in the noble gas (For example N, Ar) to 773 K for 3 hours. The hydrothermal method can produce high quality of TMDC material with uniform size (from several nanometers to several microns) but the thickness of TMDC wafer cannot be controlled.

2.2.2. Chemical vapor deposition

Chemical vapor deposition: CVD is an employed technique currently used in the synthesis of graphene and 2D TMDC materials that are exploited in the study of nanostructures including thin films and nanotubes. It is also another bottom-up method that realizes chemical reaction between vapors to form TMDC wafers at high temperature. It is one of the supreme well-organized ways to gain and grow large films on a substrate [40]. In order to form a two-dimensional material, normally, precursor is applied (S, Se and MoO_3, WO_3) on high temperature region and substrate (sapphire or SiO_2/Si substrate) on low temperature region in a pipe furnace. **Figure 1(d)** shows the procedure for 2D TMDCs [14] synthesis depends on evaporating of Sulfur;Se and MoO_3;WO_3.

The vaporized solid was then deposit on substrate to form a 2D TMDC crystal. It is the most operative way to form large area, high quality TMDCs and achievable method in thickness controlling. Other CVD methods entails the Mo sulfurization or a reagent rich in sulfur atmosphere on the substrate are used for the synthesis of 2D MoS_2 materials. The thickness and homogeneity of these 2D MoS_2 is regulated having more restrictions on the large area is not achieved and not controlled precisely. Therefore, it is preferred alternative for synthesis of 2D TMDCs material for electronic and optical device fabrication.

3. 2D transition metal dichalcogenides

Transition Metal Dichalcogenides; TMDCs; as 2D semiconductors are proposed to be a layered periodic part of elements consists of transition metal (Mo or W or Re) and chalcogen (S or Se or Te) atoms frequently represents as MX_2, where M is transition metal (usually group V/VI element) and X is Chalcogen [15, 41–44]. TMDCs provide novel and outstanding phenomena varying from indirect to direct band gap transition, large exciton binding energy, photoluminescence and high ON/OFF ratio which are particularly changed from those of their bulk materials.

The bulk material is significantly different from the nano-scaled particles that are divided into numerous categories based on dimensional topological spacing difference such as OD, 1D, 2D & 3D nano materials. 2D materials are crystalline materials containing of layered arranged atoms/molecules which has strong covalent bonds within each layer but only weak Vander Waals forces between layers that novelties application such as electrode, semiconductors and photovoltaics [9, 45].

In 2004, the production of thin carbon layer by mechanical exfoliation hailed the milestone of 2D material, so called Graphene [46, 47] that illustrates the structure of a two dimensional through a thickness less than 1 nm. The unique structure of Graphene (**Figure 2(a)**) illustrates outstanding performance such as extremely high carrier mobility (15,000 cm^2/VS) high thermal conductivity (5300 W/m.K) and high transmittance (97.7%) [48]. Those outstanding performance allures substantial attention on graphene for opening up a responsible fields in Physics (Fermi Dirac), and chemistry with many scientific publications. The electrical, optical and magnetic properties of graphene and its 2D nature has good current integration in the circuit technology, however it limits its application where the electronic band gap is fundamental in optoelectronics. The absence of band gap in graphene provides little ability for cutting off

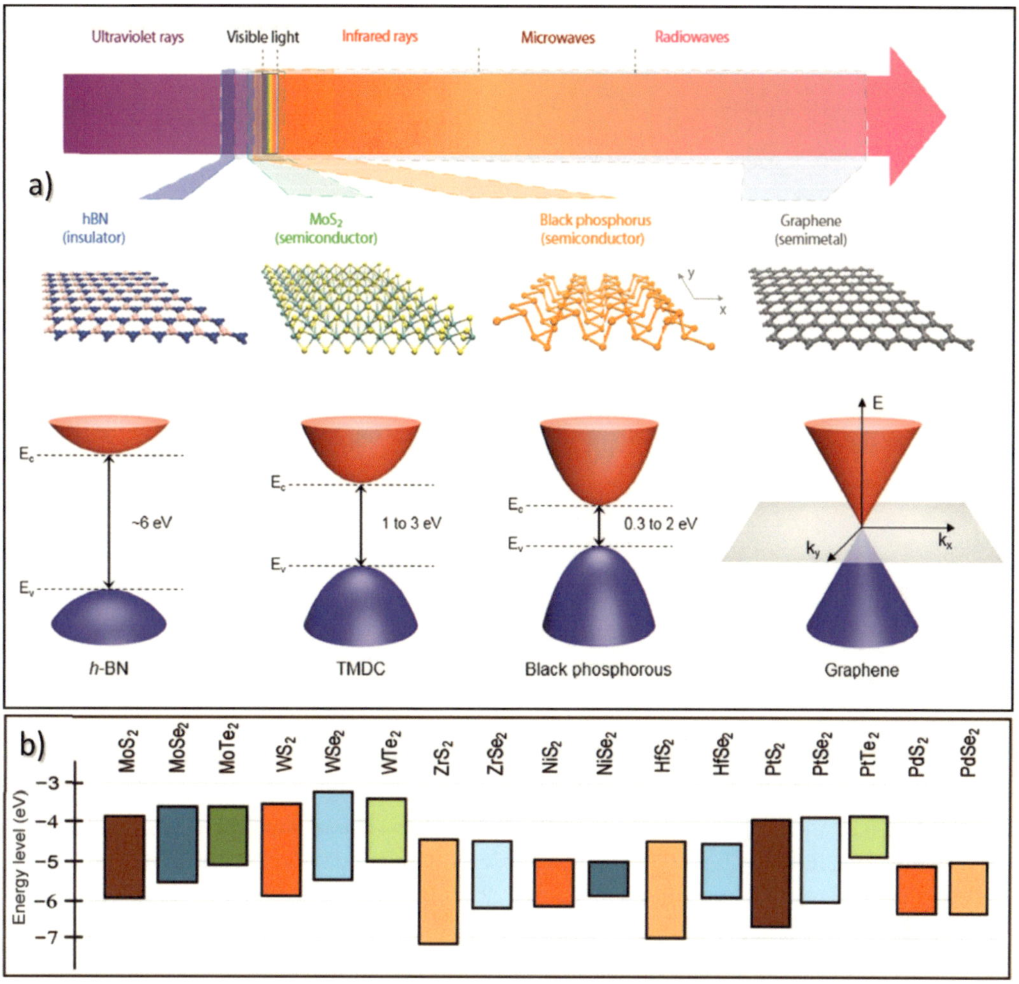

Figure 2. (a) 2D TMDC materials sweeping the energetic region of the electromagnetic spectrum. (b) Energy spectrum of various two-dimensional (2D) materials and their atomic crystal. Adapted from Kourosh et al. [8], Xia et al. [44], Lee et al. [15].

the current while negative bias is applied and it is incapable for exciton disassociation like semiconductors [49]. Now a days the interest for science is extended to other 2D engineering with physical and chemical modifications to introduce novel performances.

Therefore this urges scientists to find out other 2D materials like graphene analogue boron nitrate (BN), transition metal oxide (TMO-like titanium oxide or perovskite), transition metal dichalcogenides (TMDCs) and V-VI/IV-VI compounds naturally have a band gap [50–52]. TMDCs are the elements of the periodic table having a combination (**Table 1**) of two elements with the chemical formula of MX_2: a transitional metal:M, of groups 4–10, and chalcogens:X such as 15 sulfur:S, selenium:Se or tellurium:Te. A single layer TMDCs similar in atomic thicknesses have one layer of atoms that are not comprised as graphene but a planar layer

Graphene family	Graphene	hBN 'White graphene'	BCN	Fluorographene	Graphene oxide
2D chalcogenides	MoS_2, WS_2, $MoSe_2$, WSe_2	Semiconducting dichalcogenides: $MoTe_2$, WTe_2,ZrS_2,$ZrSe_2$ and so on		Metallic dichalcogenides: $NbSe_2$, NbS_2,TaS_2, TiS_2, $NiSe_2$ and so on	
				Layered semiconductors: GaSe, GaTe, InSe, Bi_2Se_3 and so on	
2D oxides	Micas, BSCCO	MoO_3, WO_3		Perovskite – type: $LaNb_2O_7$, $(Ca,Sr)_2Nb_3O_{10}$, $Bi_4Ti_3O_{12}$, $Ca_2Ta_2TiO_{10}$ and so on	Hydroxides: $Ni(OH)_2$, $Eu(OH)_2$ and so on
	Layered Cu oxides	TiO_2, MnO_2, V_2O_5, TaO_3, RuO_2 and so on			Others

Table 1. Current 2D library. Adapted from Geim and Grigorieva [56].

TMDCs involve a metal atom layer sandwiched between layers of chalcogen atoms. They are strongly held together by covalent bonding and the layers are feebly bound together with van der Waals forces allowing it to separate in the bulk sheets. The layered materials combination of Transition metals and Chalcogen family (**Table 1**) have almost 40 different types of TMDCs crystal phases. The different chemical nature of metal and chalcogenide atoms makes the sublattices inequivalent, producing a band gap.

The importance of band-gap engineering in the semiconducting nature of TMDCs is crucial for their potential use in photovoltaic/−catalytic applications, as the ability to tune the material to the correct spectrum or to particular donor/acceptor states in a system is vital in such applications [53, 54] (electronic & photonic). Unlike graphene the semiconducting TMDCs are with a band gap varying from 1 to 3 eV and the band gap (**Figure 2(b)** energy levels of semiconducting TMDCs are sized by changing the number of layers [27]. The TMDCs research was just at their beginning in recent years that built the foundations having a very rapid advances in the scientific field of 2D TMDCs. Obviously, all these methods which effort on gapless graphene are switched for its use on TMDCs. In 2011 with the help of quantum confinement on electronic structure [55], the indirect band gap of bulk TMDCs materials chances to direct band gap in 2D TMDC material of single layer for the coupling of spin and valley physics. The effect of strain allow TMDCs to be a promising use in transistors, optoelectronic and valleytronic devices. The predicted research of the few to monolayer TMDCs focus on identified unique compounds of the TMDCs [7, 56, 57] like MoS_2, $MoSe_2$ & WS_2 etc. exhibit a direct band gap and are of specific interest for new kinds of optoelectronic devices and the bulk crystals are indirect band gap semiconductors.

4. Photodetector technologies and performance

A photo-detector is a key principle device that rely in detecting the process by which light is converted into electrical signals by absorbing the photons [58]. The general action of a photo-detector comprises fundamentally three process, carrier generation, carrier transport and extraction of carriers as terminal output signal current. The schematics of the 2D photo-detectors are shown in **Figure 3(a)**. The fundamental principle of the photo-detector has 3 different stages as Light harvest, photocarrier (e-h) separation and charge transport. It is thus more advantages in the Image defining areas, display technology, networking devices, opto-electronic sensors, and fundamental science applications.

Different mechanisms used in the photodetectors have been identified in literature [59–61] which are crucial process in processing alterations of absorbed photos into an electrical charge. The several possible mechanisms exists, these effects are: photoconductive *(PCE)*, photovoltaic *(PVE)* and photo-thermoelectric *(PTE)* that are discussed below.

4.1. Photoconductive effect (PCE)

This effect is based on the absorption of a photon by a material results in a generation, which can enhance the electrical conductance of the 2D materials [62]. In dark when the device was under an applied bias there will be small current between the two electrodes (without illumination) **Figure 3(b)**. In light when the device was exposed to light photons electron-hole pairs are generated with a higher energy than the band gap. The drift in the electron and holes in different directions lead to an increase in current (I_{photo}) between the metal leads. This photo

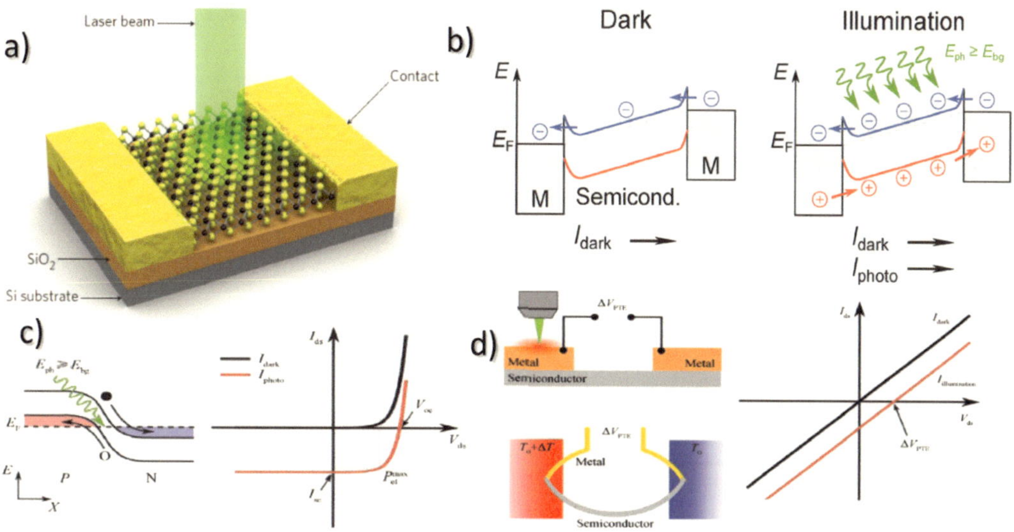

Figure 3. (a) Schematic illustration of 2D photodetector device fabricated on SiO_2/Si substrate. (b) Schematic of the photoconductive effect: in the dark and under illumination. (c) Schematic of the photovoltaic effect: band alignment in a PN junction and I-V curves in the dark & under illumination. (d) Schematic of the photo-thermoelectric effect: the thermal circuit corresponding to the field-effect transistor device and I_{ds}–V_{ds} characteristics in the dark & under illumination. Adapted from Koppens et al. [59], Buscema et al. [66].

generated current increased the conductivity of the device. Due to the large difference between the electron and hole motilities it is important to study the transit time (T) between the electron and holes [63]

$$T_{transit} = \frac{L^2}{\mu V_{ds}} \tag{2}$$

where L, V_{ds} and μ have their regular notations. Many electrons can take part in the photocurrent before whole extraction or recombination which generates the photoconductive gain (G).

4.2. Photovoltaic effect (PVE)

This effect is based on the device when exposed to light and generating voltage. This novel quantum effect for photons for its excitation of an electron hole (e-h) pair parted by an internal electric field which contributes to the photocurrent [64]. Some interface can built the internal electric field, such as Schottky barrier or a PN Junction as shown in **Figure 3(c)**. In dark the detector exhibits nonlinear I-V properties the PN junction is formed, I_{ds} α exp. $V_{ds} - 1$ as I_{ds} is exponential with the V_{ds}. When the device is under illumination the photo generated electron-hole pairs can be separated by internal electric field without external voltage leading to a large photocurrent. When the device was under illumination and reverse bias, the photogenerated carriers are wiped out in opposite directions with an increase in the reverse current and the photons energy can be converted into electrical energy by the photovoltaic effect. The voltage ΔV in the process of G-Au interface can be described as:

$$\Delta V = \phi_M - \phi_G - \Delta E_F + \text{sgn}\left(V_{gs} - V_{gs}^{Dirac}\right)\hbar v_F \sqrt{\pi\alpha}\sqrt{V_{gs} - V_{gs}^{Dirac}} \tag{3}$$

where φ_M is the work function of metal and φ_G is the work function of graphene; ΔE_F is the shift of the Fermi level of graphene by doping; V_{gs}: gate bias; V_{gs}^{Dirac}: gate bias in electrical neutral point; $\alpha = 7.2 \times 10^{10}\ cm^{-2}\ V^{-1}$ is the carrier concentration with 300 nm SiO_2; $\hbar vF = 5.52$ eVÅ. A bias voltage (external) is to be applied to improve the performance of the device to extract the charge carriers and use the phenomenon of photoelectric effect. The photocurrent can be described in an expanded area in the whole device by applying an external bias voltage that is given by [62, 65]:

$$I_{Ph} = AVe\mu\Delta n \tag{4}$$

the symbols have their usual meaning as the cross section (A), the applied voltage (V), charge (e), the mobility (μ) and the carrier density (Δn). This mechanism can offer great internal gain, which means that the device can detect with extremely low power.

4.3. Photo thermoelectric (PTE) effect

This effect is based on the thermoelectric effect caused by light illumination. When the light energy get absorbed it is sensed by Thermal detectors and the temperature gets raised [66]. **Figure 3(d)** shows the light induced heating leads to a temperature gradient through a semiconducting channel. Due to Seebeck effect the two ends of the semiconducting channel has a temperature difference ΔT that can convert into a voltage difference ΔV = S. ΔT where ΔV is linearly proportional to ΔT.

Graphene is extensively studied with this mechanism and later the MoS_2 photo-detectors [66–70] having thermoelectric effect by Buscema et al. [66] on the MoS_2 monolayer. The photoinduced current is developed with 750 nm photo source that is less than the band gap of monolayer that is caused by the heating the local junctions between MoS_2 and Gold. The temperature difference between the junctions is due to the absorption of light, creates a generation of PTE current and the voltage difference (ΔV_{PTE}) at the edge of the metal contacts is given by [71]:

$$\Delta V_{PTE} = (\text{S semiconductor : MoS}_2 - \text{S metal : TiAu})\Delta T \approx \text{S semiconductor : MoS}_2\Delta T \quad (5)$$

where S is the Seebeck coefficient and ΔT is the difference of temperature (Heated gold electrode & MoS_2) [72]

$$S = \frac{\pi^2 k_B^2 T}{3e} \cdot \frac{dln\,(\sigma(E))}{dE} \mid E = E_F \quad (6)$$

where e; the electron charge, k_B; Boltzmann constant, σ(E) is the conductivity, E_F the Fermi energy and T is the temperature.

4.4. Photo-device performance

The key terms of merit that are at high priority to characterize and assess the photo-detector performances are as listed below:

Photo gain (G_{ph})

Photo response time:

On/Off ratio

External Quantum Efficiency (EQE) and

Internal Quantum Efficiency (IQE)

Photoresponsivity (R_λ) and

Detectivity (D)

Photo gain:

The ratio of the detected charge carriers per single incident photons, given by G_{ph} and is defined by [63]

$$G_{ph} = \frac{\tau_{tr}}{\tau_{transit}} = \frac{\tau_{tr}\mu V_{bias}}{L^2} \quad (7)$$

Here, τ_{tr} is the life time of the charge carriers, $\tau_{transit}$ is the drift transit time, μ is the mobility of the charges and L is the source—drain separation distance.

Photoresponse time:

In the ON state from dark, the time required for the current change until saturation is considered. This time varies from μs to minutes in different 2D materials and related devices that can limit the application of the photo-detector.

On/Off ratio:

The ON/OFF in the photodetector is the fraction of the current with light ON and the current (I) in Dark. It is not very similar to the device ON-OFF ratio in the other FET devices.

External Quantum Efficiency (EQE):

In order to produce the photocurrent I_{ph} the fraction of the extracted free charge carriers to the photo flux φ_{in} collected at a given energy E_{ph} is the External Quantum Efficiency, defined by [73]:

$$\mathrm{EQE} = \frac{I_{ph}}{e\phi_{in}} = \frac{h\nu\, I_{ph}}{e\ P_{in}} \tag{8}$$

where P_{in} is the incident power.

Internal Quantum Efficiency (IQE):

In order to produce the photocurrent I_{ph} the sharp ratio of the number of free charge carriers collected is slightly different from external. The energy E_{ph} is given by the absorbed count of photons is the External Quantum Efficiency, described by [73]:

$$\mathrm{IQE} = \frac{I_{ph}}{e\varphi_{in} A_{abs}} = \frac{h\nu}{e}\frac{I_{ph}}{P_{in} A_{abs}} \tag{9}$$

here A_{abs} mentions the fraction of absorption.

Photoresponsivity:

Photoresponsivity is defined by R_λ on the effective area of the photocurrent I_{ph} generated per unit incident power *P* of the photo-detector (S) [5, 74, 75]:

$$\frac{I_{ph}}{P_{in}} = \frac{I_{ph}}{P \times S} = \frac{\eta\, q}{hv} \tag{10}$$

where η; is quantum efficiency, and the responsivity is termed as [67]

$$R = \frac{\eta\lambda q}{hc} = \frac{\eta\lambda\ (\mu m)}{1.24}\left(\frac{A}{W}\right) \tag{11}$$

where λ is the incident wavelength.

Quantum efficiency:

Quantum efficiency (QE) (η%) *of a photodetector* is the ratio of the electron generation rate to the photon incidence rate. QE is connected to the responsivity of the photo detector by the below equation [58, 76]

$$\eta = \frac{\frac{I_{ph}}{q}}{\frac{P_{inc}}{hv}} = \frac{I_{ph}}{q} \cdot \frac{hv}{P_{inc}} \tag{12}$$

Detectivity:

The ability of the Detector D represents the detector to distinguish between signal/noise [77, 78]:

$$D = \frac{R_{\lambda}\sqrt{A}}{\sqrt{2eI_{dark}}} \tag{13}$$

where I_{dark} is the dark current.

In 1873 the birth of photo-detectors was traced in the discovery of photoconduction of selenium [79] and was first fabricated in 2009. The Graphene as the prototype 2D material started working without any bias voltage showing steady state photo-response up to a frequency of 40 GHz. The photodetectors of graphene have broader absorption spectrum, fast response time and a carrier mobility that are energetic elements and the exfoliated graphene has the highest photoresponsivity (6.1 mA/W) and weak absorbance which is a limitation for exfoliated graphene photo-detectors [80]. Now a days, the 2D transition metal dichalcogenide semiconductors (TMDCs) have successfully fabricated for larger absorption that has raised their limitation to improve the performance of photodetector. The monolayer exfoliated MoS_2 has a photoresponsivity of 880 A/W at 561 nm and the energy band gap of TMDCs layers is tuned between direct and indirect which is crucial in designing efficient photo-detectors [81].

5. 2D photo devices

2D material devices are fabricated with the contacts and are patterned by means of physical mask & lithography (both electron beam and optical lithography) In addition, metallization, lift off process and vacuum annealing are the hardcore essential steps for a final material device fabrication. Extensive discussion is carried on the experimental techniques of material device fabrication by using physical mask and optical lithography respectively. The three parameters like metal evaporation, lift off and vacuum annealing are necessarily of more interest to be discussed.

5.1. Fabrication

Physical mask: The physical stencil mask is the modest device fabrication method that is fast and clean. The device performance is fictional with the lithography process having good transparency in the 2D material homo-structures. The fabrication of the device is carried with physical mask and prepared first on silicon substrate. The precised mask is aligned and stuck tight on the substrate by making contacts having the metals get evaporated. The 2D device is at last fabricated by eliminating the mask.

Lithography: Lithography is an effective process used in the fabrication of the device and is a planning tool to prepare the device contacts and to overcome some of the problems in the application of 2D materials. Electron beam lithography and Optical lithography differ technically but sounds good in the applications of universal devices. The E-Beam lithography can be of more advantage in device application as the precision can reach nm scale and the design of the mask is done with requirement of the samples. **Figure 4(a)** & **(b)** depicts the schematic of electron beam lithography and at first the resist is spin coated onto the substrate (for optical lithography) or sample (for electron beam lithography) and dried in a hot plate. Secondly, the UV light or electron beam exposes the patterns of the mask and starts to change the resist properties. The pattern obtained is developed after the exposure. Thirdly, the patterned samples are deposited with the metals (Ti/Au), and finally the metals/resist is removed in the lift off process to advance the device.

Metal evaporation: The metals get evaporated by arranging the devices and substrates, alignment of stencil mark or the lithography by using gold, titanium and carbon and chrome metals. The development compute the process to have normal contacts, evaporation of 5 nm Cr-Ti followed by 70 nm Au with careful consideration.

Lift off (only made by lithography): Lift off is a simple, easy method for exposing a pattern into photoresist depositing a thin film and washing to leave behind only the film in the patterned area. The resist is lifted off the substrate on the top of the metal that gets evaporated and lithography is done. This process is carried in a hot acetone of temperature around 50–60°C, in 2 h time the metal on the top gets lifted off after the resist gets dissolved.

Annealing (only made by lithography): This is the important process that is very important to fabricate the devices. It is important to remove the residual of resist/impurities in vacuum. In order to maintain good contact among the sample and the metal its crucial to input certain

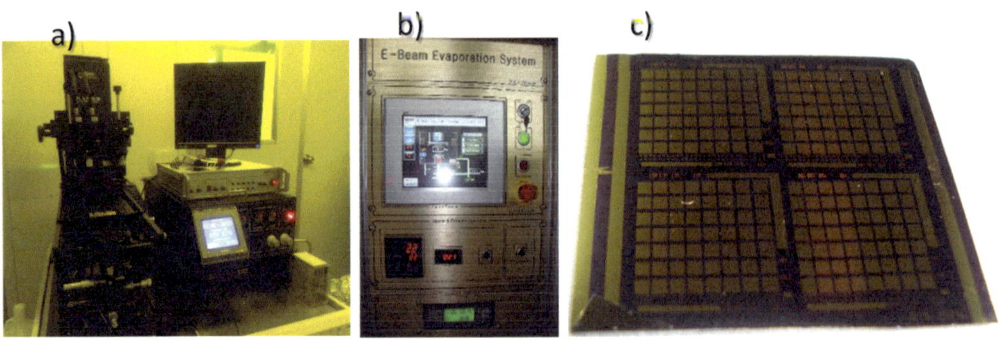

Figure 4. (a) Optical lithography. (b) E-Beam system. (c) Schematic of 2D final device fabrication with grids as a shadow mask with SiO_2 dielectric and Ti/Au electrodes.

device current to remove critical contaminants in the optimization of metal:sample interface, which is the most critical step in device fabrication (**Figure 4(c)**).

The different TMDC devices on 2D layered semiconductors is comprehensively reviewed and discussed about the photodetection applications beyond graphene. **Figure 5** [17] shows the timeline development of the applications of graphene and other 2D layered semiconductors in photodetectors based on different principles.

Most of the fabricated single/few layer TMDC FETs exhibits an n-type semi conducting behavior which might approach the saturation current. The threshold voltage indicates the natural n-type doping that arises due to the impurities like halogen (Br/Cl) atom that likely to replace S atoms. The gradual increase of the concentration of the electrons of TMDCs giving n-type doping results. The field effect device mobility of the single/few layer devices can be estimated with the below equation [82]

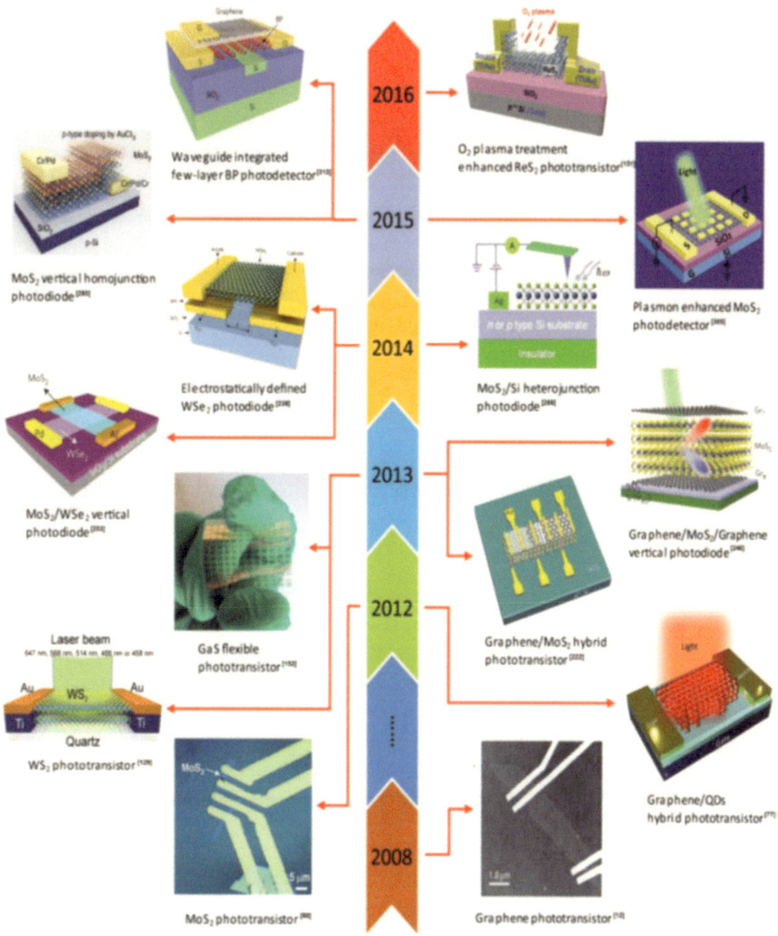

Figure 5. Timeline showing the development of the applications of graphene and other 2D layered semiconductors in photodetectors based on different principles. Adapted from Xie et al. [17].

$$\mu = \left(\frac{L}{WC_{ox}V_{ds}}\right)\left(\frac{\partial I_{ds}}{\partial V_{gs}}\right) \tag{14}$$

where L, W, C_{ox} have their usual meaning like the device channel length, width and gate capacitance per unit area of 1.15×10^{-8} F/cm^2. It is well known that the carrier mobility of semiconductors tend to decrease with increasing band gap. The electron mobility trend from III to V compounds reduces from 77,000 cm^2/VS for the narrow band gap to 3000 cm^2/VS for the wider band gap [83]. The gapless graphene show very high mobility of up to 200,000 cm^2/VS [84]. The electron mobilities among 2D materials beyond graphene shows a high field effect mobility $\mu_{eff} > 100$ cm^2/VS for MoS_2 and an average mobility of ~50 cm^2/VS/~100 cm^2/VS for single/bulk $MoSe_2$ [85, 86]. With various 2D blocks in hands the key parameters provide large collection of mobility with strong light absorption in literature [87, 88] has transfer ($I_{ds} - V_{gs}$) and output ($I_{ds} - V_{ds}$) characteristics of different TMDC synthesis.

5.2. Photodetector parameters

Choosing to study the optoelectronic properties of the semiconducting two dimensional materials like the TMDCs family consisting of transition metal dichalcogenides have shown a great deal of attention towards strong light-matter interaction for optoelectronic applications. The selection of the photodetector is based on the wavelength of light detected and the basic photodetectors are semiconducting optical devices that adapt the incident light to an electrical light that is focused as photocurrent [89]

$$I_{ph} = BP^m \tag{15}$$

where I_{Ph} is the photocurrent, B is a constant, m is an exponent determining the photoconduction mechanism and P is the illumination intensity. In order to understand the photodetection high responsivity is required and essential for its potential application in TMDCs materials which is a good parameter for the performance of an improved photodetector. It is defined by [90]

$$R_\lambda = \frac{I_P}{P_{light}} \tag{16}$$

where $I_P = I_{illumination} - I_{dark}$ and P_{light} is the power of incident light.

The detector response should be great at the detected wavelength and different laser beams are used at several optical power and applied gate voltage. The laser beams are chosen with a laser wavelength of 532 nm is used in producing a photocurrent by focusing the light with a spot diameter less than 1 μm. The I & V parameters tend to increase with the increasing light intensity that strongly suggest the photocarriers having electric field separated at the interface of 2D and Si. To know the potential application of TMDC materials in photodetection, high responsivity in various 2D TMDCs have been used for high performance photo-detectors such as, MoS_2, $MoSe_2$, WS_2, $MoTe_2$, & $ReSe_2$ and so on [91–94]. These TMDCs in **Figure 6** can provide additional advantages over graphene based photodetectors that have larger photoresponsivity

http://dx.doi.org/10.5772/intechopen.72295

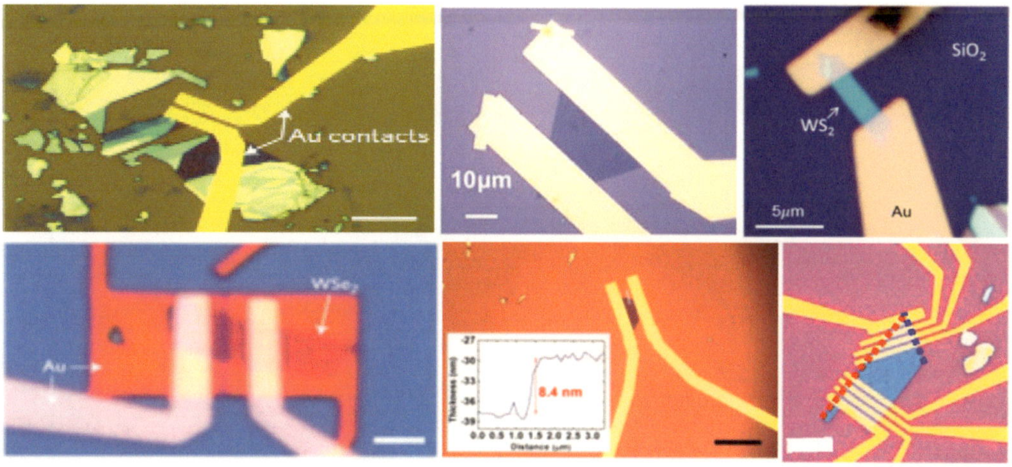

Figure 6. Optical and AFM images of different 2D devices gated using a degenerately doped silicon substrate with two gold contacts. Adapted from Lopez-Sanchez et al. [23], Xia et al. [101], Liu et al. [18], Baugher et al. [19], Huang et al. [20], Liu et al. [112].

of 10 mA/W to date in pure graphene photo-detector in visible region. Compared to classical direct band gap semiconductors TMDCs are more advantages due to its transparency, easy processing, and flexibility has ability to detect light at different wavelengths by tuning their optical gap with various number of layers.

Graphene cannot be used as a competent photo-detector, with great dark current & small absorption and photoresponsivity. MoS_2 is a representative member of the transition metal dichalcogenide (TMDC) family that consists of two-dimensional material planar sheets that stack on each other together using weak van der Waals interlayer interactions. MoS_2 is a said to replace the graphene having promising alternative properties to enhance photocurrent response because it has a band gap of 1.8 eV and high mobility of above 100 cm^2/VS [95]. MoS_2 based photodetectors are of two kinds; (a) direct band gap on single-layer MoS_2 (b) indirect band gap on few-layer MoS_2. In 2011 the first photo-detector is based on single layer MoS_2 was fabricated and developed [32], having a photoresponse of 7.5 mA/W, response time of 50 ms which can measure 750 nm light that is comparable to graphene based photo-detectors. As in **Figure 7** in ambient condition Lopez-Sanchez with his group [24] developed a device based on monolayer MoS_2 yielding a highest photoresponsivity of 880 A/W and on the other hand in high vacuum condition a CVD monolayer MoS_2 based photo-detector arrive a photoresponse and photogain limiting to 2200 and 5000 A/W respectively [96]. This implies that the environment has a great influence on both the electronic and the optoelectronic properties. Beside single layer MoS_2, few-layer MoS_2 photodetectors have unique properties, such as large absorption, band gap reduction and lower photoresponsivity than single layer photodetectors. The reduction in responsivity is caused by the indirect band gap and the trap states dominate the photodetection mechanism in few layer MoS_2. In 2012 W. Choi et al. [20] fabricated a photo-detector based on multilayer MoS_2 having a wide ($\sim$900 nm) spectral response with a high photoresponse (>100 mA/W) and photo-detector based on a double layer

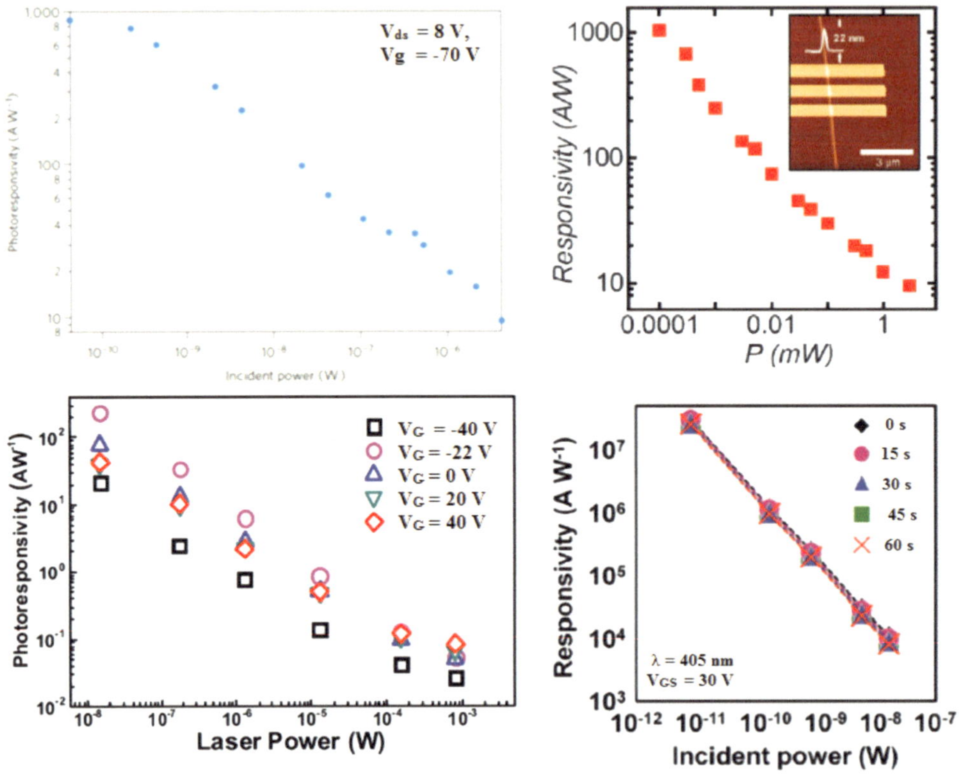

Figure 7. Photoresponsivity of 2D TMDC devices under illumination wavelength. The TMDC devices shows an increasing response of the photodetectors used for a high broad range of wavelengths. Adapted from Lopez-Sanchez et al. [23], Ko et al. [21], Shim et al. [113], Buscema et al. [66].

MoS_2 device giving a faster photoresponse of (~110 μs; decay ~70 μs; rise) 570 mA/W of photoresponsivity of 10 V and had high thermal stability In a different study top-gated photodetectors depending on three different layers like single: double: triple 2D MoS_2 [33–35] are compared and there outcomes indicate that the triple-layer device exhibits a very improved detection with red source, and the others like double-layers and single layer were more valuable for an energy of higher green detection. Since decade's significant advances have been achieved to synthesize large continues films or triangular flakes or hexagonal flakes on different substrates [57, 97–99]. Like MoS_2 the properties of monolayer $MoSe_2$ exhibits a direct band gap of 1.6 eV and changes into indirect band gap of 1.1 eV for bulk or multilayer $MoSe_2$. $MoSe_2$ Photodetectors based on single/multilayer devices shows a remarkable light response, speed response (15 ms–8 s) remains too low for practical applications [100, 101].

The CVD grown $MoSe_2$ Photodetectors exhibit high photoresponsivity between 0.26 and 13 mA/W [101] and Duan et al. [102] demonstrated the epitaxial growth of $MoSe_2/MoS_2$ heterojunctions which showed a pronounced photoresponse characteristics. The exfoliated flake based photodetector shows a responsivity of 10^6 was transferred on Ti electrodes up to 97.1 A/W and the heterojunction photodetector like Gr/$MoSe_2$/Si display a photoresponse in the wide range of 350–1310 nm having an extreme fast response speed of 270 ns. The $MoSe_2$ devices show a fast response speed within a few tens of milliseconds for high performance optoelectronic devices. Among the 2D materials the tungsten disulfate (WS_2) has been investigated for its use in photodetectors under the influence of gaseous environment. Recently Huo and his group [81, 103] carried the measurements under low excitation powers and the photoresponsivity of the devices ranges from 13 A/W in vacuum to 884 A/W in NH_3 atmosphere. The WS_2 photodetector has good improvement in the responsively and sensing performances. Besides WS_2, tungsten diselenide (WSe_2) can also be used as sensing material for photodetectors and a high responsively of 180 and 30 A/W is reported on the influence of metal electrodes on photoresponse [104, 105].

The $MoTe_2$ 2D material shows the electrical properties of a phototransistor based on mechanically exfoliated few layer $MoTe_2$ device that reached a responsivity of 2560 A/W [106]. Xu and coworkers successfully isolated the single and few layer HfS_2 flakes by mechanical exfoliation that has an indirect band gap of 2 eV shows a responsivity of $\sim$ 890 A/W [107]. $ReSe_2$ based photodevices on single layer reports an extraordinary responsivity of 95 A/W with a fast response times on the order of tens of ms. ReS_2 has a direct band gap of 1.5 eV and the CVD grown and exfoliated few layer ReS_2 shows a responsivities ranging from 16.14 A/W to 8.86×10^5 A/W [108] with a response time of several ms to hundreds of s. Few layer ReS_2 based transistors under the oxygen plasma treatment can advance the optoelectronic properties with a maximum responsivity of 16 A/W, 88,600 A/W & 2.5×10^7 A/W [109–112] and a response time of several tens of seconds. The GaSe based devices show a photoresponse of 2.8 A/W and a response time of 0.02 s while GaS based devices show responsivities of up to 4.2 A/W and 19.2 A/W and the devices with GaTe exhibit increased responsivities between 274.3 A/W and 10^4 A/W having a response speed down to several ms [113]. The GaSe and GaS has photoresponsivity of 4 A/W on flexible substrates in the UV Blue region and the direct band gap GaTe and In_2Se_3 exhibits large responsivities of 1×10^4 A/W at short light illumination and response time varies from ms to s [114–116]. A thicker InSe flake has a great photoresponse of 160 A/W and a fast response speed of few seconds that indicates a great influence of trap states. SnS_2 flakes has a great attention used in photodetectors & transistors synthesized by the exfoliation method and CVD grown flakes can reach up to a high responsivity of 8 mA/W with a response time of 5 ms [117]. To enhance the performance of photodetector devices at faster response times in the absence of gain and responsivity on the order of mA/W can achieve the highest sensitivity ($D^* \geq 7.7 \times 10^{11}$ jones) in TMDC photodetectors [17]. Combining the critical parameters presented in table [17] like photoresponse time, Detectivity and EQE a high speed and highly sensitive photodetector could be realized by stacking of TMDCs and other two dimensional materials. The photodetectors on different semiconducting TMDCs beyond MoWSeS (MoS_2:R_{ph} = 0.57 A/W, GaS: R_{ph} = 19 A/W, GaSe: R_{ph} = 2.8 A/W, In_2Se_3: R_{ph} = 3.95 A/W) materials have their use in the flexible electronics of noble metals (PdS_2, PtS_2…) and the early transition metals (e.g. TiS_2, ZrS_2, HfS_2…) [118–122]. The growth of the devices between their in-plane and out-of-plane modes have recently been discovered as nanodiscs. These scientific discoveries are a frontline portion of the technological evolving field and these 2D materials beyond graphene are blooming with new devices based on MoWSeS materials, and investigation on the layered TMDCs research is still in early stages.

6. Conclusions

This chapter is discussed on the Photodetectors based on 2D Transition metal Dichalcogenides (TMDCs) that are rapidly established themselves as fascinating blocks for optoelectronics focusing on the photoresponsivity. After the graphene discovery in 2004 thin film TMDCs are used in the demonstration of the novel nanoelectronic and optoelectronic devices that are becoming more eminent. The topdown & bottomup synthesis routes including Chemical vapor deposition, hydrothermal method, electrical and mechanical exfoliation methods are the promising methods to improve photocurrent response and discussion is carried on 2D TMDC materials for a strategy to obtain large area with controlled thickness. The 2D TMDCs with wide range of material properties can exhibit an ironic physical behavior extending from an insulator to a narrow gap semiconductor to a semi metal or metal. The 2D TMDC semiconductors can have strong interaction with light and enhance the photon absorption with the number of layers and evident with the band gap structures to create electron hole creation in the merit of photodetection. The large band gap of TMDCs has higher carrier lifetimes allowing them as a promising candidate for high sensitivity photodetectors. The direct band gap semiconductors are suitable for application in optoelectronic devices generating electron-hole pair with high absorption coefficient for effective photoresponse characteristics. The key principle in the photo detection is explained and several different mechanism have been discussed to enhance the photodetection efficiency. The theoretical key parameters of photo performance is well explained to improve the light-matter interaction. The phototransistor device preparation is shown and the Phototransistor devices based on TMDCs monolayer/multilayers show and report the device photoresponsivity reaching 880–2570 A/W and having a high photodetectivity under the light intensity. The improvement of TMDC photodetectors performance like the parameter photoresponsivity is explained clearly in all the different TMDC materials which is favorable for photodetection current that can greatly enhance the photoresponse speed. The predictions of 2D materials in the commercialization depends on the ability to offer at large scale depend not only on their performance but also on the ability low cost integration to the existing photonic and electronic devices.

Acknowledgements

The authors would like to thank the financial support from National Research Foundation of Korea (grant NRF – **2017035339**).

Author details

Inturu Omkaram*, Young Ki Hong and Sunkook Kim*

*Address all correspondence to: inturuomkar@gmail.com and kimskcnt@gmail.com

School of Advanced Materials Sciences and Engineering, Sungkyunkwan University, Gyeong-gi, South Korea

References

[1] Huang X, Yin ZY, Wu SX, Qi XY, He QY, Zhang QC, Yan QY, Boey F, Zhang H. Graphene-based materials: Synthesis, characterization, properties and applications. Small. 2011;**7**:1876-1902

[2] Sze SM, Ng KK. Physics of Semiconductor Devices. 3rd ed. United States of America: Wiley; 2007

[3] Kolobov AV, Tominaga J. Two-Dimensional Transition-Metal Dichalcogenides. Switzerland: Springer; 2016. p. 239

[4] Jiang HJ. Chemical preparation of graphene-based nanomaterials and their applications in chemical and biological sensors. Small. 2011;**7**:2413-2427

[5] Koppens FHL, Mueller T, Avouris P, et al. Photodetectors based on graphene, other two-dimensional materials and hybrid systems. Nature Nanotechnology. 2014;**9**(10):780-787

[6] Neto AC, Guinea F, Peres N, Novoselov KS, Geim AK. The electronic properties of graphene. Reviews of Modern Physics. 2009;**81**:109-162

[7] Mak KF, Lee C, Hone J, Shan J, Heinz TF. Atomically thin MoS_2: A new direct-gap semiconductor. Physical Review Letters. 2010;**105**:136805 (1-4)

[8] Kourosh KZ, Jian ZO, Torben D, Michael SS, Martin P, Sally LG. Two-dimensional transition metal dichalcogenides in biosystems. Advanced Functional Materials. 2015; **25**:5086-5099

[9] Wang QH, Kalantar-Zadeh K, Kis A, Coleman JN, Strano MS. Electronics and optoelectronics of two-dimensional transition metal dichalcogenides. Nature Nanotechnology. 2012;**7**:699-712

[10] Bultler SZ, Hollen SM, Cao L, et al. Progress, challenges and opportunities in two-dimensional materials beyond graphene. ACS Nano. 2013;**7**(4):2898-2926

[11] Wang H, Yu L, Lee YH, Shi Y, Hsu A, Chin ML, Li L-J, Dubey M, Kong J, Palacios T. Integrated circuits based on bilayer MoS_2 transistors. Nano Letters. 2012;**12**:4674-4680

[12] Novoselov KS, Castro Neto AH. Two-dimensional crystals-based heterostructures: Materials with tailored properties. Physica Scripta. 2012;**T146**:014006 (6pp)

[13] Zhang W, Zhang P, Su Z, Wei G. Synthesis and sensor applications of MoS_2-based nanocomposites. Nanoscale. 2015;**7**:18364-18378

[14] Shi Y, Li H, Li LJ. Recent advances in controlled synthesis of two dimensional transition metal dichalcogenides via vapour deposition techniques. Chemical Society Reviews. 2015;**44**:2744-2756

[15] Lee JY, Shin J-H, Lee G-H, Lee C-H. Two-dimensional semiconductor optoelectronics based on van der Waals heterostructures. Nanomaterials. 2016;**6**:193-211

[16] Zheng L, Zhongzhu L, Guozhen S. Photodetectors based on two dimensional materials. Journal of Semiconductors. 2016;**37**:091001 (1-11)

[17] Xie C, Mak C, Tao X, Yan F. Photodetectors based on two-dimensional layered materials beyond graphene. Advanced Functional Materials. 2017;**27**:1603886-1603927

[18] Liu X, Hu J, Yue C, Fera ND, Ling Y, Mao Z, Wei J. High performance field-effect transistor based on multilayer tungsten disulfide. ACS Nano. 2014;**8**:10396-10402

[19] Baugher BWH, Churchill HOH, Yang Y, Jarillo-Herrero P. Optoelectronic devices based on electrically tunable p–n diodes in a monolayer dichalcogenide. Nature Nanotechnology. 2014;**9**:262-267

[20] Huang H, Wang J, Hu W, Liao L, Wang P, Wang X, Gong F, Chen Y, Wu G, Luo W. Highly sensitive visible to infrared $MoTe_2$ photodetectors enhanced by the photogating effect. Nanotechnology. 2016;**27**:445201-445208

[21] Ko PJ, Abderrahmane A, Kim N-H, Sandhu A. High-performance near-infrared photodetector based on nano-layered $MoSe_2$. Semiconductor Science and Technology. 2017;**36**: 065015-065022 (8pp)

[22] Jiang FK, Lee GB, Tai YC, et al. A flexible micromachine-based shear-stress sensor array and its application to separation-point detection. Sensors and Actuators A: Physical. 2000;**79**:194-203

[23] Lopez-Sanchez O, Lembke D, Kayci M, Radenovic A, Kis A. Ultrasensitive photodetectors based on monolayer MoS_2. Nature Nanotechnology. 2013;**8**:497-501

[24] Jo S, Ubrig N, Berger H, Kuzmenko AB, Morpurgo AF. Mono- and bilayer WS_2 light-emitting transistors. Nano Letters. 2014;**14**:2019-2025

[25] Bae S, Kim H, Lee Y, Xu XF, Park JS, Zheng Y, Balakrishnan J, Lei T, Kim HR, Song YI, Kim YJ, Kim KS, Ozyilmaz B, Ahn JH, Hong BH, Iijima S. Roll-to-roll production of 30-inch graphene films for transparent electrodes. Nature Nanotechnology. 2010;**5**:574-578

[26] Bonaccorso F, Sun Z, Hasan T, Ferrari AC. Graphene photonics and optoelectronics. Nature Photonics. 2010;**4**:611-622

[27] Splendiani A et al. Emerging photoluminescence in monolayer MoS_2. Nano Letters. 2010;**10**:1271-1275

[28] Sze S. Semiconductor Devices, Physics and Technology. 2nd ed. New York, USA: John Wiley & Sons; 2002

[29] Tran QT, Le TD, Subramanian R, Nae-Eung L. Transparent, stretchable, and rapid-response humidity sensor for body-attachable wearable electronics. Nano Research. 2017;**10**:2021-2033

[30] Wu C, Kim TW, Li F, Guo T. Wearable electricity generators fabricated utilizing transparent electronic textiles based on polyester/Ag nanowires/graphene core–shell nanocomposites. ACS Nano. 2016;**10**(7):6449-6457

[31] Wang X, Yusheng C, Xinlong W, Lei L, Xinhua O, Ziyi G, Hui H. MoS_2 quantum dots with a tunable work function for high-performance organic solar cells. ACS Applied Materials & Interfaces. 2016;**8**(40):26916-26923

[32] Yin Z et al. Single-layer MoS_2 phototransistors. ACS Nano. 2011;**6**:74-80

[33] Kwon J, Hong YK, Han G, Omkaram I, Choi W, Kim S, Yoon Y. Giant photoamplification in indirect-bandgap multilayer MoS_2 phototransistors with local bottom-gate structures. Advanced Materials. 2015;**27**:2224-2230

[34] Lee HS, Min S-W, Chang Y-G, Park MK, Nam T, Kim H, Kim JH, Ryu S, Im S. MoS_2 nanosheet phototransistors with thickness-modulated optical energy gap. Nano Letters. 2012;**12**:3695-3700

[35] Choi W, Cho MY, Konar A, Lee JH, Cha G-B, Hong SC, Kim S, Kim J, Jena D, Joo J, Kim S. High-detectivity multilayer MoS_2 phototransistors with spectral response from ultraviolet to infrared. Advanced Materials. 2012;**24**:5832-5836

[36] Britnell L, Ribeiro R, Eckmann A, Jalil R, Belle B, Mishchenko A, Kim Y-J, Gorbachev R, Georgiou T, Morozov S. Strong light-matter interactions in heterostructures of atomically thin films. Science. 2013;**340**:1311-1314

[37] Geim AK, Novoselov KS. The rise of graphene. Nature Materials. 2007;**6**:183-191

[38] Zheng J, Zhang H, Dong S, Liu Y, Tai Nai C, Suk Shin H, et al. High yield exfoliation of two-dimensional chalcogenides using sodium naphthalenide. Nature Communications. 2014;**5**:2995 (1-7)

[39] Guo X, Wang Y, Wu F, Ni Y, Kokot S. A colorimetric method of analysis for trace amounts of hydrogen peroxide with the use of the nano-properties of molybdenum disulfide. The Analyst. 2015;**140**:1119-1126

[40] Lee YH, Zhang XQ, Zhang W, Chang MT, Lin CT, Chang KD, et al. Synthesis of large-area MoS_2 atomic layers with chemical vapor deposition. Advanced Materials. 2012;**24**: 2320e5

[41] Wilson JA, Yoffe AD. Transition metal dichalcogenides discussion and interpretation of observed optical, electrical, and structural properties. Advances in Physics. 1969;**18**:193-335

[42] Novoselov KS, Jiang D, Schedin F, Booth TJ, Khotkevich VV, Morozov SV, et al. Two-dimensional atomic crystals. Proceedings of the National Academy of Sciences of the United States of America. 2005;**102**:10451-10453

[43] Chalcogenides fill the gap, Nature Materials. 2014;**13**:1073-1073

[44] Xia F, Wang H, Xiao D, Dubey M, Ramasubramaniam A. Two-dimensional material nano-photonics. Nature Photonics. 2014;**8**:899-907

[45] Radisavljevic B, Kis A. Mobility engineering and a metal–insulator transition in mono-layer MoS_2. Nature Materials. 2013;**12**:815-820

[46] Novoselov KS, Geim AK, Morozov SV, Jiang D, Zhang Y, Dubonos SV, Grigorieva IV, Firsov AA. Electric field effect in atomically thin carbon films. Science. 2004;**306**(5696):666-669

[47] Tang Q, Zhou Z. Graphene-analogous low-dimensional materials. Progress in Materials Science. 2013;**58**:1244-1315

[48] Fuhrer MS. Graphene: Ribbons piece-by-piece. Nature Materials. 2010;**9**:611-612

[49] Giovannetti G, Khomyakov PA, Brocks G, Kelly PJ, van den Brink J. Substrate-induced band gap in graphene on hexagonal boron nitride: Ab initio density functional calculations. Physical Review B. 2007;**76**:073103

[50] Jariwala D, Sangwan VK, Lauhon LJ, Marks TJ, Hersam MC. Emerging device applications for semiconducting two-dimensional transition metal dichalcogenides. ACS Nano. 2014;**8**:1102-1120

[51] Kuc A, Zibouche N, Heine T. Influence of quantum confinement on the electronic structure of the transition metal sulfide TS_2. Physical Review B. 2011;**83**:245213

[52] Kappera R. Electrical properties and phase engineering of two dimensional MoS_2 [thesis]; United States: The state University of New Jersey; 2015

[53] Tonndorf P, Schmidt R, Böttger P, Zhang X, Börner J, Liebig A, Albrecht M, Kloc C, Gordan O, Zahn DRT, et al. Photoluminescence emission and Raman response of monolayer MoS_2, $MoSe_2$, and WSe_2. Optics Express. 2013;**21**:4908-4916

[54] Andreas P, Thomas M. Optoelectronic devices based on atomically thin transition metal dichalcogenides. Applied Sciences. 2016;**6**:78. DOI: 10.3390/app6030078

[55] Chernikov A, van der Zande AM, Hill HM, Rigosi AF, Velauthapillai A, Hone J, Heinz TF. Electrical tuning of exciton binding energies in monolayer WS_2. Physical Review Letters. 2015;**115**:126802

[56] Geim AK, Grigorieva IV. Van der Waals heterostructures. Nature. 2013;**499**:419-425

[57] Rhyee J-S, Kwon J, Dak P, Kim J-H, Kim SM, Park J, Hong YK, Song W, Omkaram I, Alam MA. High mobility transistors based on large area and highly crystalline CVD grown $MoSe_2$ films on insulating substrates. Advanced Materials. 2016;**28**:2316-2321

[58] Sze SM, Ng KK. Physics of Semiconductor Devices. Hoboken, NJ: John Wiley & Sons, Inc.; 2006

[59] Koppens FHL, Mueller T, Avouris P, Ferrari AC, Vitiello MS, Polini M. Photodetectors based on graphene, other two-dimensional materials and hybrid systems. Nature Nanotechnology. 2014;**9**:2571-2574

[60] Lee EJH, Balasubramanian K, Weitz RT, Burghard M, Kern K. Contact and edge effects in graphene devices. Nature Nanotechnology. 2008;**3**:486-490

[61] Lou Z, Liang Z, Shen G. Photodetectors based on two dimensional materials. Journal of Semiconductors. 2016;**37**:9

[62] Freitag M, Low T, Xia F, Avouris P. Photoconductivity of biased graphene. Nature Photonics. 2013;**7**:53-59

[63] Saleh BEA, Teich MC, Saleh BE. Fundamentals of Photonics. New York: Wiley; 1991

[64] Rappaport P. The photovoltaic effect and its utilization. Solar Energy. 1959;**3**:8-18

[65] Furchi MM, Polyushkin DK, Pospischil A, Mueller T. Mechanisms of photoconductivity in atomically thin MoS_2. Nano Letters. 2014;**14**(11):6165-6170

[66] Buscema M, Barkelid M, Zwiller V, van der Zant HSJ, Steele GA, Castellanos-Gomez A. Large and tunable photothermoelectric effect in single-layer MoS_2. Nano Letters. 2013;**13**:358-363

[67] Wei P, Bao WZ, Pu Y, Lau CN, Shi J. Anomalous thermoelectric transport of Dirac particles in graphene. Physical Review Letters. 2009;**102**:166808

[68] Zuev YM, Chang W, Kim P. Thermoelectric and magnetothermoelectric transport measurements of graphene. Physical Review Letters. 2009;**102**:096807

[69] Li Z, Bae MH, Pop E. Substrate-supported thermometry platform for nanomaterials like graphene, nanotubes and nanowires. Applied Physics Letters. 2014;**105**(2):023107

[70] Wu J, Schmidt H, Amara KK, et al. Large thermoelectricity via variable range hopping in chemical vapour deposition grown single-layer MoS_2. Nano Letters. 2014;**14**(5):2730-2734

[71] Bhushan B. Springer Handbook of Nanotechnology. New York: Springer; 2007

[72] Nolas GS, Sharp J, Goldsmid J. Thermoelectrics: Basic Principles and New Materials Developments. New York: Springer; 2001

[73] Jyh MW, Wei EC. Ultrahigh responsivity and external quantum efficiency of an ultraviolet-light photodetector based on a single VO_2 microwire. ACS Applied Materials & Interfaces. 2014;**6**:14286-14292

[74] Choi MS, Qu D, Lee D, Liu X, Watanable K, Taniguchi T, Yoo WJ. Lateral MoS_2 p–n junction formed by chemical doping for use in high-performance optoelectronics. ACS Nano. 2014;**8**:9332-9340

[75] Mao J, Yu Y, Wang L, Zhang X, Wang Y, Shao Z, Jie J. Ultrafast, broadband photodetector based on $MoSe_2$/silicon heterojunction with vertically standing layered structure using graphene as transparent electrode. Advancement of Science. 2016;**3**:1600018

[76] Schroder DK. Semiconductor Material and Device Characterization. 3rd ed. New Jersey: John Wiley & Sons; 2006

[77] Jha AR. Infrared Technology. New York: Wiley; 2000. pp. 245-359

[78] Bhattacharya P. Semiconductor Optoelectronics Device. Upper Saddle River, NJ: Prentice-Hall; 1997. pp. 345-367

[79] Bube RH. Photoconductivity of Solids. New York: John Wiley; 1960. p. 1

[80] Jinhua L, Liyong N, Zijian Z, Feng Y. Photosensitive graphene transistors. Advanced Materials. 2014;**26**:5239-5273

[81] Perea-López N, Elías AL, Berkdemir A, Castro-Beltran A, Gutiérrez HR, Feng S, Lv R, Hayashi T, López-Urías F, Ghosh S, Muchharla B, Talapatra S, Tereones H, Terrones M. Photosensor device based on few-layered WS_2 films. Advanced Functional Materials. 2013; **23**(44):5511-5517

[82] Radisavljevic B, Radenovic A, Brivio J, Giacometti V, Kis A. Single-layer MoS_2 transistors. Nature Nanotechnology. 2011;**6**:147-150

[83] Riikonen J, Tuomi T, Lankinen A, Sormunen J, Saynatjoki A, Knuuttila L, Lipsanen H, McNally PJ, O'Reilly L, Danilewsky A, Sipila H, Vaijarvi S, Lumb D, Owens A. Synchrotron X-ray topography study of defects in indium antimonide P-I-N structures grown by

metal organic vapour phase epitaxy. Journal of Materials Science: Materials in Electronics. 2005;**116**:449-453

[84] Bolotin KI, Sikes KJ, Jiang Z, Klima M, Fudenberg G, Hone J, Kim P, Stormer HL. Ultrahigh electron mobility in suspended graphene. Solid State Communications. 2008; **146**:351-355

[85] Wang X, Gong Y, Shi G, Chow WL, Keyshar K, Ye G, Vajtai R, Lou J, Liu Z, Ringe E. Chemical vapor deposition growth of crystalline monolayer $MoSe_2$. ACS Nano. 2014; **8**:5125-5131

[86] Kumar N, Cui Q, Ceballous F, He D, Wang Y, Zhao H. Exciton diffusion in monolayer and bulk $MoSe_2$. Nanoscale. 2014;**6**:4915-4919

[87] Dean CR, Young AF, Meric I, Lee C, Wang L, Sorgenfrei S, Watanabe K, Taniguchi T, Kim P, Shepard K, et al. Boron nitride substrates for high-quality graphene electronics. Nature Nanotechnology. 2010;**5**:722-726

[88] Lee G-H, Yu Y-J, Cui X, Petrone N, Lee C-H, Choi MS, Lee D-Y, Lee C, Yoo WJ, Watanabe K, et al. Flexible and transparent MoS_2 field-effect transistors on hexagonal boron nitride-graphene heterostructures. ACS Nano. 2013;**7**:7931-7936

[89] Mekki A, Dere A, Mensah-Darkwa K, Al-Ghamdi A, Gupta RK, Harrabi K, Farooq WA, El-Tantawy F, Yakuphanoglu F. Graphene controlled organic photodetectors. Synthetic Metals. 2016;**217**:43-56

[90] Liu E, Long M, Zeng J, Luo W, Wang Y, Pan Y, Zhou W, Wang B, Hu W, Ni Z, You Y, Zhang X, Qin S, Shi Y, Watanable K, Taniguchi T, Yuan H, Hwang HY, Cui Y, Miao F, Xing D. High responsivity phototransistors based on few-layer ReS_2 for weak signal detection. Advanced Functional Materials. 2016;**26**:1938-1944

[91] Tosun M, Chuang S, Fang H, Sachid AB, Hettick M, Lin Y, Zeng Y, Javey A. High-gain inverters based on WSe_2 complementary field-effect transistors. ACS Nano. 2014;**8**:4948-4953

[92] Hwang WS, Remskar M, Yan R, Protasenko V, Tahy K, Chae SD, Zhao P, Konar A, Xing H, Seabaugh A, Jena D. Transistors with chemically synthesized layered semiconductor WS2 exhibiting 105 room temperature modulation and ambipolar behaviour. Applied Physics Letters. 2012;**101**:013107

[93] Larentis S, Fallahazad B, Tutuc E. Field-effect transistors and intrinsic mobility in ultra-thin $MoSe_2$ layers. Applied Physics Letters. 2012;**101**:223104

[94] Fathipour S, Ma N, Hwang WS, Protasenko V, Vishwanath S, Xing HG, Xu H, Jena D, Appenzeller J, Seabaugh A. Exfoliated multilayer $MoTe_2$ field-effect transistors. Applied Physics Letters. 2014;**105**:192101

[95] Ayari A, Cobas E, Ogundadegbe O, et al. Realization and electrical characterization of ultrathin crystals of layered transition-metal dichalcogenides. Journal of Applied Physics. 2007;**101**(1):14507

[96] Zhang W, Huang JK, Chen CH, et al. High-gain photo transistors based CVD MoS_2 monolayer. Advanced Materials. 2013;**25**(25):3456-3461

[97] Zhan Y, Liu Z, Najmaei S, Ajayan PM, Lou J. Large-area vapor-phase growth and characterization of MoS_2 atomic layers on a SiO_2 substrate. Small. 2012;**8**:966-971

[98] Liu KK, Zhang W, Lee YH, Lin YC, Chang MT, Su CY, Chang CS, Li H, Shi Y, Zhang H, Lai CS, Li LJ. Growth of large-area and highly crystalline MoS_2 thin layers on insulating substrates. Nano Letters. 2012;**12**:1538-1544

[99] Jung C, Kim SM, Moon H, Han J, Kwon J, Hong YK, Omkaram I, Yoon Y, Kim S, Park J. Highly crystalline CVD-grown multilayer $MoSe_2$ thin film transistor for fast photodetector. Scientific Reports. 2015;**5**:15313

[100] Chang Y-H, Zhang W, Zhu Y, Han Y, Pu J, Chang J-K, Hsu W-T, Huang J-K, Hsu C-L, Chiu M-H, Takenobu T, Li H, Wu C-I, Chang W-H, Wee ATS, Li L-J. Monolayer $MoSe_2$ grown by chemical vapor deposition for fast photodetection. ACS Nano. 2014;**8**:8582-8590

[101] Xia J, Huang X, Liu L-Z, Wang M, Wang L, Huang B, Zhu D-D, Li J-J, Gu C-Z, Meng X-M. CVD synthesis of large-area, highly crystalline $MoSe_2$ atomic layers on diverse substrates and application to photodetectors. Nanoscale. 2014;**6**:8949-8955

[102] Duan X, Wang C, Shaw JC, Cheng R, Chen Y, Li H, Wu X, Tang Y, Zhang Q, Pan A. Lateral epitaxial growth of two-dimensional layered semiconductor heterojunctions. Nature Nanotechnology. 2014;**9**:1024-1030

[103] Huo N, Yang S, Wei Z, et al. Photoresponsive and gas sensing field-effect transistors based on multilayer WS_2 nanoflakes. Scientific Reports. 2014;**4**:5209

[104] Zhang W, Chiu MH, Chen CH, et al. Role of metal contacts in high-performance phototransistors based on WSe_2 monolayers. ACS Nano. 2014;**8**(8):8653-8661

[105] Hu P, Wen Z, Wang L, Tan P, Xiao K. Synthesis of few-layer GaSe nanosheets for high performance photodetectors. ACS Nano. 2012;**6**:5988-5994

[106] Yin L, Zhan X, Xu K, Wang F, Wang Z, Huang Y, Wang Q, Jiang C, He J. Ultrahigh sensitive $MoTe_2$ phototransistors driven by carrier tunnelling. Applied Physics Letters. 2016;**108**:043503

[107] Xu K, Wang Z, Wang F, Huang Y, Wang F, Yin L, Jiang C, He J. Ultrasensitive phototransistors based on few-layered HfS_2. Advanced Materials. 2015;**27**:7881-7887

[108] Yang S, Tongay S, Li Y, Yue Q, Xia J-B, Li S-S, Li J, Wei S-H. Layer-dependent electrical and optoelectronic responses of $ReSe_2$ nanosheet transistors. Nanoscale. 2014;**6**:7226-7231

[109] Zhang E, Jin Y, Yuan X, Wang W, Zhang C, Tang L, Liu S, Zhou P, Hu W, Xiu F. ReS_2-based field-effect transistors and photodetectors. Advanced Functional Materials. 2015;**25**:4076-4082

[110] Liu F, Zheng S, He X, Chaturvedi A, He J, Chow WL, Mion TR, Wang X, Zhou J, Fu Q, Fan HJ, Tay BK, Song L, He R-H, Kloc C, Ajayan PM, Liu Z. Highly sensitive detection of polarized light using anisotropic 2D ReS_2. Advanced Functional Materials. 2016;**26**: 1169-1177

[111] Hafeez M, Gan L, Li H, Ma Y, Zhai T. Large-area bilayer ReS_2 film/multilayer ReS_2 flakes synthesized by chemical vapor deposition for high performance photodetectors. Advanced Functional Materials. 2016;**26**:4551-4560

[112] Liu E, Long M, Zeng J, Luo W, Wang Y, Pan Y, Zhou W, Wang B, Hu W, Ni Z, You Y, Zhang X, Qin S, Shi Y, Watanabe K, Taniguchi T, Yuan H, Hwang HY, Cui Y, Miao F, Xing D. High responsivity phototransistors based on few-layer ReS_2 for weak signal detection. Advanced Functional Materials. 2016;**26**:1938-1944

[113] Shim J, Oh A, Kang DH, Oh S, Jang SK, Jeon J, Jeon MH, Kim M, Choi C, Lee J, Lee S, Yeom GY, Song YJ, Park J-H. High-performance 2D rhenium disulfide (ReS_2) transistors and photodetectors by oxygen plasma treatment. Advanced Materials. 2016;**28**:6985-6992

[114] Hu P, Zhang J, Yoon M, Qiao X-F, Zhang X, Feng W, Tan P, Zheng W, Liu J, Wang X, Idrobo JC, Geohegan DB, Xiao K. Highly sensitive phototransistors based on two-dimensional GaTe nanosheets with direct band gap. Nano Research. 2014;**7**:694-703

[115] Liu F, Shimotani H, Shang H, Kanagasekaran T, Zólyomi V, Drummond N, Fal'ko VI, Tanigaki K. High-sensitivity photodetectors based on multilayer GaTe flakes. ACS Nano. 2014;**8**:752-760

[116] Hu PA, Wang LF, Yoon M, Zhang J, Feng W, Wang XN, Wen ZZ, Idrobo JC, Miyamoto Y, Geohegan DB, Xiao K. Highly responsive ultrathin GaS nanosheet photodetectors on rigid and flexible substrates. Nano Letters. 2013;**13**:1649-1654

[117] Tao Y, Wu X, Wang, et al. Flexible photodetector from ultraviolet to infrared based on SnS_2 nonosheet microsphere film. Journal of Materials Chemistry C. 2015;**3**(6):1347-1353

[118] Tao Y-R, Wu X-C, Xiong W-W. Flexible visible-light photodetectors with broad photoresponse based on ZrS_3 nanobelt films. Small. 2014;**10**:4905-4911

[119] Tao Y-R, Wu J-J, Wu X-C. Enhanced ultraviolet-visible light responses of phototransistors based on single and a few ZrS_3 nanobelts. Nanoscale. 2015;**7**:14292-14298

[120] Xiong W-W, Chen J-Q, Wu X-C, Zhu J-J. Individual HfS_3 nanobelt for field-effect transistor and high performance visible-light detector. Journal of Materials Chemistry C. 2014;**2**:7392-7395

[121] Xiong W-W, Chen J-Q, Wu X-C, Zhu J-J. Visible light detectors based on individual $ZrSe_3$ and $HfSe_3$ nanobelts. Journal of Materials Chemistry C. 2015;**3**:1929-1934

[122] Xie LM. Two-dimensional transition metal dichalcogenide alloys: Preparation, characterization and applications. Nanoscale. 2015;**7**:18392-18401

Chapter 3

Nanoscale III-V Semiconductor Photodetectors for High-Speed Optical Communications

Jack Jia-Sheng Huang, Yu-Heng Jan, H.S. Chang,
Chih-Jui Ni, Emin Chou, Shih-Kai Lee,
Horng-Shyang Chen and Jin-Wei Shi

Additional information is available at the end of the chapter

http://dx.doi.org/10.5772/intechopen.73054

Abstract

Nanophotonics involves the study of the behavior of light on nanometer scale. Modern nanoscale semiconductor photodetectors are important building blocks for high-speed optical communications. In this chapter, we review the state-of-the-art 2.5G, 10G, and 25G avalanche photodiodes (APDs) that are available in commercial applications. We discuss the key device parameters, including avalanche breakdown voltage, dark current, temperature dependence, bandwidth, and sensitivity. We also present reliability analysis on wear-out degradation and optical/electrical overload stress. We discuss the reliability challenges of nanoscale photodetectors associated with device miniaturization for the future. The reliability aspects in terms of high electric field, Joule heating, and geometry inhomogeneity are highlighted.

Keywords: semiconductor photodetectors, avalanche photodiodes, APD photodetectors, InGaAs/InAlAs APD, III-V photodetectors, nanophotonics, reliability, temperature dependence, device miniaturization

1. Introduction

Recently, two-dimensional (2D) materials have drawn great interest in the field of nanophotonics. The 2D material sometimes is referred to as single-layer material, composed of a single layer of atoms. One notable example of the single-layer material is graphene that was discovered by Nobel Laureates Andre Geim and Konstantin Novoselov in 2004 [1]. Since then, a vast amount of research has been directed at 2D materials because of their novel characteristics and potential use in the applications of photovoltaics, semiconductors, electrodes, and water purification.

Generally, 2D materials can be classified into two categories. The first is called 2D allotrope, a material consisting of single elements. Among them, the most widely studied 2D allotrope is graphene. Graphene is a 2D single-atom-thick material consisting of a monolayer of carbon atoms in a honeycomb array. It exhibits unusual electronic [2] and magnetic [3] properties such as high carrier mobility and ambipolar effect. Graphene is considered as a revolutionary material for future generation of high-speed electronic, radio frequency logic devices, thermally/electrically conductive reinforced composites, sensors, transparent electrodes, and so on [2–6].

The second category is a 2D compound composed of two or more covalently bonding elements [7]. Layered combinations of different 2D materials are generally called van der Waals heterostructures. However, the efficient integration of 2D functional layers with three-dimensional (3D) systems remains a significant challenge, limiting device performance and circuit design.

Since the first discovery of 2D graphene, there have been about 700 2D materials predicted to be stable, and many of them remain to be explored and synthesized [8]. The global market for 2D materials is expected to reach US$390 million within a decade, mostly for graphene in the semiconductor, electronics, battery energy, and composites markets [9–12].

2. Nanoscale photonic and electronic devices

Although 2D materials exhibit great technological potential, modern commercial photonic and electronic devices are mainly based on 3D nanoscale structures. Several 2D materials such as graphene photodetectors [13] and InAs quantum membranes [14] have been demonstrated in research laboratories with the promising performance. Nevertheless, efficient integration of 2D functional layers with 3D systems remains a significant challenge, hence limiting its commercial use. Moreover, high-volume manufacturing is still immature, and reliability issues remain unknown.

Three-dimensional nanostructures are the main building blocks of modern photonic and electronic components because of their proven technology and established field reliability. **Table 1** lists the smallest feature size of photonic and electronic devices that are commercially fabricated by using 3D nanostructures. For the electronics sector, integrated circuit (IC) has been deployed in vast majority of semiconductor products [15]. In the IC, the feature size refers to the smallest features on an IC where the feature size is usually gated by the length of the transistor channel [16]. Over the past few decades, the IC has followed Moore's law where the number of transistors has roughly doubled in about every 18 months [17–19]. The recent examples of IC's device miniaturization include the 32 nm technology used in Apple's A5/A6 chips for iPad Mini/iPhone 5, the 28 nm technology in Apple's A7 chip for iPad Air, the 20 nm technology in Apple's A8 chip for iPhone 6, the 16 nm technology in Apple's A9/A10 chips for iPhone 6S/iPhone 7, and the 10 nm in A11 chip for iPhone 8 [20, 21].

For the photonics sector, avalanche photodiodes (APDs) have followed similar device scaling with less restrictions in design rule and technology roadmap. The APD structure is typically composed of a thick absorption layer, a charge control layer, and a multiplication layer (M-layer) for avalanche device operation. Among the various layers of APD, the charge control

	Photonic sector: APD photodetector	Electronic sector: IC transistor
Recent technology since 2010	2.5G APD: • M-layer about 500 nm • Charge control layer~100 nm 10G APD: • M-layer about 160–200 nm • Charge control layer~50 nm 25G APD: • M-layer about 80–100 nm • Charge control layer~30 nm	32 nm node: • Transistor length~32 nm 28 nm node: • Transistor length~28 nm 20 nm node: • Transistor length~20 nm 16 nm node: • Transistor length~16 nm 10 nm node: • Transistor length~10 nm
Future technology	50G APD: • M-layer < 80 nm • Charge control layer < 30 nm 100G APD • M-layer < 50 nm • Charge control layer < 20 nm	7 nm node • Transistor length~7 nm 5 nm node • Transistor length~5 nm 3 nm node • Transistor length~3 nm

Table 1. List of minimum feature size of photonic and electronic devices based on 3D nanostructures. The devices that are commercially fabricated at the present and in the future are shown.

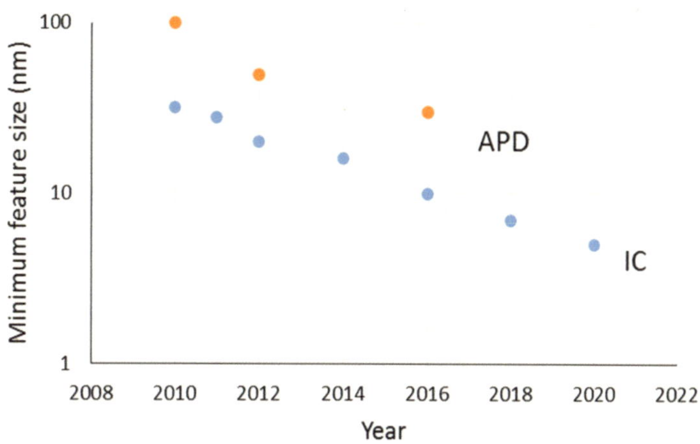

Figure 1. Evolution of minimum feature size for APD and IC. Device miniaturization occurs for both photonic and electronic devices to attain higher speed.

layer is denoted as the minimum feature size. Semiconductor photodetectors have evolved from 2.5G APD to 10G APD and recently advanced to 25G APD [22, 23]. Driven by 100G data-center demand, the development and manufacturing of 25G APD have been accelerated with rapid pace and growth [24–26]. Comparing the charge control layer, the feature size roughly reduced from about 100 nm in 2.5G APD to 50 nm in 10G APD and continued to shrink to about 30 nm in 25G APD.

Figure 1 shows the evolution of the minimum feature size for photonic and electronic devices where APD and IC are taken as the respective examples of photonics and electronics devices. The year indicates roughly the time of commercial manufacturing. For APD, the charge control layer scales with device speed in a similar fashion compared to IC. For IC, device scaling has rigorously followed Moore's law for decades.

3. Nanoscale III-V semiconductor photodetectors

Semiconductor photodiodes are important components for high sensitivity, low-noise receivers, and detectors deployed in the optical communication systems such as passive optical network (PON) [27]. Among the photodiode portfolio, APDs are attractive devices due to their significant improvement in photoreceiver sensitivity compared with traditional p-i-n (PIN) photodiodes [28]. By adding the multiplication layer, the avalanche photodiodes combine the detection and amplification properties simultaneously.

Recently, 10G and 25G APDs have drawn great interest in commercial and military applications due to their high bandwidth and low-noise performance advantages. In order to achieve high bandwidth, a mesa structure with coplanar P and N electrodes is typically employed [29–32]. For APD, there are two critical device parameters for the reverse bias operation. The first parameter is the avalanche breakdown voltage associated with the multiplication layer. The breakdown voltage is typically measured at reverse current of 10 μA. The avalanche breakdown has been formulated to study the impact ionization coefficients of electrons and holes [33–35]. The second parameter is the dark current that is typically measured at reverse bias below the breakdown voltage. Since these two parameters strongly influence the device performance of the APD, it is important to understand the temperature dependence of these two parameters. A good knowledge of temperature dependence is critical for the design of robust APD that can maintain stable performance when subject to temperature fluctuations. In Sections 3.1–3.4, we discuss the avalanche breakdown voltage, dark current, and temperature dependence in detail.

Table 2 summarizes the device cross-sectional schematics and key features of 2.5G APD, 10G APD, and 25G APD. There are several differences between 2.5G APD and 10G/25G APD. First, planar-type design is usually employed in 2.5G APD due to the advantages of processing simplicity and low cost. On the other hand, mesa-type structure has been incorporated in the high-speed APD such as 10G and 25G. Second, the p-InP region of 2.5G APD is typically formed by Zn diffusion in the intrinsic InP layer [35], while the p-InP of 10G and 25G APDs is grown by molecular beam epitaxy (MBE) to attain more precise control [23, 24].

For 2.5G APD, the device shown in Table II is composed of p-InP/i-InP/n-InP/i-InGaAs/n-InP. From top to bottom, the first layer is the p-InP formed by Zn diffusion. Guard ring is formed to avoid the edge breakdown at the p-InP diffused region. The intrinsic InP underneath the p-InP serves as the multiplication layer. Next, a thin layer (~100 nm) of n-type InP

	Device structure	Key feature
2.5G APD	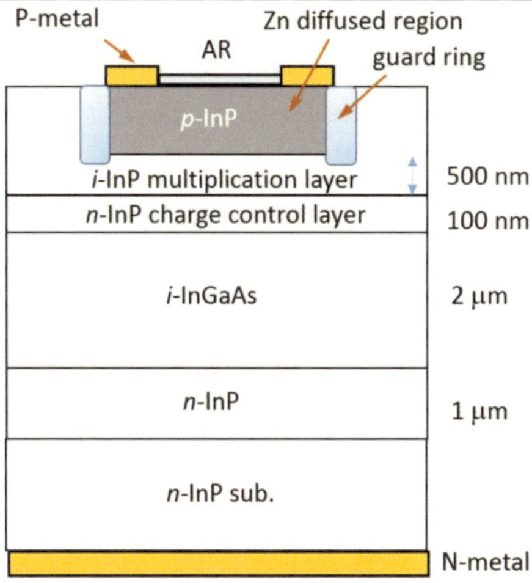	• Planar-type • Hole carrier for multiplication • InP multiplication layer (~500 nm) determined by Zn diffusion • Charge control layer (~100 nm) by epitaxial growth
10G APD	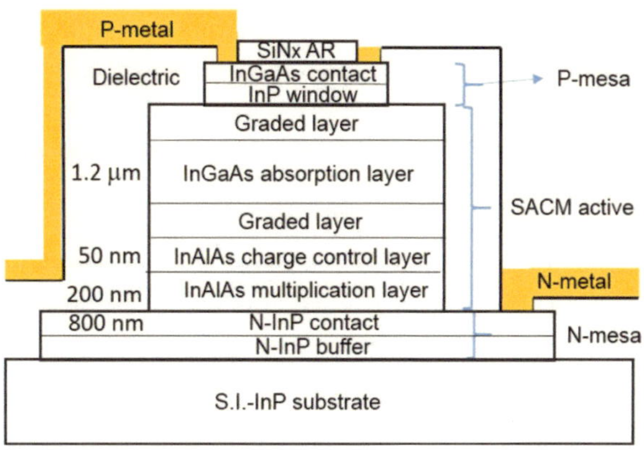	• Mesa-type • Electron carrier for multiplication • InAlAs multiplication layer (160–200 nm) grown by MBE • Charge control layer (50 nm) grown by MBE
25G APD	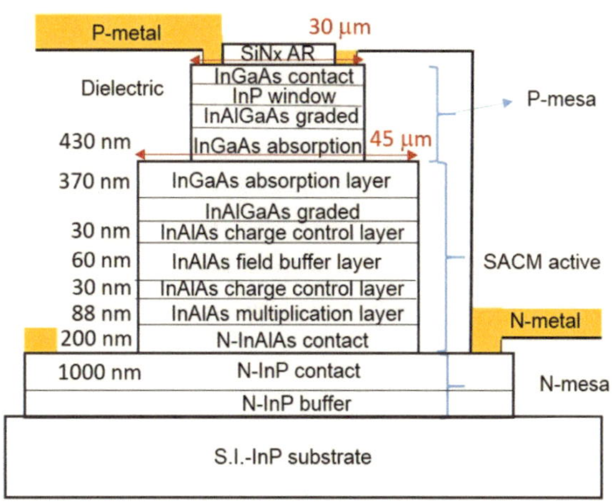	• Mesa-type • Dual charge layer • Electron carrier for multiplication • InAlAs multiplication layer (80–100 nm) grown by MBE • Charge control layer (30 nm) grown by MBE

Table 2. Structure cross-sectional schematics and key features of 2.5G APD, 10G APD, and 25G APD. The schematics are not drawn in scale [23, 24, 35].

is used as charge control. The intrinsic i-InGaAs of 2 μm in thickness is for absorption. Below the i-InGaAs, there are n-InP buffer and n-InP substrates.

For 10G APD, mesa-type structure with coplanar p- and n-metal contacts is employed to enhance the speed [23, 36]. The active region is sandwiched between the P-mesa at the top and the N-mesa at the bottom. The P-mesa consists of the p-InP window and p-InGaAs contact layers. The p-contact is usually made by forming a metal ring immediately outside the antireflective (AR) window. The p-ring is connected to the outside p-pad by the metal bridge. The active region consists of the InAlAs multiplication, InAlAs charge control, graded InGaAs/InAlAs, and InGaAs absorption layers. Such active structure is also called separate absorption, charge, and multiplication (SACM). The InGaAs absorption layer is undoped with a thickness of about 1200 nm. The InAlAs charge control layer is p-type doped (~$1.0–1.1\times10^{18}$ cm^{-3}). The InAlAs multiplication layer is undoped with a thickness of about 160 nm. The N-mesa, consisting of the N-InP buffer and contact layers, is grown on a semi-insulating (S.I.) InP substrate. The n-metal contact is connected to the N-mesa. For the passivation, the low-k dielectric material such as polyimide is used to reduce the capacitance.

For 25G APD, more sophisticated mesa structure such as dual charge layers has been designed to improve the bandwidth-gain product [22, 24]. Table II shows the conceptual cross-sectional views of a top-illuminated 25G APD device structure. The epitaxial layers are grown by molecular beam epitaxy (MBE). From top to down, it is composed of the p^+-InGaAs contact layer, p^+-InP window layer, p-type partially depleted InGaAs absorber, two p-type InAlAs charge layers, one intrinsic InAlAs field buffer layer, one intrinsic InAlAs multiplication (M-) layer, and n^+-InAlAs/InP contact layers. The partially depleted p-type absorber, which has a graded doping profile (top: 5×10^{19} to bottom: 1×10^{17} cm^{-3}) is used to shorten the hole transit time, accelerate the electron diffusion process, and increase the high-power and linearity performances [36]. The total thickness of InGaAs absorber is 0.8 μm and the ratio of depleted versus p-doped region is chosen to balance the RC delay and internal carrier transit/avalanche-delay time under low gain operation ($M_G < 5$) [24]. In order to shorten the avalanche delay time, a thin M-layer (around 90 nm) is chosen in our device structure [22, 36].

3.1. IV characteristics of APD

Figure 2 shows the typical reverse current-voltage (IV) curve of a 10G mesa-type SACM APD photodetector. On IV curve, there are two transitions along the reverse voltage at about 10 and 32 V, respectively. When the reverse voltage is applied to the device, the InAlAs multiplication layer begins to be depleted first. The first transition at 10 V corresponds to the punch-through voltage at which electric field depletes the InAlAs multiplication and *i*-InGaAs absorption layers [23, 37]. The punch-through voltage related to the absorption layer is denoted as V_{pt}. The second transition represents the avalanche breakdown voltage (V_{br}) for APD. The breakdown is typically determined by the InAlAs multiplication layer. By properly controlling the charge density of the charge control layer, sufficient electric field can be reached to achieve a good avalanche gain while keeping the tunneling and impact ionization away from the InGaAs absorption layer. In this case, the breakdown voltage taken at 10 μA is estimated to be 31.7 V. Between the two transitions, the reverse current is of technological importance for optical detection. The dark current is usually referred to the reverse current at 0.9 V_{br}. In this case, the dark current is estimated to be about 12.1 nA at 20°C.

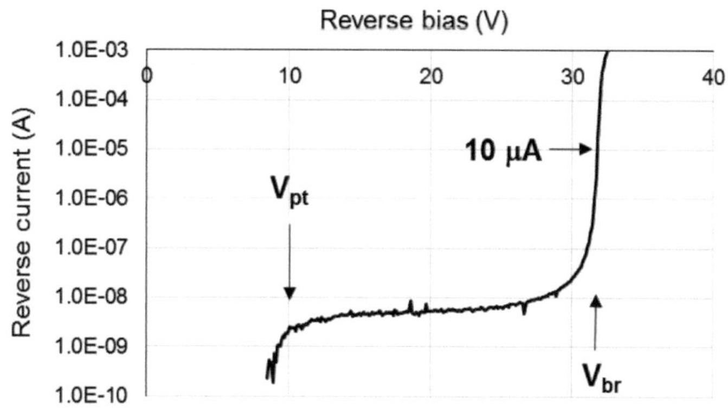

Figure 2. Typical IV characteristics of 10G InAlAs mesa-type APD. The IV is measured in reverse bias at 20°C [23].

Figure 3 shows the typical reverse IV of a 25G mesa-type SACM APD photodetector [38]. On the IV curve, there are two transitions along the reverse voltage that occur at 3 and 26 V, respectively. Again, the first transition at 3 V corresponds to V_{pt} at which the electric field depletes both the InAlAs multiplication and i-InGaAs absorption layers [23]. The second transition represents V_{br} determined by the InAlAs multiplication layer. The avalanche breakdown voltage taken at 10 μA is estimated to be about 26 V for this device. The typical range of avalanche breakdown voltage is 24–26 V. The dark current refers to the reverse current at 0.9 V_{br} is estimated to be about 13.4 nA at 25°C.

3.2. Avalanche breakdown mechanism

Figure 4 shows the schematic cross-section and the internal electric field profile of a SACM APD [39]. A top illumination is illustrated here where the absorption layer is next to the charge control and multiplication layers.

There are two major types of junction breakdown in semiconductor diodes. The first type is associated with the tunneling breakdown where the tunneling mechanism is dominant for the devices with lower breakdown voltage. The tunneling breakdown process exhibits a negative temperature coefficient [40, 41]. The second is the avalanche breakdown that is the dominant mechanism for large breakdown voltage, as illustrated in **Figure 5**. At very high electric field (~10^5–10^6 V/cm), some electrons within the diffusion distance near the depletion layer gain enough energy to create the secondary electron–hole pair by raising the electron from the valence band into the conduction band. This excitation process creates an electron-hole pair due to impact ionization. The electrons and holes created by the impact ionization are accelerated by high electric field. Consequently, the secondary electron-hole pair can create even more carriers, leading to a snowball avalanche effect. The avalanche breakdown process typically shows a positive temperature coefficient.

3.3. Temperature dependence

Figure 6 shows the reverse IV as a function of temperature for the mesa-type APD. The breakdown of APD is due to the avalanche mechanism since the device is operating at high electric field. At higher temperature, the avalanche breakdown voltage is expected to increase due to

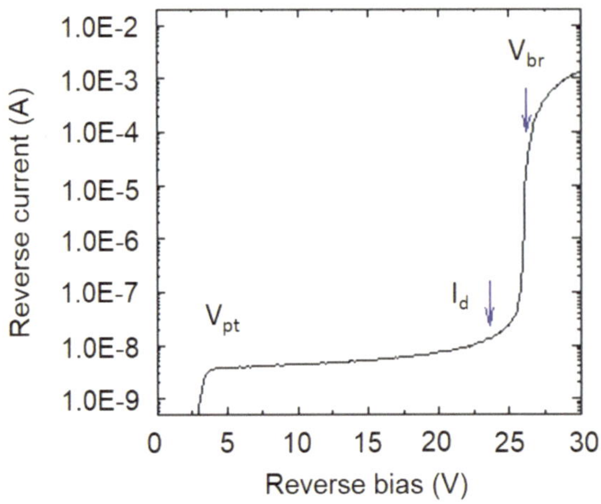

Figure 3. The reverse IV curve of the 25G mesa-type APD photodetector where the first transition is related to the punch-through (V_{pt}) and the second related to the avalanche breakdown (V_{br}) [38].

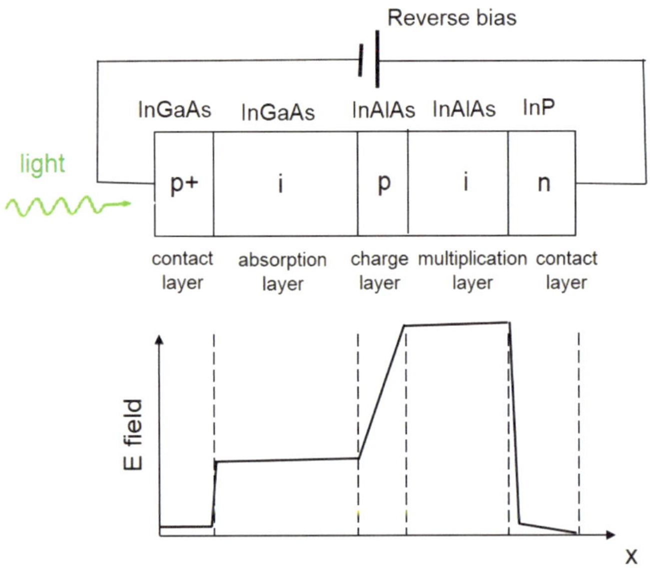

Figure 4. Schematic cross-section and internal electric field profile of the SACM APD structure. In the illustration, the InGaAs absorption layer is next to the InAlAs charge control and multiplication layers [39].

the effect of phonon [42]. As the temperature increases, the population of phonons increases. Thus, a higher electric field and applied voltage are required to reach breakdown in order to overcome the increased carrier cooling caused by phonon scattering [43].

The positive temperature coefficient of the APD has been experimentally found and theoretically described where the breakdown voltage of the Si APD followed a linear expression with temperature [44]. Here, it is assumed that the avalanche breakdown voltage increased linearly with increasing temperature as shown in Eq. (1).

$$V_{br}(T) = V_{br}(T_0)[1 + \alpha(T - T_0)] \quad (1)$$

where $V_{br}(T)$ is the avalanche breakdown voltage at temperature T, $V_{br}(T_0)$ is the avalanche breakdown voltage at reference temperature T_0, and α is the normalized temperature coefficient. For the sake of comparison, Eq. (1) can be rewritten as Eq. (2) where the second term shows the temperature coefficient of the breakdown voltage.

$$V_{br}(T) = V_{br}(T_0) + \alpha V_{br}(T_0)(T - T_0) \quad (2)$$

Figure 7 shows the avalanche breakdown voltage as a function of temperature based on two APD wafers with similar structures but from different processing runs. There are several

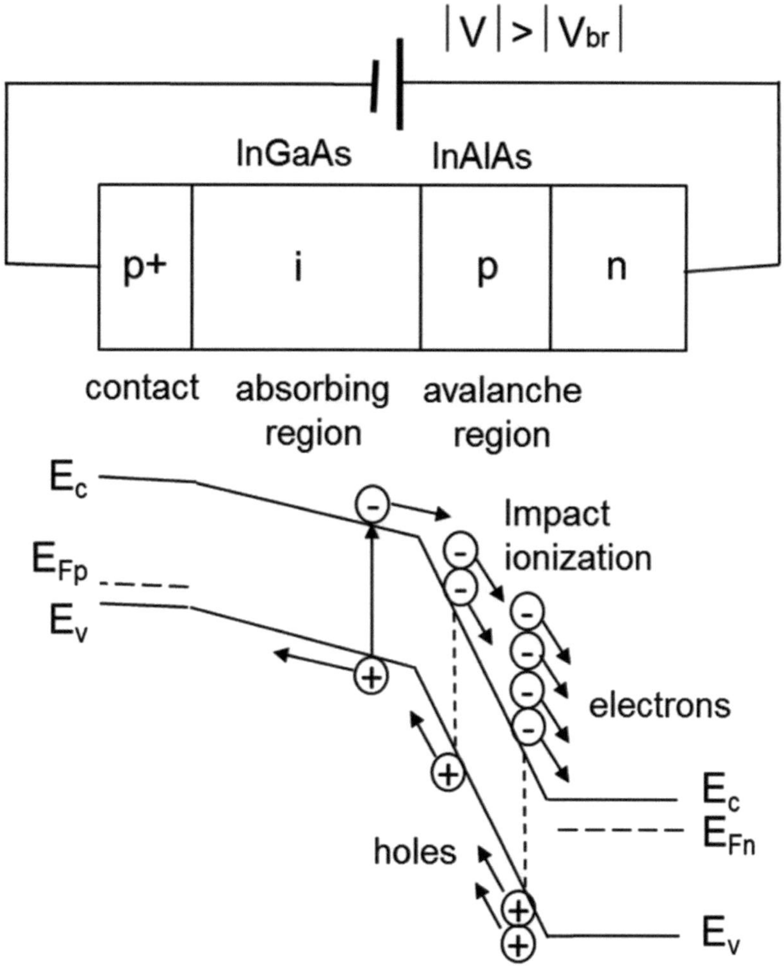

Figure 5. Schematics of avalanche breakdown mechanism of APD with simplified energy band diagram at the reverse voltage of$|V|>|V_{br}|$. The electron with high enough kinetic energy drifts to the avalanche region and gets accelerated to create a secondary electron-hole pair. The secondary electron-hole pair in turn generates another electron-hole pair, leading to snowball avalanche process [23].

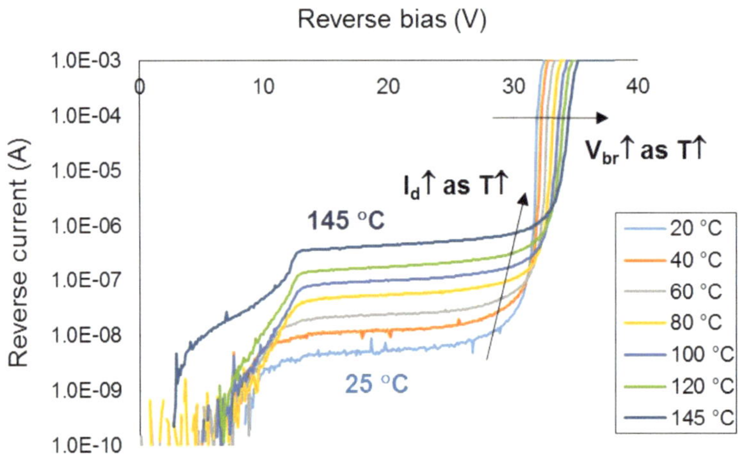

Figure 6. The reverse IV of 10G mesa-type APD as a function of temperature over a wide range of 20–145°C [23].

interesting features worth mentioning. First, the two wafers show similar breakdown voltage. The intercepts at the reference temperature (T_0 = 20°C) for wafers A and B are 32.8 and 31.7 V, respectively. The similarity between the wafers from two different processing runs suggests that the breakdown voltage is largely determined by the epitaxial structure. Second, both wafers show the same temperature coefficient of breakdown voltage where αV_{br} taken from the slope in **Figure 6** is equal to 0.017 V/°C. Third, the normalized temperature coefficient α is determined to be 5.1 × 10^{-4} and 5.5 × 10^{-4}°C^{-1}, as shown in **Table 3**. The measured value of the normalized temperature coefficient in the mesa-type APD is excellent, lower than the reported value of 7.2 × 10^{-4}°C^{-1} [45]. The superior temperature dependence of our mesa-type InGaAs/InAlAs APD can bring performance advantage to maintain the gain when the device is subject to temperature fluctuations. The superior temperature dependence of the avalanche

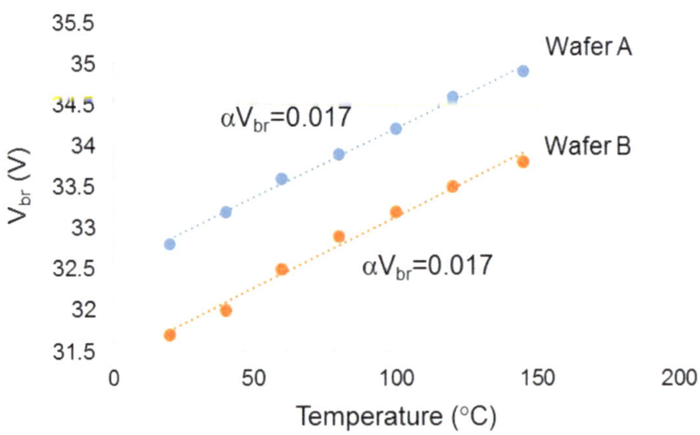

Figure 7. The breakdown voltage of mesa-type APD as a function of temperature from 20 to 145°C. The symbols are experimental data, and the dashed lines are fitting from Eq. (2) [23].

	$V_{br}(T_0)$	$\alpha V_{br}(T_0)$	α
Wafer A	32.8 V	0.017 V/°C	5.1×10^{-4} /°C
Wafer B	31.7 V	0.017 V/°C	5.5×10^{-4} /°C

Table 3. The temperature dependence of the mesa-type APD measured from 20 to 145°C [23].

breakdown voltage can be attributed to the optimized design of the InAlAs multiplication layer thickness and InAlAs charge control layer doping.

3.4. Dark current, Id

Another key device parameter of APD is dark current [46, 47]. As shown in **Figure 6**, the dark current is typically referred to the reverse current below the breakdown voltage. For different temperature ranging from 20 to 145°C, the dark current is taken from the reverse voltage of 0.9 V_{br} where V_{br} increases with the temperature. **Figure 8** illustrates the transport

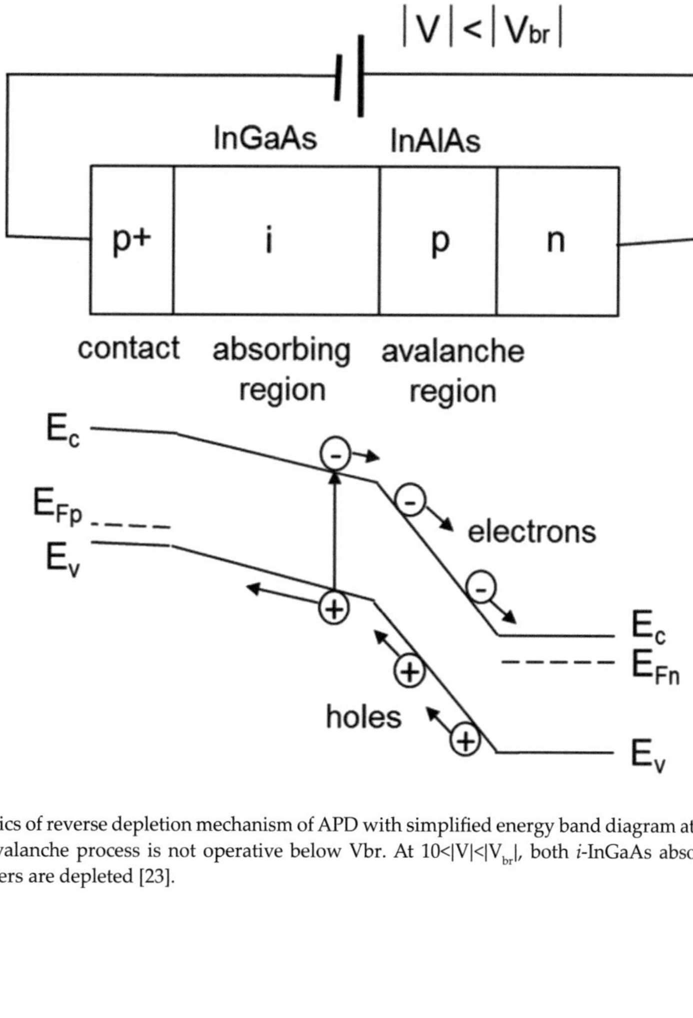

Figure 8. Schematics of reverse depletion mechanism of APD with simplified energy band diagram at the reverse voltage of$|V|<|V_{br}|$. The avalanche process is not operative below Vbr. At $10<|V|<|V_{br}|$, both *i*-InGaAs absorption and InAlAs multiplication layers are depleted [23].

of electrons and holes in the energy diagram at the reverse voltage below the avalanche breakdown ($|V|<|V_{br}|$) and above the punch-through voltage ($|V|>10$). In this regime, both *i*-InGaAs absorption and InAlAs multiplication layers are depleted, but the avalanche breakdown in the multiplication layer is not activated.

It has been found that the dark current increases with increasing temperature due to the influence of bandgap energy [48]. We note that the generation-recombination component of the dark current has temperature dependence. Therefore, the bandgap could be estimated based on the temperature dependence of the dark current. The generation current can be expressed in Eq. (3) where A_{pn} is the area of the p-n junction, W is the width of the depletion region, q is the carrier charge, σ is the carrier capture cross-section, υ_{th} is the carrier thermal velocity, N_t is the trap density, and ni is the intrinsic carrier concentration.

$$I_{dg} = A_{pn} W q \sigma \upsilon_{th} N_t n_i \tag{3}$$

The intrinsic carrier concentration can be expanded in Eq. (4) where N_c is the effective density of state in the conduction band, N_v is the effective density of state in the valence band, $E_g(T)$ is the energy bandgap, k is the Boltzmann's constant, and T is the temperature. The energy bandgap, $E_g(T)$, shows a temperature dependence where the bandgap value decreases with increasing temperature [49].

$$I_{dg} = A_{pn} W q \sigma \upsilon_{th} N_t (N_c N_v)^{1/2} \exp\left(-\frac{E_g(T)}{2kT}\right) \tag{4}$$

Based on Eq. (4), the energy bandgap can be extracted from the plot of dark current versus temperature. Taking the natural logarithm of Eq. (4), the temperature dependence of the dark current can be expressed in Eq. (5). The slope in the plot of I_d versus 1/kT shown in **Figure 9** is equal to $-E_g(T)/2$. The energy bandgap is estimated to be about 0.71 eV, close to the value of InGaAs where the bandgap energy of $In_{0.53}Ga_{0.47}As$ is reported to be 0.75 eV at 295 K [50, 51]. The slight difference between our value and others is likely attributable to the variance from junction temperature and test measurement.

$$Ln(I_d) \sim \left(-\frac{E_g(T)}{2}\right)\left(\frac{1}{kT}\right) \tag{5}$$

3.5. Bandwidth

Figure 10 shows the bandwidth plot of a 25G APD based on small-signal modulation response at 25°C. The bandwidth curves measured at reverse voltage of −17 and −21 V are shown. At the reverse bias of −17 V, the bandwidth taken by the 3 dB roll-off can reach 20 GHz. Such bandwidth is adequate for meeting the requirement of 4 × 25G ER4 Ethernet [24]. The corresponding multiplication factor or gain at −17 V is about 2.2 (M = 2.2). As the reverse bias was adjusted to −21 V for high gain operation (M = 5), the 3 dB bandwidth can be maintained at around 15 GHz.

3.6. Sensitivity

Figure 11 shows the bit error rate (BER) of a 25G APD photodetector as a function of input optical power. At the reverse bias near V_{br}, the 25G APD can achieve error-free bit error

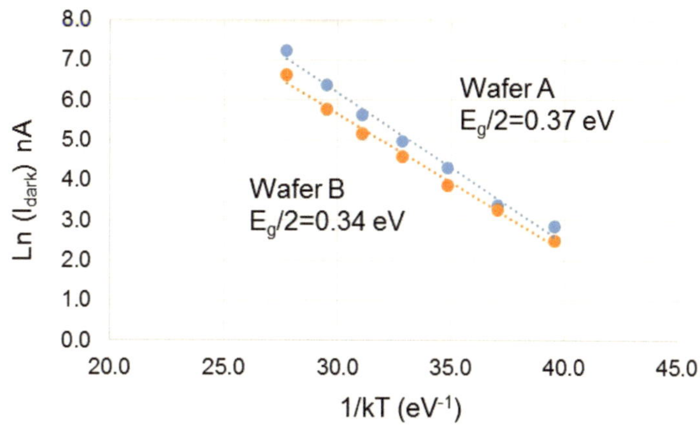

Figure 9. The natural logarithm of dark current versus reciprocal of temperature for mesa-type APD. The dark current was measured from 20 to 145°C [23].

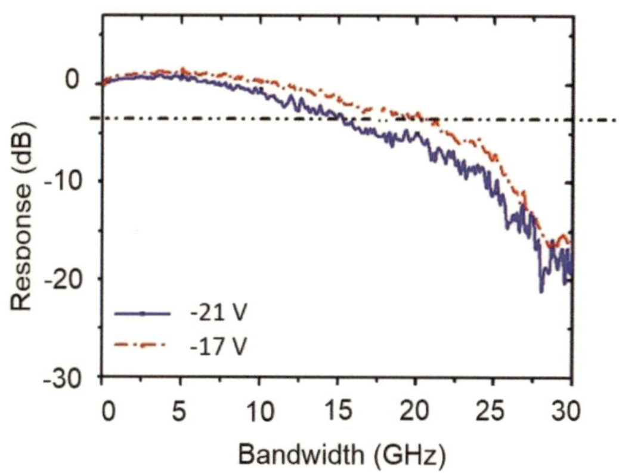

Figure 10. Bandwidth plot of 25G APD photodetector measured at 25°C with the reverse bias at −17 and −21 V. For a gain of 2, the BW can reach 20 GHz [38].

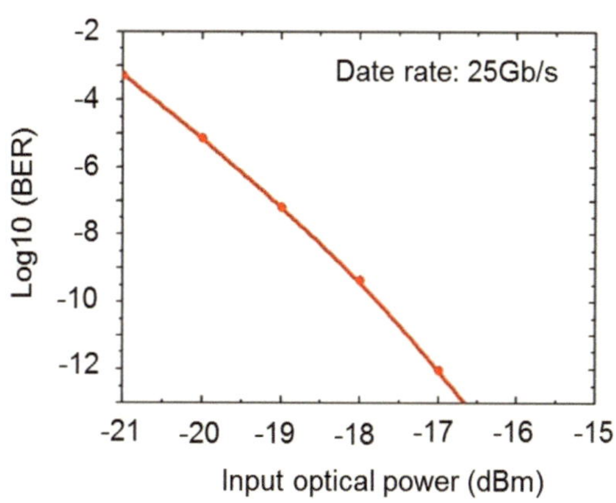

Figure 11. BER of a 25G APD device showing that the sensitivity of −17 dBm can be achieved for BER of 10^{-12} [38].

rate (BER) < 10^{-12} and the sensitivity of −17 dBm with a gain of 6. Such sensitivity level can provide a good margin for photoreceiver detection over the 40 km transmission over fiber [52, 53].

4. Reliability

4.1. Optical and electrical overload stress

In order to verify photodetector's robustness against the simultaneous electrical and optical stresses, fiber optic component manufacturers typically perform overload stress in burst and continuous modes, as illustrated in **Figure 12**. To determine the damage threshold of overload, optical stress is ramped up from −4 to +4 dBm when APD is subjected to electrical stress at $V_{br} - 2$ V.

In the burst mode, the optical stress of 1% pulsed duty cycle is applied to the APD for 60 s. With robust design and process, the APD devices can achieve good survival rate under the harsh overload stress. **Table 4** summarizes the overload results indicating that no failure occurs after being overload stressed up to +4 dBm. In the continuous mode, the damage threshold can also sustain the stress level of +1 dBm, well exceeding the requirement for −6 dBm.

For the study of reliability physics, it is important to identify the failure location and morphology of APD device after the overload stress. **Figure 13** shows an example of the damage morphology of the mesa-type APD sample stressed with an overload test of +4 dBm at burst mode. After the stress, the APD device shows functional failure due to short circuit, likely related to the InAlAs window. The short failure can be correlated with the metal shorting that shunts between the top InGaAs contact layer and the InAlAs multiplication layer, punching through

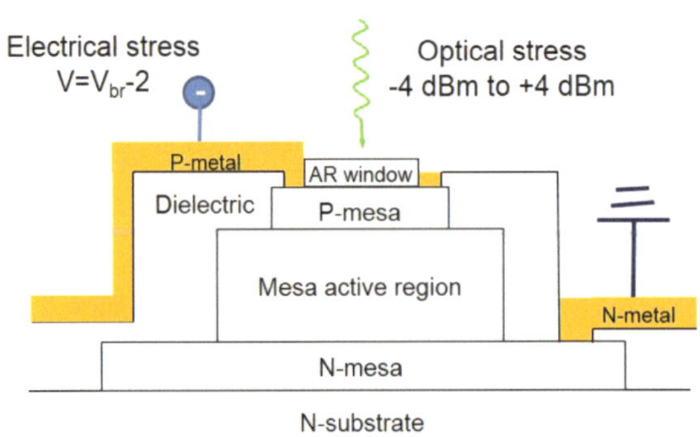

Figure 12. Schematic of electrical and optical overload stress applied to the APD for both burst and CW modes [38].

	Burst mode	CW mode
Damage threshold	No failure up to +4 dBm	No failure up to +1 dBm

Table 4. Damage threshold of optical and electrical overload stresses of 25G APD for burst and CW modes [38].

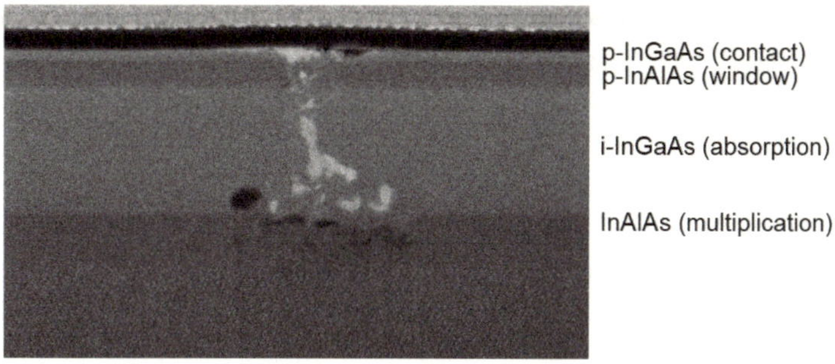

Figure 13. Failure morphology of a damaged mesa-type APD device after electrical and optical overload stress.

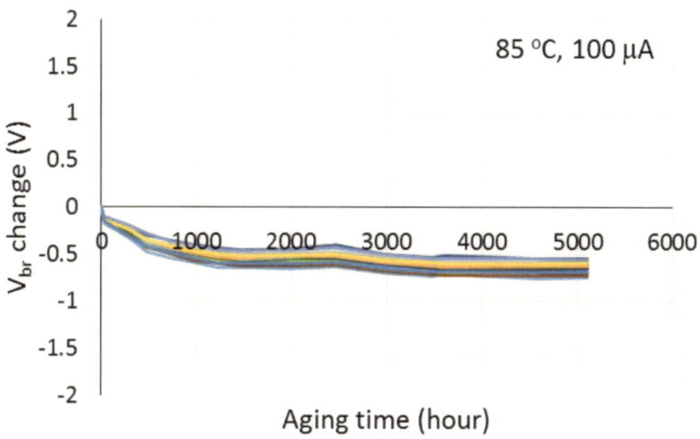

Figure 14. The relative breakdown voltage change as a function of aging time based on the stress condition of 85°C, 100 μA [38].

the InGaAs absorption layer. In comparison, the short failure can be suppressed by employing an InP window structure. The failure morphology reveals the spatial location of the weak spot and provides useful information about the robust overload design for the photodetector.

4.2. Reliability aging data

To establish long-term reliability performance, the APD devices are tested with temperature and current stresses to monitor the degradation. [54]. **Figure 14** shows the relative change in avalanche breakdown voltage as a function of aging time where the mesa-type APD samples are stressed at 85°C under a reverse current of 100 μA. The failure criterion is defined as 1 V change in V_{br}. All APD photodetectors pass reliability test with excellent reliability margin after 5100 h aging.

Due to the small degradation at regular aging condition of 85°C, mesa-type APD devices are also stressed at highly elevated temperature experimentally determine the failure times. **Figure 15** shows the failure time distributions of the APD from the aging groups of 165 and 175°C. The failure time at 50% cumulative probability shows the statistical mean-time-to-failure (MTTF). The MTTF values for 165 and 175°C are 896 and 446 h, respectively.

The device failure time (t_f) follows the modified Black's Eq. [55–57] which provides a good empirical description of device degradation over time as a function of stress current and temperature as shown in Eq. (6).

$$t_f = \frac{A}{I^N}\exp\left(\frac{E_a}{kT}\right) \tag{6}$$

In Eq. (6), the first term represents the current acceleration factor where A is a constant, I is the stress current, and N is the current exponent; the second term represents the temperature acceleration where E_a is the activation energy, k is the Boltzmann's constant, and T is the temperature.

For the sake of activation energy study, Eq. (6) can be rewritten in the form of natural logarithm as shown in Eq. (7) where the third term can readily determine the activation energy.

$$Ln(MTTF) = Ln(A) - N \bullet Ln(I) + \frac{E_a}{kT} \tag{7}$$

Figure 16 shows the plot of Ln(MTTF) versus 1/(kT) where the slope is equal to Ea. Based on the experimental aging data of 165 and 175°C, the E_a of APD is estimated to be 1.18 eV, which is in close agreement with other reported values [27, 58].

With the E_a establishment, the failure times from the aging test can be extrapolated to project the device lifetimes at the operating condition by using Eq. (6). For the operating condition of 50°C, the device lifetime of the mesa-type APD is estimated to be around 6900 years as shown in **Table 5**, which represents great reliability margin for the 20 year stringent requirement per Telcordia.

4.3. Future reliability challenges

Nowadays, semiconductor devices often incorporate design-in reliability in the early development phase [21, 59, 60]. During device miniaturization, several aspects including high electric field, Joule heating, and geometry inhomogeneity may impose reliability challenges.

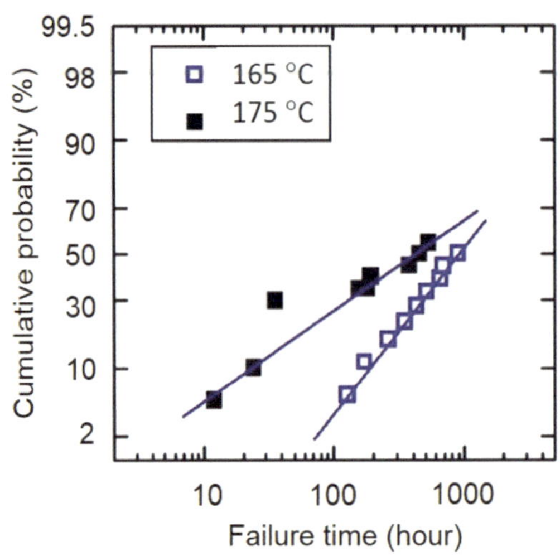

Figure 15. The failure time distributions of APD photodetectors based on the stress condition of 165 and 175°C [38].

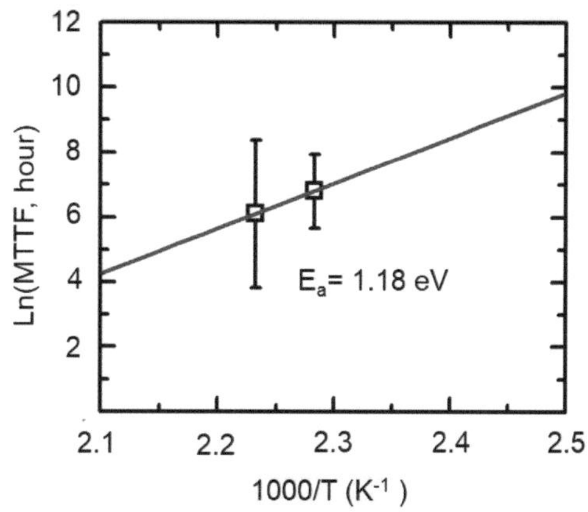

Figure 16. The plot of Ln(MTTF) versus 1/(kT) for the determination of E_a measured by mesa-type APD devices. The E_a value was 1.18 eV [38].

Aging stress temperature	Failure time at stress temperature	Device operating lifetime at 50°C
165°C	896 h	6912 years
175°C	446 h	6909 years

Table 5. Projected device lifetime of mesa-type APD at 50°C operating condition based on extrapolation from the stress condition [38].

(1) High electric field

The breakdown electric field of APD generally increases with decreasing thickness of multiplication layer. The increase in electric field with feature size reduction can accelerate device degradation over long-term field use [61–63].

(2) Joule heating

As device size shrinks, Joule heating is expected to be increased at a given bias current [64–66]. The increased Joule heating would raise device junction temperature that may degrade functional performance and long-term reliability.

(3) Geometry inhomogeneity

For electronic IC, the geometry inhomogeneity may induce current crowding and cause early reliability failure [67–69]. For APD, the inhomogeneous structure is typically generated by mesa etch [70–72]. The etched mesa interface may cause an increase in leakage current due to the generation of surface state [70, 71].

Among the three factors mentioned above, high electric field may impose the most reliability challenge for APD. This is because the effect of Joule heating can be mitigated by the use of mesa-type structure. For example, the resistance of the mesa-type APD is typically lower than that of the planar-type due to the shorter conduction path [73]. The lower resistance would result in lower Joule heating. Regarding geometry inhomogeneity, the surface leakage from mesa structure is expected to be similar from design standpoint. With careful processing control, the dark current of modern mesa-type APD devices can be confined within a few tens of nanoampere [23, 38].

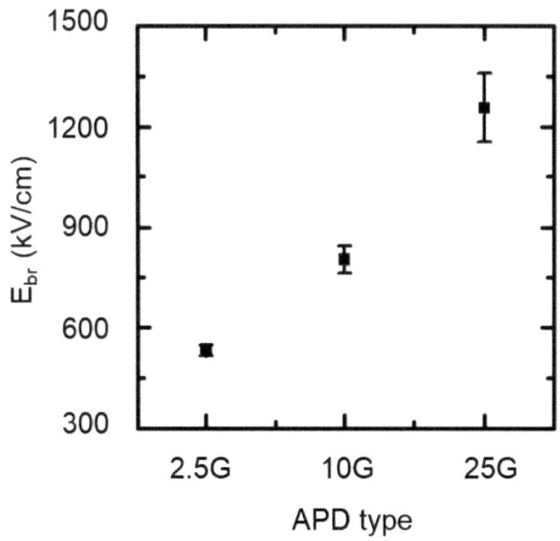

Figure 17. The breakdown electric field of the multiplication layer for various APD generations. Generally, the breakdown field increases with increasing device speed [39].

Figure 17 shows the estimated breakdown electric field in the multiplication layer of commercial 2.5G, 10G, and 25G APDs. At the breakdown voltage for avalanche operation, majority of the voltage drop typically occurs at the absorption layer and multiplication layer [22]. Since the absorption layer is very thick, the electric field of the absorber is usually low enough to prevent breakdown. Due to the thin layer, the electric field is much higher in the multiplication layer as shown in the electric field profile in **Figure 4**. Assuming half of the voltage drop is across the multiplication layer, we can estimate the electric breakdown field at the multiplication layer [74]. It is shown that the breakdown field may reach about 800 and 1300 kV/cm for 10G APD and 25G APD, respectively.

Although the breakdown field depends on the design of epitaxial doping and thickness [38], the ballpark estimate shown in **Figure 17** suggests that reliability issue may need to be extensively studied for 25G APD and beyond. As illustrated in our recent studies [26, 75, 76], the two seemingly different fields of electronics and photonics share some interesting similarities in reliability physics. As device scaling continues to occur to attain higher speed and greater performance, 25G APD and beyond may mark another new frontier of reliability studies. With effort as relentless as IC industry, the reliability robustness of APD is expected to be achieved and matured in the forthcoming.

5. Conclusions

In this chapter, we review the state-of-the-art commercial APD photodetectors. A 2.5G APD is typically based on the planar-type structure where the p-InP is formed by Zn diffusion with guard ring on the edge of diffused region to avoid edge breakdown. The advantages of planar-type APD include processing simplicity and low cost. On the other hand, 10G and 25G APDs are based on more sophisticated mesa-type structures in order to meet the speed requirements. The common mesa-type APD photodetectors are composed of SACM where the absorption, charge control, and multiplication layers are precisely grown and controlled by MBE.

The important device parameters of APD include: (1) avalanche breakdown voltage, (2) dark current, (3) temperature coefficient, (4) bandwidth, and (5) sensitivity.

1. For the avalanche breakdown voltage, the typical V_{br} of 10G and 25G APD photodetectors are about 31 and 26 V, respectively. The V_{br} value may vary depending on the design and process of component vendors.

2. The dark current of modern planar-type and mesa-type APDs is typically in the range of a few tens of nanoampere. By plotting the dark current versus temperature, the activation energy close to the energy bandgap of InGaAs (~0.75 eV) can be found.

3. The temperature dependence of avalanche breakdown voltage shows a positive coefficient due to the effect of phonon. The temperature coefficient of the breakdown voltage is determined to be 0.017 V/°C, corresponding to a normalized temperature coefficient of 5.1×10^{-4}°C^{-1}. The superior temperature dependence of the avalanche breakdown voltage is attributable to the optimized design of the InAlAs multiplication layer thickness and InAlAs charge control doping.

4. The typical 3 dB bandwidths of 10G and 25G APDs can reach around 8 and 20 GHz, respectively. The bandwidth is a function of voltage bias and multiplication gain.

5. The sensitivity and BER are paramount to high-speed photodetection. For the BER of 1×10^{-12}, a sensitivity of −17 dBm or better can be achieved.

The main reliability assessment of APD photodetector includes: (1) life test aging and (2) optical/electrical overload.

1. For the life test aging, mesa-type APDs can pass 5000 h aging with V_{br} change of <0.5 V at the stress condition of 85°C, 100 μA. Based on the aging data of 165 and 175°C, the activation energy of the APD is estimated to be 1.18 eV. The projected device lifetime of the APD is extrapolated to be about 6900 years at the operating condition of 50°C.

2. For the overload, APD is tested for its durability against simultaneous electrical and optical stresses. In the burst mode, the APD usually shows higher level of optical stress durability with a threshold up to +4 dBm. In the CW mode, the damage threshold is around +1 dBm, also exceeding the requirement of −6 dBm.

Author details

Jack Jia-Sheng Huang[1,2]*, Yu-Heng Jan[1,2], H.S. Chang[2], Chih-Jui Ni[2], Emin Chou[2], Shih-Kai Lee[2], Horng-Shyang Chen[2] and Jin-Wei Shi[3]

*Address all correspondence to: jack.huang@sourcephotonics.com

1 Source Photonics, West Hills, CA, USA

2 Source Photonics, Hsinchu, Taiwan

3 Department of Electrical Engineering, National Central University, Zhongli, Taiwan

References

[1] Novoselov KS, Geim AK, Morozov SV, Jiang D, Zhang Y, Dubonos SV, Grigorieva IV, Firsov AA. Electric field effect in atomically thin carbon films. Science. 2004;**306**:666-669

[2] Castro Neto AH, Guinea F, Peres NMR, Novoselov KS, Geim AK. The electronic properties of graphene. Reviews of Modern Physics. 2009;**81**:109-162

[3] Stajic J. Hydrogen atom makes graphene magnetic. Science. 2016;**352**(6284):424

[4] Avouris P, Chen ZH, Perebeinos V. Carbon-based electronics. Nature Nanotechnology. 2007;**2**:605-615

[5] Dragoman M, Dragoman D. Graphene-based quantum electronics. Progress in Quantum Electronics. 2009;**33**:165-214

[6] Castro Neto AH. The carbon new age. Materials Today. 2010;**13**:12-17

[7] Garcia JC, de Lima DB, Assali LVC, Justo JF. Group IV graphene- and graphene-like nanosheets. Journal of Physical Chemistry C. 2011;**115**(27):13242-13246

[8] Ashton M, Paul J, Sinnott SB, Hennig RG. Topology-scaling identification of layered solids and stable exfoliated 2D materials. Physical Review Letters. 2017;**118**:106101

[9] MaterialsWeb.org–Databases of Structural, Electronic, and Thermodynamic data for 2D and 3D Materials

[10] Graphene-Info Market Report. Graphene-info; June 2015

[11] https://www.wired.com/2010/10/graphene/

[12] Global Demand for Graphene after Commercial Production to be Enormous. AZONANO.com; 28 February 2014

[13] Koppens FHL, Mueller T, Avouris P, Ferrari AC, Vitiello MS, Polini M. Photodetectors based on graphene, other two dimensional materials and hybrid systems. Nature Nanotechnology. 2014;**9**:780-793

[14] Fang H, Bechtel HA, Plis E, Martin MC, Krishna S, Yablonovitch E, Javey A. Quantum of optical absorption in two-dimensional semiconductors. Proceeding National Academy Science. 2013;**110**(29):11688-11691

[15] https://www.semiconductors.org/

[16] https://en.wikipedia.org/wiki/Integrated_circuit

[17] Moore GE. Cramming more components onto integrated circuits. Electronics Magazine. 1965;**38**(8):4

[18] Arden W, Brillouet M, Cogez P, Graef M, Huizing B, Manhkopf R. More-than-Moore White Paper

[19] Waldrop MM. The chips are down for Moore's law. Nature. 2016;**530**:144-147

[20] https://en.wikipedia.org/wiki/Apple_mobile_application_processors

[21] Huang JS, Jan YH. Environmental Engineering of Photonic and Electronic Reliabilities: From Technology and Energy Efficiency Perspectives. Saarbrücken, Germany: Scholars' Press; 2017

[22] Campbell JC. Recent advances in avalanche photodiodes. Journal of Lightwave Technology. 2016;**34**(2):278-285

[23] Huang JS, Chang HS, Jan YH, Chen HS, Ni CJ, Chou E. Temperature dependence study of mesa-type InGaAs/InAlAs avalanche photodiode characteristics. Advances in OptoElectronics. 2017:1-5. Article ID 2084621

[24] Chen YH, Wun JM, Wu SL, Chao RL, Huang JS, Jan YH, Chen HS, Ni CJ, Chang HS, Chou E, Shi JW. Top-illuminated $In_{0.52}Al_{0.48}As$-based avalanche photodiode with dual charge layers for high-speed and low dark current performances. IEEE Journal of Selected Topics in Quantum Electronics. March/April, 2018;**24**(2):3800208

[25] Nada M, Muramoto Y, Yokoyama H, Ishibashi T, Kodama S. InAlAs APD with high multiplied responsivity-bandwidth product (MR-bandwidth product) of 168 A/W·GHz for 25 Gbit/s high-speed operations. Electronics Letters. 2012;**48**(7):397-399

[26] Huang M, Cai P, Li S, Wang L, Su T-I, Zhao L, Chen W, Hong C-Y, Pan D. Breakthrough of 25Gb/s Germanium on Silicon Avalanche Photodiode. Optical Fiber Communications Conf., Technical Digest, paper Tu2D.2, (OFC, Anaheim, CA, 2016)

[27] Ishimura E, Yagyu E, Kaji MN, Ihara S, Yoshiara K, Aoyaji T, Tokuda Y, Ishikawa T. Degradation mode analysis on highly reliable guardring-free planar InAlAs avalanche photodiode. Journal of Lightwave Technology. 2007;**25**(12):3686-3693

[28] Achouche M, Glastre G, Caillaud C, Lahrichi M, Chitoui M, Carpentier D. InGaAs communication photodiodes: From low- to high-power-level designs. IEEE Photonics Journal. 2010;**2**(3):460-468

[29] Kim HS, Choi JH, Bang HM, Jee Y, Yun SW, Burm J, Kim MD, Choo AG. Dark current reduction in APD with BCB passivation. Electronics Letters. 2001;**37**(7):455-457

[30] Takeshita T, Hirota Y, Ishibashi T, Muramoto Y, Ito T, Tohmori Y, Ito H. Degradation behavior of avalanche photodiodes with a mesa structure observed using a digital OBIC monitor. IEEE Transactions on Electron Devices. 2006;**53**(7):1567-1574

[31] Watanabe I, Tsuji M, Hayashi M, Makita K, Taguchi K. Reliability of mesa-structure InAlGaAs-InAlAs superlattice avalanche photodiodes. IEEE Photonics Technology Letters. 1996;**8**(6):824-826

[32] Smith GM, McIntosh KA, Donnely JP, Funk JE, Mahoney LJ, Verghese S. Reliable InP-based Geiger-mode avalanche photodiode array. Proceedings of SPIE. 2009;**7320**:1-10

[33] Pearsall TP. Impact ionization rates for electrons and holes in GaInAs. Applied Physics Letters. 1980;**36**:218-220

[34] Oasaka F, Mikawa T. Excess noise design of InP/InGaAsP/InGaAs avalanche photodiodes. IEEE Journal of Quantum Electronics. 1986;**QE-22**:471-478

[35] Neitzert HC, Cappa V, Crovato R. Influence of the Device Geometry and Inhomogeneity on the Electrostatic Discharge Sensitivity of InGaAs/InP Avalanche Photodiode. EOS/ESD Symposium. CA: Santa Clara; 1997

[36] Nada M, Yoshimatsu T, Muramoto Y, Yokoyama H, Matsuzaki H. Design and performance of high-speed avalanche photodiodes for 100-Gb/s systems and beyond. IEEE/OSA Journal of Lightwave Technology. 2015;**33**(5):984-990

[37] Ma CLF, Deen MJ, Tarof LE, Yu J. Modelling of breakdown voltage and its temperature dependence in SAGCM InP/InGaAs avalanche photodiodes. IEEE International Electron Devices Meeting (IEDM, Dec., 1994, San Francisco, CA), pp. 22.5.1-22.5.4

[38] Huang JS, Chang HS, Jan YH, Chen HS, Ni CJ, Chou E, Lee SK, Shi J-W. Highly Reliable, Cost-Effective and Temperature-Stable Top-Illuminated Avalanche Photodiode (APD) for 100G Inter-Datacenter ER4-Lite Applications Photoptics. Portugal: Funchal; 2018 in press

[39] Huang JS, Chang HS, Jan YH. Reliability challenges of nanoscale avalanche photodiodes. Open Access Journal of Photoenergy. 2017;**1**(3):00015

[40] Viswanathan CR. Physical Principles of Semiconductor Devices. UCLA: EE Class Note; 1993. p. 172-208

[41] Hu C. PN and Metal-semiconductor Junctions. UC, Berkeley, EE Class Note, Chapter 4. 2009. pp. 89-156

[42] Bendib T, Pancheri L, Dieffal F, Betta GFD. Impact of temperature and doping concentration on avalanche photodiode characteristics. Proceed. World Congress Eng., vol.I, (WCE, July, 2014, London, UK), pp. 5-8

[43] Massey DJ, David JPR, Rees GJ. Temperature dependence of impact ionization in submicronmeter silicon devices. IEEE Transactions on Electron Devices. 2006;**53**:2328-2334

[44] Tyagi MS. Zener and avalanche breakdown in silicon alloyed p-n junctions. Solid State Electronics. 1968;**11**:99-128

[45] Tan LLJ, Ong DSG, Ng JS, Tan CH, Jone SK, Qian Y, David JPR. Temperature dependence of avalanche breakdown in InP and InAlAs. IEEE Journal of Quantum Electronics. 2010;**46**(8):1153-1157

[46] Ohnaka K, Kubo M, Shibata J. A low dark current InGaAs/InP p-i-n photodiode with covered mesa structure. IEEE Transactions on Electron Devices. 1987;**ED-34**(2):199-204

[47] Fukuda M. Optical semiconductor devices. Chapter 4 Photodiodes. 1999. Wiley, New York, NY

[48] Forrest SR, Leheny RF, Nahory RE, Pollack MA. $In_{0.53}Ga_{0.47}As$ photodiodes with dark current limited by generation-recombination and tunneling. Applied Physics Letters. 1980;**37**:322-324

[49] Thomson JD, Summers HD, Smowton PM, Herrmann E, Blood P, Hopkinson M. Temperature dependence of the lasing wavelength of InGaAs quantum dot lasers. Journal of Applied Physics. 2001;**90**(9):4859-4861

[50] Pearsall TP. $Ga_{0.47}In_{0.53}As$: A ternary semiconductor for photodetector applications. IEEE Journal of Quantum Electronics. 1980;**16**(7):709-720

[51] Pearsall TP, Eaves L, Portal JC. Photoluminescence and impurity concentration in $Ga_xIn_{1-x}As_yP_{1-y}$ alloys lattice-matched to InP. Journal of Applied Physics. 1983;**54**(2):1037-1047

[52] Laird JS, Hirao T, Onoda S, Ohyama H, Kamiya T. Heavy-ion induced single-event transients in high-speed InP-InGaAs avalanche photodiodes. IEEE Transactions on Nuclear Science. 2003;**50**(6):2225-2232

[53] Alpert A. High-Speed Jitter Testing of XFP Transceivers. Viavi White Paper; 2015

[54] Generic reliability assurance requirements for optoelectronic devices used in telecommunication equipment. Telcordia, GR-468-CORE; 2004

[55] Black JR. Electromigration failure modes in aluminum metallization for semiconductor devices. Proceedings of the IEEE. 1969;**57**(9):1587-1594

[56] Huang JS. Temperature and current dependences of reliability degradation of buried heterostructure semiconductor lasers. IEEE Transactions on Device and Materials Reliability. 2005;**5**(1):150-154

[57] Huang JS, Nguyen T, Hsin W, Aeby I, Ceballo R, Krogen J. Reliability of etch-mesa buried-heterostructure semiconductor lasers. IEEE Transactions on Device and Materials Reliability. 2005;**5**(4):665-674

[58] Watanabe I, Tsuji M, Hayashi M, Makita K, Taguchi K. Reliability of mesa-structure InAlGaAs-InAlAs superlattice avalanche photodiodes. IEEE Photonics Technology Letters. 1996;**8**(6):824-826

[59] Huang JS. Design-in reliability of modern wavelength-division multiplex (WDM) distributed feedback (DFB) lasers. Applied Physics Research. 2012;**4**(2):15-28

[60] Bandi UR, Dasaka M, Kumar PK. Design in Reliability for Communication Designs. 43rd ACM/IEEE Design Automation Conference (San Francisco, CA, 2006)

[61] Fukuda M. Reliability and Degradation of Semiconductor Lasers and LEDs. Artech House: Norwood, MA; 1994

[62] Oates AS, Lin MH. Electromigration failure distributions of cu/low-k dual-damascene vias: Impact of the critical current density and a new reliability extrapolation methodology. IEEE Transactions on Device and Materials Reliability. 2009;**9**(2):244-254

[63] Huang JS, Shofner TL, Zhao J. Direct observation of void morphology in step-like electromigration resistance behavior and its correlation with critical current density. Journal of Applied Physics. 2001;**89**(4):2130-2133

[64] Efremov AA, Bochkareva NI, Gorbunov RI, Lavrinovich DA, Tarkhin YTRDV, Shreter YG. Effect of the joule heating on the quantum efficiency and choice of thermal conditions for high-power blue InGaN/GaN LEDs. Physics of Semiconductor Devices. 2006;**40**(5):605-610

[65] Wachutka GK. Rigorous thermodynamic treatment of heat generation and conduction in semiconductor device modeling. IEEE Transactions on Computer-Aided Design of Integrated Circuits and Systems. 1990;**9**(11):1141-1149

[66] Huang JS. "Reliability of Optoelectronics", Chapter 6 in Reliability Characterisation of Electrical and Electronic Systems. Cambridge, UK: Woodhead Publishing; 2015

[67] Tu KN, Yeh CC, Liu CY, Chen C. Effect of current crowding on vacancy diffusion and void formation in electromigration. Applied Physics Letters. 2000;**76**(8):988-990

[68] Yeh ECC, Choi WJ, Tu KN. Current crowding-induced electromigration in flip chip solder joints. Applied Physics Letters. 2001;**80**(4):580-282

[69] Huang JS, Yeh ECC, Zhang ZB, Tu KN. The effect of contact resistance on current crowding and electromigration in ULSI interconnects. Materials Chemistry and Physics. 2002;**77**(2):377-383

[70] Ma Y, Zhang Y, Gu Y, Chen X, Shi Y, Ji W, Xi S, Du B, Li X, Tang H, Li Y, Fang J. Impact of etching on the surface leakage generation in mesa-type InGaAs/InAlAs avalanche photodetectors. Optics Express. 2016;**24**(7):7823-7834

[71] Sudo H, Suzuki M. Surface degradation mechanism of InP/InGaAs APD. Journal of Lightwave Technology. 1988;**6**(10):1496-1501

[72] Huang JS, Jan YH, Chen HS, Chang HS, Ni CJ, Chou E. Predictive reliability model of 10G/25G mesa-type avalanche photodiode degradation. Applied Physics Research. 2016;**8**(3):66-74

[73] Huang JS, Jan YH, Chang HS, Chang J, Chang R, Liu G, Ren D, Chou E. ESD polarity effect study of monolithic, integrated DFB-EAM EML for 100/400G optical networks. CLEO-PR (Singapore, July31-Aug.4, 2017), Paper#1018

[74] Saleh MA, Hayat MM, Kwon OH, Holmes AL, Campbell JC, Saleh BEA, Teich MC. Breakdown voltage in thin III-V avalanche photodiodes. Applied Physics Letters. 2001;**79**(24):4037-4039

[75] Huang JS, Yu-Heng J. Interconnected analogy and resemblance: from classical music to modern semiconductor technology. International Journal of Electronics and Electric al Engineering. 2016;**5**(2):1-8

[76] Huang JS, Jan YH. A complexity science study: Effect of critical current density on semiconductor laser diodes and integrated circuits. Journal of Systemics, Cybernetics Informatics. 2017;**15**(1):77-83

Chapter 4

Two-Dimensional Halide Perovskites for Emerging New-Generation Photodetectors

Yingying Tang, Xianyi Cao and Qijin Chi

Additional information is available at the end of the chapter

http://dx.doi.org/10.5772/intechopen.71032

Abstract

Compared to their conventional three-dimensional (3D) counterparts, two-dimensional (2D) halide perovskites have attracted more interests recently in a variety of areas related to optoelectronics because of their unique structural characteristics and enhanced performances. In general, there are two distinct types of 2D halide perovskites. One represents those perovskites with an intrinsic layered crystal structure (i.e. MX_6 layers, M = metal and X = Cl, Br, I), the other defines the perovskites with a 2D nanostructured morphology such as nanoplatelets and nanosheets. Recent studies have shown that 2D halide perovskites hold promising potential for the development of new-generation photodetectors, mainly arising from their highly efficient photoluminescence and absorbance, color tunability in the visible-light range and relatively high stability. In this chapter, we present the summary and highlights of latest researches on these two types of 2D halide perovskites for developing photodetectors, with an emphasis on synthesis methods, structural characterization, optoelectronic properties, and theoretical analysis and simulations. We also discuss the current challenging issues and future perspective. We hope this chapter would add new elements for understanding halide perovskite-based 2D materials and for developing their more efficient optoelectronic devices.

Keywords: photodetector, halide perovskite, two-dimensional layered structure, nanoplatelet, and optoelectronic device

1. Introduction to photodetectors and halide perovskites

Photodetection techniques have demonstrated many important applications in a variety of fields or occasions, including medical X-ray imaging, ubiquitous visible light cameras, near-infrared medical imaging and short-wave infrared surveillance and machine vision [1–5]. Current photodetection relies primarily on conventional semiconductor materials with an appropriate bandgap, capable of transducing photons with different energies into electrical signals for subsequent processing, image reconstruction and storage. Historically, silicon and III–V semiconductors have been most commonly used as photodiodes and phototransistors (**Figure 1**) [6–8]. However, these conventional inorganic semiconductors are broadband absorbers. Thus, they produce a broad spectral photoresponse that could complicate the architecture and fabrication of devices and makes it difficult in pure color replication. Since the discovery of electrical conductivity in doped polypyrrole in the 1960s [9] and myriads of small-molecule and polymeric conductors and semiconductors during the 1970s and 1980s, organic semiconductors have shown promising applications in solar cells, field-effect transistors (FET) and photodetectors [10–12]. However, engineering organic semiconductors with an optical bandgap below 1 eV remains a challenge, as well as their carrier mobility is normally low. In addition, most organic planar photodetectors suffer from slow response time (≥10 ms), high operation voltage (≥10 V) and lack of stability to UV light (190–350 nm) [13]. This has driven the development of colloidal quantum dot (CQD)-based photodetectors, such as CdS, CdSe and PbS [14–16], as a possible option to expand the range of optical path difference (OPD). In most CQD-based photodetectors, however, a compromise between carrier extraction and absorption efficiency is required in order to reduce the electrical volume of the photoactive material but without sacrificing absorption. More recently, metal-halide perovskites have emerged as new candidate materials for optoelectronic devices (solar cells in particular) with the conversion efficiency reaching over 20% [17]. It is thus expected that using metal-halide perovskites as the key material component could construct new-generation photodetectors. In general, perovskites are optically tunable from 390 to 790 nm via halide-exchange or quantum confinement effect (**Figure 1**), so that they possess an appropriate bandgap exhibiting a steep absorption edge and demonstrate remarkable efficiency in photo-generated charge collection [18–22]. These unique optical properties make halide perovskites play key roles in the recent rapid development of solar cells, which has also spawned the exploration of diverse applications of halide perovskites across the domain of photodetectors. Metal halide perovskite-based photodetectors could

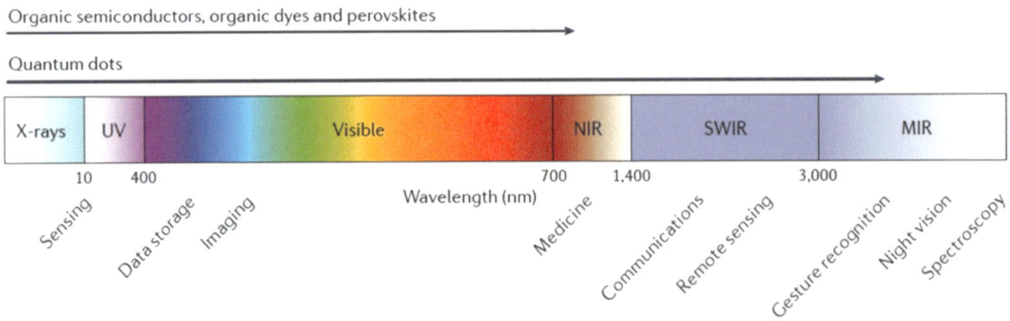

Figure 1. Schematic illustration of the wavelength region and potential application of semiconducting materials including organic semiconductors, organic dyes, halide perovskites and quantum dots. These semiconducting materials having the different applicable spectra region [8]. Abbreviation: UV, ultraviolet; NIR, near-infrared; MIR, mid-infrared; SWIR for short-wave infrared.

have the notable advantages such as ease of processing, tailorable optical property, good compatibility with flexible substrates and synergistic integration with complementary metal oxide semiconductors [23–26].

Compared to three-dimensional (3D) materials, two-dimensional (2D) materials, defined as crystals with very high aspect ratios and thicknesses corresponding to a few atomic layers, usually exhibit unique properties [27–30]. Graphene is the first example discovered with a true 2D atomic crystal, which is known by its high chemical stability [31] and large specific surface area [32]. In particular, graphene has a very high electronic conductivity [33] and absorbs the incoming radiation almost independent of wavelength [34], which can perform synergistic functionality in optoelectronic devices. Besides, other newly emerged 2D materials, such as hexagonal boron nitride (h-BN), [35] transition-metal dichalcogenides (TMDs) [36], layered metal oxides [37], and Venes (phosphorene and arsenene) [38, 39] have also attracted broad interests in various fields. Unlike above-mentioned 2D materials, however, 2D perovskites in the form of nanoplatelet (NPL) or nanosheet and layered-structure perovskites are ionic materials. This chemical bonding nature could endow them special properties distinguished from their 3D structures. For example, these 2D perovskites have an increased exciton binding energy, can reduce fluorescence decay time, and enhance absorption cross sections with respect to the bulk as well as having a notable optical nonlinearity [40–42]. Moreover, energy transfer in the stack of 2D perovskite sheets or platelets is significantly enhanced due to the large contact area between individual platelets. Note that the construction of 3D perovskites needs to match some critical parameters such as Goldschmidt's Tolerance Factor, ionic radii, charge balance and bonding/coordination preference, in order to produce stable structures. In 2D perovskites, in contrast, a remarkable structural tunability is feasible especially for the interlayer "A" cation length. This character provides a high flexibility for the design and synthesis of desirable 2D perovskite crystal structures with new physicochemical properties.

This chapter aims at summarizing recent advances and discussing current challenges and ongoing efforts in the development of 2D perovskite-based photodetectors. We *first* briefly overview recent research progress in 2D perovskites before focusing on the optoelectronic properties of 2D layered perovskite structures or perovskite nanoplatelets and nanosheets and photodetectors. We summarize a number of popular methods for the preparation of 2D perovskites, their structural characterizations and fabrication and tests of photodetectors. Also included is a brief discussion on theoretical simulations of electronic structures of halide perovskites for their perspective applications in future optoelectronic devices.

2. Design and preparation of 2D halide perovskites

To obtain desirable optical properties, the design and controlled synthesis of 2D metal-halide perovskites is a critical step. In this section, we focus on summarizing the recent progress in the synthesis of 2D halide perovskites.

2.1. Solution-processed methods

Some 2D halide perovskites can be prepared via solution-processed methods, which represent low-cost, flexible and efficient approach in many cases. For example, Sun et al. recently reported the synthesis of $CsPbBr_3$ nanostructures with different morphologies by adjusting different groups of acids and amines (**Figure 2a**) [43]. In the presence of hexanoic acid and octylamine, the nanocrystals can be tailored into spherical quantum dots. Nanocubes were generated under the coordination effects of oleic acid and dodecylamine, while nanoplatelets formed with the help of oleic acid and octylamine. The average size for these three types of nanostructures is 4.3, 9 and 100 nm, respectively. Their photoluminescence (PL) decay lifetime was determined to be in the range from several tens to hundreds of nanoseconds [43]. In another approach, Bekenstein et al. prepared colloidal nanoplatelets (NPLs) of $CsPbBr_3$ at

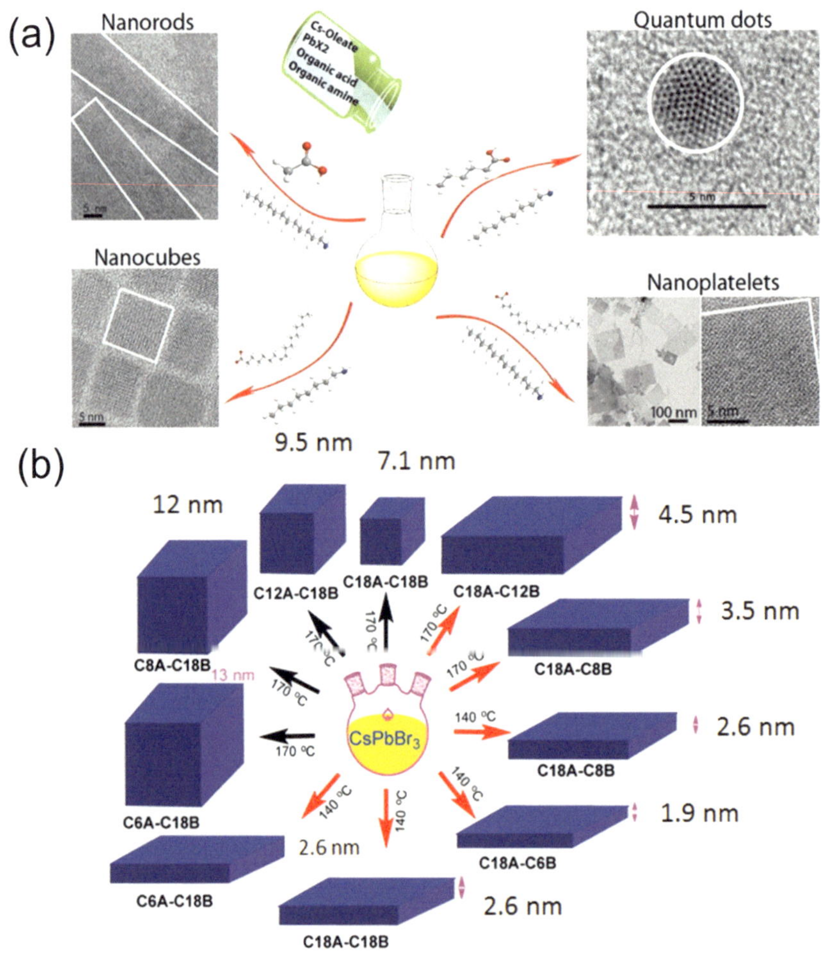

Figure 2. Summary of the synthesis methods for halide perovskite nanostructures. (a) Schematic view of the formation process of $CsPbX_3$ (X = Cl, Br, I) nanocrystals with different morphology [43]. (b) Morphology and size control of $CsPbBr_3$ nanocrystals through varying the temperature and surfactants [45].

elevated temperatures, in which hot Cs-oleate solution was quickly injected into the $PbBr_2$ solution containing octadecene, oleylamine and oleic acid [44]. The shape and thickness of $CsPbBr_3$ NPLs was roughly controllable by adjusting the injection temperature of Cs-oleate solution. Thin NPLs, thicker NPLs, nanocubes and large nanocubes together with nanowires were obtained under 90, 130, 150, 170–200°C, respectively. However, no nanocrystals are generated below 70°C. $CsPbBr_3$ NPLs obtained at 130°C have a lateral dimension of ~20 nm and a thickness of ~3 nm. PL spectra were tuned from 512 nm (5 unit cell thickness) to 405 nm (1 unit cell thickness). For those NPLs with 5, 4, and 3 unit cell thickness, the photoluminescence quantum yield (PLQY) is 84.4 ± 1.8, 44.7 ± 2.6 and 10 ± 0.5%, respectively. The perovskite NPLs can be further assembled into stacked columnar phases by concentrating NPLs solutions and large 2D sheets by diluting samples [44]. Pan et al. studied the effects of organic acid, base and cesium precursors on the morphology of $CsPbBr_3$ perovskite nanocrystals (**Figure 2b**) [45]. The use of shorter chain amines resulted in the formation of thin nanoplatelets, while using shorter chain carboxylic acids led to large-size nanocubes at high temperatures. Furthermore, using CsOAc as the Cs source can overcome the solubility problem that existed in Cs_2CO_3 precursor, showing superior performance, which allows the use of low boiling point reactants, such as acid or amine with chain length shorter than eight carbons. These findings have demonstrated the feasibility to tune the shape, size and optoelectronic properties of halide perovskites [45].

2.2. Vapor-phase methods

Vapor-deposition methods including chemical vapor deposition (CVD), van der Waals (VDW) epitaxy and other related methods are effective approaches to the preparation of high quality 2D nanocrystals especially for conventional semiconductors [46, 47]. These methods have now been attempted for the synthesis of 2D perovskites. Following is a brief summary with some representative examples illustrated.

Ha et al. first reported the preparation of $CH_3NH_3PbX_3$ nanoplatelets via a gas-solid heterophase reaction, [48] as illustrated in **Figure 3a**. By means of VDW epitaxy, lead halide nanoplatelets were initially grown on muscovite mica and then were converted into perovskites via gas-solid reaction with CH_3NH_3X. The resulting nanoplatelets had the lateral dimension in the range of 5–30 μm and the thickness from several atomic layers to several 100 nm. $CH_3NH_3PbI_3$ nanoplatelets exhibited a large diffusion length of ~200 nm (**Figure 3b**) [48]. Adding phenyl-C61-butyric acid methyl ester (PCBM) layer onto the $CH_3NH_3PbI_3$ platelets resulted in the reduction of PL lifetime from 6.8 ± 0.4 to 0.278 ± 0.004 ns. Besides, the electron-diffusion length was estimated to be 210 ± 50 nm, [49] which can be attributed to the high quality perovskites prepared from vapor-phase method. Wang et al. prepared crystalline 2D $MAPbCl_3$ perovskite thin platelets on muscovite mica using VDW epitaxy method, with the thickness of sub-10 nm and a few tens of micrometers in the lateral dimension (**Figure 3c**) [50]. Weak VDW interaction and delocalized bonding characters of ionic and metallic crystals play key roles in forming ultrathin and large-scale $MAPbCl_3$ platelets.

A single-step CVD approach was developed by Wang et al. to prepare all-inorganic halide perovskite, $CsPbX_3$ microplatelets [51]. The two types of precursor powders (PbX_2 and CsX)

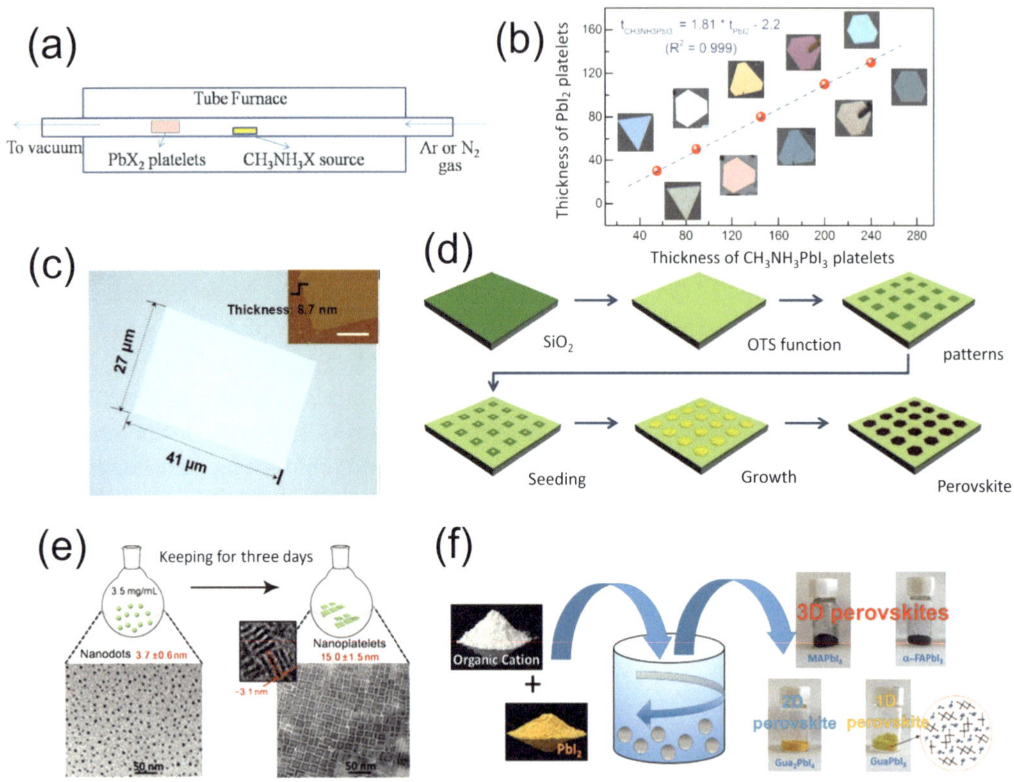

Figure 3. (a) Diagram of vapor-phase-transport system setup [48]. (b) Correlation of the thickness of PbI_2 and $CH_3NH_3PbI_3$ platelets [48]. The inset represents the optical image of $CH_3NH_3PbI_3$ platelets. (c) Optical image of an individual large-size ultrathin sheet [50]. The inset shows the thickness of the sheet by AFM measurement. (d) The procedure for preparing $CH_3NH_3PbI_3$ plates on a patterned substrate [52]. (e) Self-transformation of $CH_3NH_3PbBr_3$ nanodots to nanoplatelets [53]. The enlarged parts represent the corresponding TEM images of $CH_3NH_3PbBr_3$ nanocrystals. (f) Scheme of the mechanochemical synthesis for the 3D, 2D and 1D perovskites [54].

were mixed in a stoichiometric ratio and reacted in a home-built CVD system under controlled temperatures [51]. A quality factor up to 2863 was achieved in $CsPbX_3$ microplatelets-based gain medium and whispering gallery mode cavity, while the photodetector based on a sandwiched structure of graphene/CsPbX3 microplatelets/vertically stacked graphene displayed a high photoresponsivity >10^5 A/W. Similarly, Wang et al. demonstrated a two-step wafer-scale growth of large arrays of $CH_3NH_3PbI_3$ microplate crystals [52]. The SiO_2/Si substrate was functionalized by (octadecyl)trichlorosilane (OTS) forming a hydrophobic surface, followed by a selective growth of PbI_2 plates on the hydrophilic region generated by the oxygen-plasma treatment. Finally, perovskite platelets were obtained via immersing PbI_2 plates into CH_3NH_3I vapor (**Figure 3d**).

2.3. Other synthesis methods

Besides solution-processed and vapor-phase methods, other methods have also been attempted to prepare 2D metal halide perovskites such as self-organization synthesis and mechanochemical approaches. Still, more alternative methods are in progress with the aim to control the morphology and thickness of 2D halide perovskites more effectively.

Liu et al. demonstrated the self-organization process for the transformation of $CH_3NH_3PbBr_3$ nanodots (NDs) to NPLs [53]. In detail, the $CH_3NH_3PbBr_3$ NDs were dissolved in a nonpolar solvent and aged in darkness (**Figure 3e**). The thickness of resulting $CH_3NH_3PbBr_3$ NPLs is similar to the size of original NDs, but the lateral dimension is about fourfold to the original NDs. Theoretical simulations show that the side-side orientation with the coordination between electrostatic and van der Waals interactions is favored during the self-organization process. The resulting NPLs loaded into a polymer matrix (4-methyl-1-pentene, TPX) displayed cos/sin intensity dependence on the polarizer angles for the PL emission. In addition, this method offers a new way for controlling the shape of perovskite-based nanomaterials with polarized emission [53]. Jodlowski et al. reported the preparation of a bi-dimensional (i.e. 2D) perovskite (Gua_2PbI_4) using a mechanochemical synthesis approach (**Figure 3f**) [54]. The precursors of PbI_2 and MAI were mixed in a 1:1 molar ratio inside a ball mill under ambient conditions to produce polycrystalline powders. This synthesis process featured solventless environment, swiftness, simplicity, reproducibility and promising potential for exploring the alternative design of various perovskite nanostructures [54].

3. Structural characteristics and analysis of 2D halide perovskites

3.1. General structural features

In general, there are two kinds of 2D metal halide perovskites. One is the intrinsically layered crystal structure, which are electronically "2D" due to the quantum confinement in the layers composed of MX_6 octahedra separated by organic cations. The other is those perovskites with 2D nanostructured morphology but intrinsically have 3D structures.

In the 2D layered perovskites, there is no restricted size confinement for the interlayer cations, as indicated in 3D perovskites. This structure feature allows organic cations such as butylammonium ion ($C_4H_9NH_3^+$) [55], phenylethylammonium ion ($C_8H_9NH_3^+$) [56], aromatic biimidazolium dications ($C_6H_8N_4^+$) to access to the interlayer space between inorganic layers [57]. In $(NBT)_2PbI_4$, for example, PbI_6 octahedra can link together to form a flat layer in the *bc*-plane. Instead of NBT cation, EDBE cation makes $(EDBE)PbI_4$ be featured as a distorted layer, that is, a <110>-oriented zigzag structure. Increasing the alkyl chain length can narrow the band gap of hybrid perovskites, which is invoked by the distortion of inorganic layers and B-I-B bond angles. Moreover, the geometrical distortion of perovskite layers can also affect the defectivity and charge relaxation dynamics. Tan et al. reported the synthesis of $(C_4H_9NH_3)_2PbBr_4$ crystals by a solution-processed method, which have a domain size of several to tens of micrometers and the thickness of several to tens of nanometers [58]. $PbBr_6$ octahedra are connected to each other via corner-sharing bromide atoms forming a 2D perovskite layer in the *bc*-plane, as shown in **Figure 4a**. Meanwhile, the space between layers is separated by $C_4H_9NH_3^+$ cations for charge balance. $(C_4H_9NH_3)_2PbBr_4$ crystal-based photodetectors assisted with the protection and electric contact of monolayer graphene was reported to exhibit very low dark current (~ 10^{-10} A),

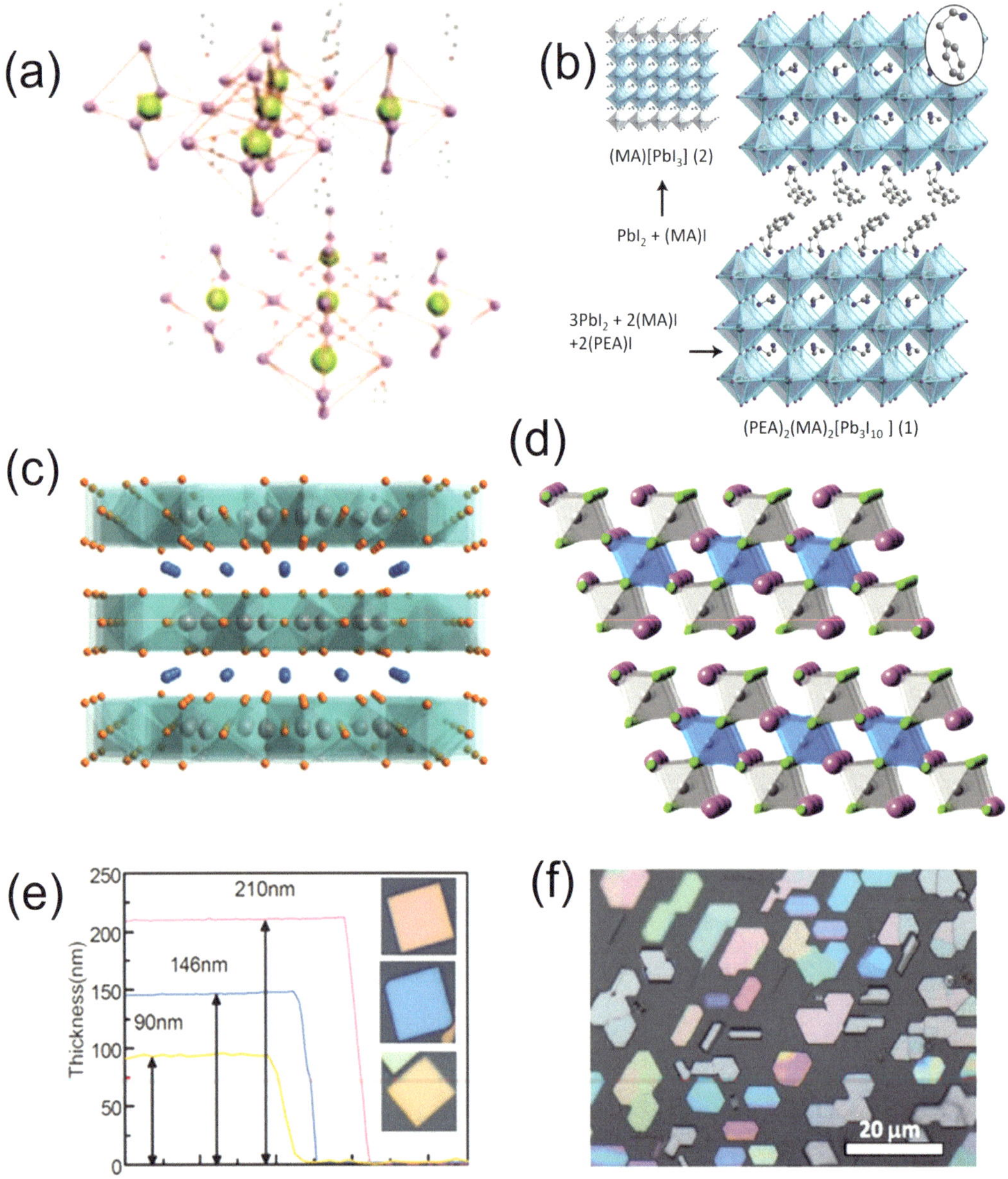

Figure 4. (a) Structural illustration of 2D layered $(C_4H_9NH_3)_2PbBr_4$ [58]. (b) 2D layered structure in $(PEA)_2(MA)_2[Pb_3I_{10}]$ [60]. (c) $[Pb_2Br_5]^-$ layer in $CsPb_2Br_5$ [62]. (d) The crystal structures of $Cs_4CuSb_2Cl_{12}$ [63]. (e) AFM characterization of individual $CsPbBr_3$ nanoplatelets with different colors [64]. (f) An optical image of $CH_3NH_3PbI_3$ nanoplatelets prepared by vapor phase method [48].

high on/off current ratio (up to 10^3) and a high responsivity of ~2100 A/W [58]. In addition, the "A" moiety can be further substituted by two types of organic cations, named as Ruddlesden-Popper family. In $(BA)_2(MA)_{n-1}Pb_nI_{3n+1}$ (n = 1, 2,3,4, ∞) perovskites, the layers are composed of tilted, corner-sharing PbI_6 octahedra, propagating in the *ac* plane, while the inorganic sheets are disconnected by the organic bilayers. For such perovskites $(BA)_2(MA)_{n-1}Pb_nI_{3n+1}$, their band gap is 2.43 (n = 1), 2.17 (n = 2), 2.03 ((n = 3) and 1.91 (n = 4) eV, respectively [59]. If BA^+ cation is substituted with aromatic cations, such as PEA^+ cation, then resulting $(PEA)_2(MA)_2Pb_3I_{10}$ (PEA = $C_6H_5(CH_2)_2NH_3^+$, MA = $CH_3NH_3^+$) features n = 3 layers with the PEA cations sandwiched between layers and MA^+ cations situated in the layers, displaying an optical band gap of 2.06 eV (**Figure 4b**) [60].

If the inorganic framework of 3D perovskites are slit along <111> axis, <111> oriented perovskites can be obtained. Leng et al. reported a collaborative solvent- and ligand-assisted re-precipitation method for the synthesis of $MA_3Bi_2X_9$ (X = Cl, Br, I) quantum dots. $(CH_3NH_3)_3Bi_2Br_9$ has a distorted layered structure composed of corner-shared $BiBr_6$ octahedra with Cs^+ ions filled between interlayers [61]. Also, other interesting types of layered perovskite structures have been investigated. For instance, the tetragonal phase of $CsPb_2Br_5$ exhibits a sandwich structure consisting of $[Pb_2Br_5]^-$ layers and intercalated Cs^+ as shown in **Figure 4c**. In the $[Pb_2Br_5]^-$ layer, one Pb^{2+} cation coordinates with four Br^- forming the elongated pentahedron. All the Pb^{2+} ions are confined in the center of the layer, while Br^- ions are located in both the bottom and top surface of the layer. The structural feature of the $CsPb_2Br_5$ crystals is similar to that of layered double hydroxides, the nanoplatelets of which can be steadily prepared by a facile precipitation process [62]. Vargas et al. prepared a 2D inorganic metal halide perovskite single crystal of $Cs_4CuSb_2Cl_{12}$ by a solution-processed method (**Figure 4d**). They found that this perovskite has a direct bandgap of 0.98 eV obtained by DFT calculations and exhibits higher conductivity than that of $MAPbI_3$ (MA = methylammonium). Besides this, perovskite displayed high photo/thermal-stability and also tolerant to humidity, which makes it a promising candidate for optoelectronic applications [63].

As for the perovskites with a 2D morphology, Levchuk et al. reported the synthesis of $CH_3NH_3PbX_3$ (X = Br, I) nanoplatelets via a ligand-assisted re-precipitation method, tailoring the thickness of nanoplatelets between 1 and 8 unit cell monolayers [64]. Thanks to the quantum confinement effects, the thickness-dependent PL properties of these nanoplatelets would offer an effective method to tune the emission color of perovskites [64]. Vapor-phase methods have also been employed for the synthesis of 2D perovskites. CVD was used to synthesize 2D $CsPbBr_3$ nanoplatelets directly on the mica substrate via the reaction between PbX_2 and CsX (**Figure 4e**). The growth mechanism of the nanoplatelets suggested that layered mica could facilitate the VDW epitaxial growth of 3D perovskites into ultrathin 2D morphology [65]. Meanwhile, they also employed a two-step CVD procedure for the preparation of $MAPbI_3$ platelets. In this method, PbI_2 was firstly deposited on the silicon substrate, followed by the vapor-phase conversion of the PbI_2 into $MAPbI_3$ [48]. Zhang et al. reported the preparation of $CH_3NH_3PbI_{3-a}X_a$ (X = I, Br, Cl) nanoplatelets grown on muscovite mica, which showed triangular or hexagonal platelets with nanoscale thickness (10–300 nm) and edge length of several to tens of micrometers (**Figure 4f**) [66]. For a typical $CH_3NH_3PbI_3$ triangular nanoplatelet, a lasing threshold of ~37 μJ cm^{-2} is deduced; while for $CH_3NH_3PbI_{3-a}Br_a$ and $CH_3NH_3PbI_{3-a}Cl_a$, a higher threshold of ~128 μJ cm^{-2} is achieved but with a larger cavity quality factor, Q ~ 900. Integrating onto other conductive substrates such as Si, Au and ITO, respectively, the spontaneous emission lifetimes are 2.6, 1.2 and 1.9 ns. These emission times are much shorter than those on mica. However, the threshold of these nanoplatelet lasers remained to be ~40 μJ cm^{-2}, because the exciton avalanche occurs in a much shorter time-scale than that of the carrier transfer/trapping at the surface/interfaces. In the family of the unique perovskite nanoplatelets, they have demonstrated strong optical-pumped room-temperature near-infrared lasing with low thresholds and wide mode-tunability [66]. Liao et al. reported the single crystalline square microdisks (MDs) of $CH_3NH_3PbBr_3$ prepared by using a one-step solution self-assembly method for the first time [67]. Square MDs of $CH_3NH_3PbBr_3$ are highly emissive with PLQY = 22% ± 5%. Q value of 430 was obtained at 555 nm presented by a transverse magnetic (TM) mode, which is smaller than the simulated one. This is most likely due to the shape deformation of $CH_3NH_3PbBr_3$ MDs derivate from perfect squares. Single mode lasing at 557.5 nm was achieved in a 2.0 × 2.0 × 0.6 μm^3 square MD with a threshold of 3.6 ± 0.5 $\mu J\ cm^{-2}$. Built-in whispering-gallery mode (WGM) microresonator in single-crystalline square MDs, ambipolar charge transport and the high carrier mobility make perovskites attractive candidates to realize electrically pumped on-chip coherent light sources [67].

3.2. High-resolution characterization tools

Grazing incidence wide angle X-ray scattering (GIWAXS) has been widely used to evaluate the lattice expansion in perovskite materials. As shown in **Figure 5a**, the *d* spacing of relative peaks is larger than that of the bulk crystals, consistent with the transmission electron microscopy (TEM) measurements of single sheets [68].

Atomic force microscopy (AFM) measurements are usually employed to detect the thickness and surface of perovskite nanosheets and nanoplatelets. Yang et al. reported that the thickness of 2D $(PEA)_2PbI_4$ NSs is measured as 2.0 ± 0.1 nm by AFM (**Figure 5b**), showing the formation of single-layer nanosheets [69]. According to the crystallographic model, the theoretical interlayer distance should be 1.64 nm. Thus, it is very close to the AFM results, given the fact that some organic molecules adsorbed on the surface of perovskite nanosheets. Besides, a quite smooth surface of the $(PEA)_2PbI_4$ nanosheet was observed from the high-magnification height profile (right, **Figure 5b**).

Transmission electron microscopy (TEM), high-resolution TEM (HRTEM) and high-angle annular dark-field scanning TEM (HAADF-STEM) are all-powerful tools for analysis of 2D perovskite structures. Akkerman et al. reported the face-to-face alignment of $CsPbBr_3$ nanoplatelets with a thickness of 3.0 ± 0.4 nm (corresponding to 5 unit cells) and the average lateral dimensions of 7.9 ± 1.2 nm × 40.9 ± 6.8 nm (**Figure 5c**). HRTEM images confirmed that $CsPbBr_3$ nanoplatelets are lying flat with respect to the substrate (**Figure 5d**). And the high crystallographic structure of these nanoplatelets can be well characterized by Fast Fourier Transform (FFT) images (**Figure 5e**) [70]. In the STEM mode, these nanoplatelets are lying flat or edge-on on the substrate (**Figure 5f**).

Single crystal X-ray diffraction is another powerful technique to analyze the structure of new materials by collecting diffraction data on single crystals. Recently, Cortecchia et al. found that a significant deformation of the Pb − X bond length and the X − Pb − X bond angle can result in broadened PL spectra for the two perovskites, $(NBT)_2PbI_4$ [71] and $(EDBE)PbI_4$ [72] (NBT = n-butylammonium and EDBE = 2,2-(ethylenedioxy)bis(ethylammonium)) [73].

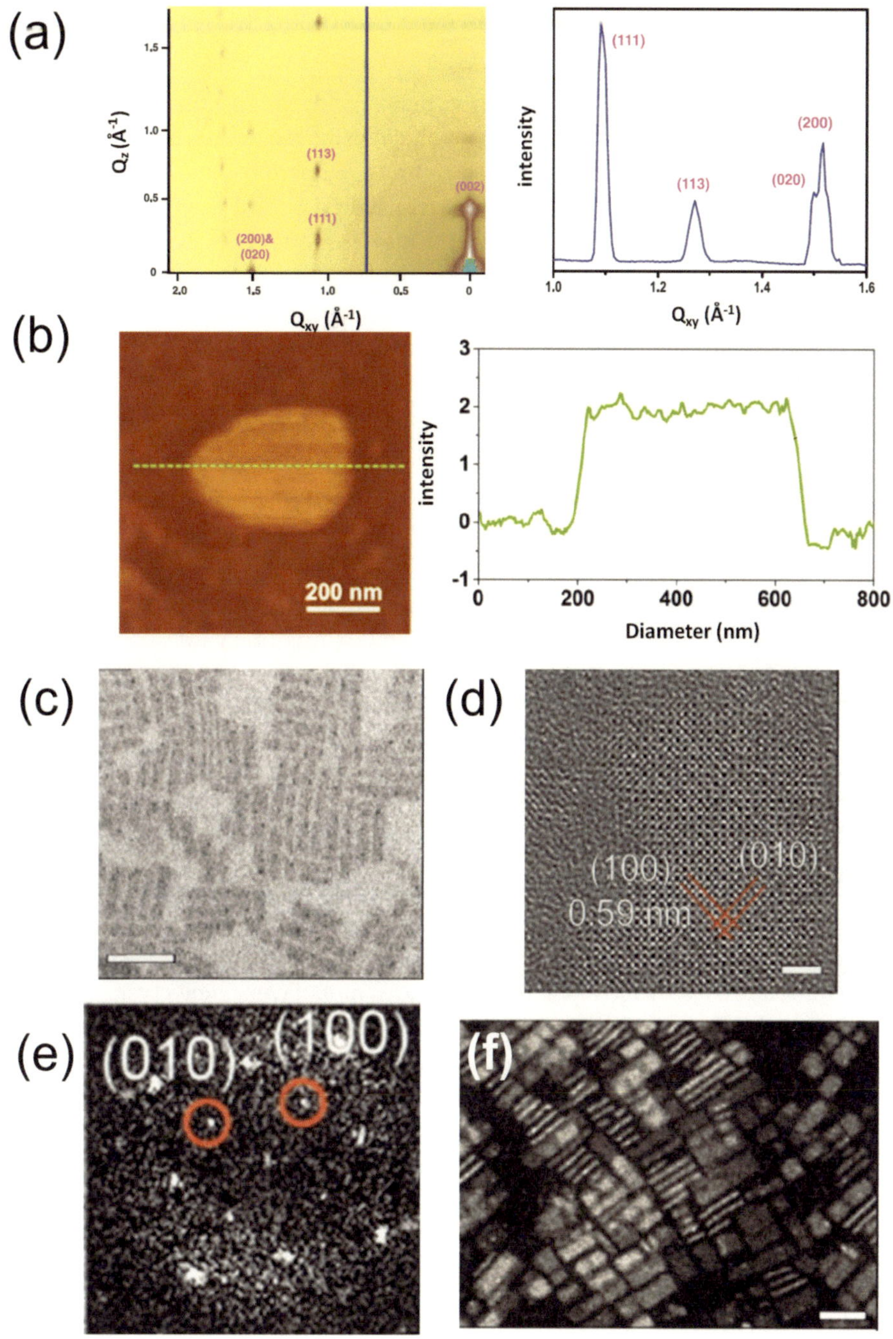

Figure 5. (a) GIWAXS image of the 2D thin sheets (left) and integrated GIWAXS spectrum of the 2D thin sheets (right) [68]. (b) AFM image (left) and height profile (right) of a typical $(PEA)_2PbI_4$ nanosheet synthesized in toluene [69]. (c) TEM, (d) HRTEM and (e) FFT images of $CsPbBr_3$ NPLs. (f) STEM dark-field image of $CsPbBr_3$ nanoplatelets [70]. Scale bars correspond to 50 nm in (c), 2 nm in (d) and 20 nm in (f).

4. Applications of 2D perovskites as photodetector materials

Responsivity, speed and sensitivity are three main factors for evaluating photodetectors. Responsivity is the ratio of photocurrent/photovoltage to the incident optical power, $R = I_{out}/P_{in}$. Speed is derived from the rise and fall response times, respectively, of a transient photocurrent. Sensitivity measures the ability to detect a weak light signal and is reported as the specific detectivity, $D^* = R\ (A{\cdot}\Delta f)1/2\ I_{noise}^{-1}$. Devices with a high responsivity and a small noise current can achieve a signal-to-noise ratio of 1, which corresponds to the noise-equivalent intensity (i.e. the detection limit) at a lower incident intensity. Specific detectivity is normalized to the device area and the frequency bandwidth of the noise measurement, enabling comparison among different device architectures, areas and measurement bandwidths.

There are three kinds of photodetectors: *photoconductors*, *phtototransistors* and *photodiodes*. Charge carriers in photoconductors are re-circulated through symmetrical contacts, in which an external voltage is applied to read out the photon-derived changes and to achieve gains through multiple carrier re-circulation. In contrast to photoconductors, a built-in potential in a junction of photodiodes is essential for extraction of the photo-generated carriers. In phototransistors, a gate-programmable semiconducting channel is optoelectronically modulated, which might assist or hinder the charge transport while providing gain.

In this section, applications of 2D layered perovskites and perovskite nanoplatelets or nanosheets as the key component in fabrication of photodetectors are summarized and discussed.

4.1. Photodetectors based on 2D layered perovskites

Compared to perovskite film-based photodetectors, photodetectors made of perovskite single crystals exhibit more efficient photoresponsivity. In 2016, Tan *et al.* reported a solution-processed synthesis of $(C_4H_9NH_3)_2PbBr_4$ crystals on SiO_2/Si substrates under room temperature and application in the fabrication of photodetectors [58]. Individual 2D perovskite crystal-based photodetectors were constructed along with large-area monolayer single-crystal graphene film used as a protection layer from dissolution into water or acetone, which has improved the stability of this device (**Figure 6a**). More importantly, water was carefully excluded during the transfer of graphene onto the perovskite crystals through a special "dry" transfer method. Interdigital graphene electrodes were designed to improve the absorption cross section of 2D perovskite crystals (**Figure 6b**). Finally, the photodetector displayed an ultrahigh photoresponsivity up to 2100 A/W and an On/Off ratio of 10^3 (**Figure 6c** and **d**). Such an efficient photoresponse is most likely attributed to the strong adsorption of 2D perovskite crystals on and enhanced carrier collection by graphene electrodes. This report represents the first time to fabricate a novel photodetector based on 2D individual perovskite crystals combining graphene-based source-drain electrodes, and the results could open a new avenue for the development perovskite crystals based optoelectronic devices [58].

Zhou et al. reported the synthesis of three different 2D layered structures, $(C_4H_9NH_3)_2PbI_4$ (n = 1, one-layered perovskite), $(C_4H_9NH_3)_2(CH_3NH_3)Pb_2I_7$ (n = 2, two-layered perovskite) and $(C_4H_9NH_3)_2(CH_3NH_3)_2Pb_3I_{10}$ (n = 3, three layered perovskite) [74]. Especially, the photodetector based on three-layered perovskite displayed higher I_{light}/I_{dark} ratio on the order of ~10^3

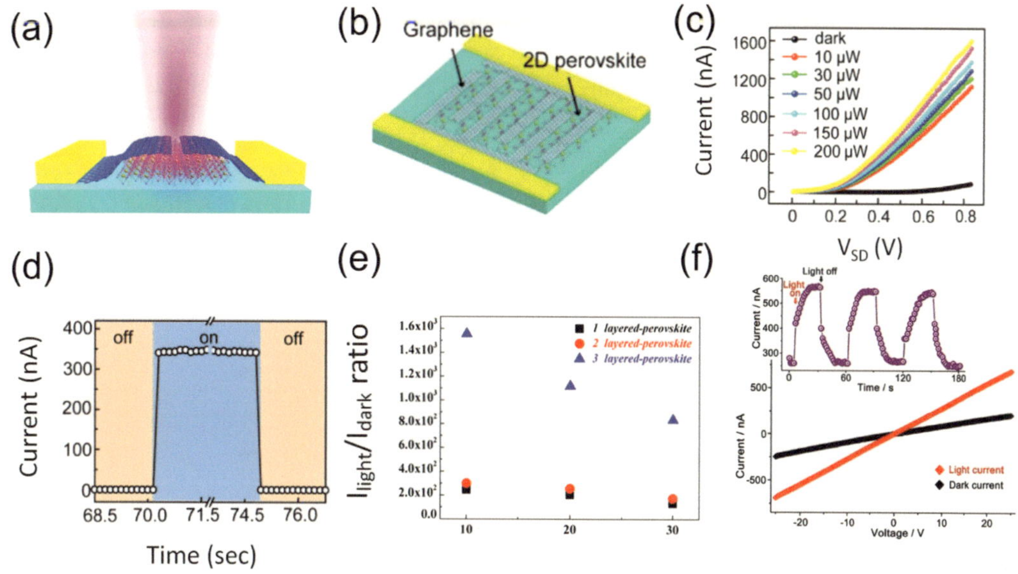

Figure 6. Schematic illustration of the structure of 2D $(C_4H_9NH_3)_2PbBr_4$-based photodetector with (a) single-crystalline graphene film connecting the source-drain top electrodes [58] and (b) interdigital graphene electrodes [58]. (c) Current-voltage curves obtained in the dark and under different illumination intensity using a 470 nm defocused laser [58]. (d) Transient photocurrent recorded [58]. (e) Dependence of I_{light}/I_{dark} ratio on applied voltage [74]. (f) Photocurrents generated by a pellet of $(TMP)_{1.5}[Bi_2I_7Cl_2]$ [75].

(**Figure 6e**). However, compared to 3D perovskites, these photodetectors showed much lower responsivities, which can be attributed to their larger optical band gap, small absorption in the UV-vis absorption spectrum as well as the limitation of charge carrier mobility from the insulting butylammonium layers. Among three photodetectors, three-layered photodetector displayed better performance than the others did. The reason might be that it has a narrow band gap and its microstructure is favorable to achieve higher output current.

More recently, the first 2D and mixed-halide anion of bismuth was synthesized by Li et al. via solution-processed method [75]. In this compound, a unique 2D honeycomb-like $[Bi_2I_7Cl_2]_n^{3n-}$ layers are isolated by TMP^{2+} cations (N,N,N′,N′-tetramethylpiperazine). The bandgap of this material is 2.10 eV according to optical measurements. For the electrical conductivities, pellets and single crystal of $(TMP)_{1.5}[Bi_2I_7Cl_2]$ displayed higher conductivity of 1.59×10^{-6} S/cm and 2.37×10^{-6} S/cm, respectively. Under the illumination of 450 W Xenon lamp and external applied voltage of 25 V, the photocurrent increased to 0.68 μA (**Figure 6f**). Furthermore, no material decomposition was observed after stored in air for 1 week, showing superior stability over commonly used $APbI_3$ [75].

4.2. Photodetectors using 2D perovskite nanoplatelets or nanosheets

Song et al. first reported the fabrication of atomically thin, 2D $CsPbBr_3$ nanosheets on ITO/PET substrates as a flexible and ultrathin photodetector via centrifugal casting method (**Figure 7a**) [76]. $CsPbBr_3$ nanosheets were synthesized with a thickness of ~3.3 nm and an edge length of ~1 μm, and their typical shape is tetragonal. In contrast to organometallic halide perovskites, the as-prepared $CsPbBr_3$ nanosheets are well dispersible in several organic solvents. UV–vis absorption spectra indicated a favorable absorption capability with a direct bandgap of ~2.32 eV for the nanosheets.

Linear and symmetrical photocurrent plots further evidenced the uniform dispersion of perovskite nanosheets on the ITO electrode [76]. A high light on/off ratio of >10^3 under 442 nm laser was demonstrated, showing good light-switching behavior. And a peak value of 0.25 A W^{-1} was obtained under 5 V at 517 nm, as shown in **Figure 7b**, comparable to that of commercial Si photodetectors (< 0.2 A W^{-1}). Besides, external quantum efficiency (EQE) was measured to be 53% at 515 nm and 10 V bias voltage (**Figure 7c**). The rise and decay time of the $CsPbBr_3$

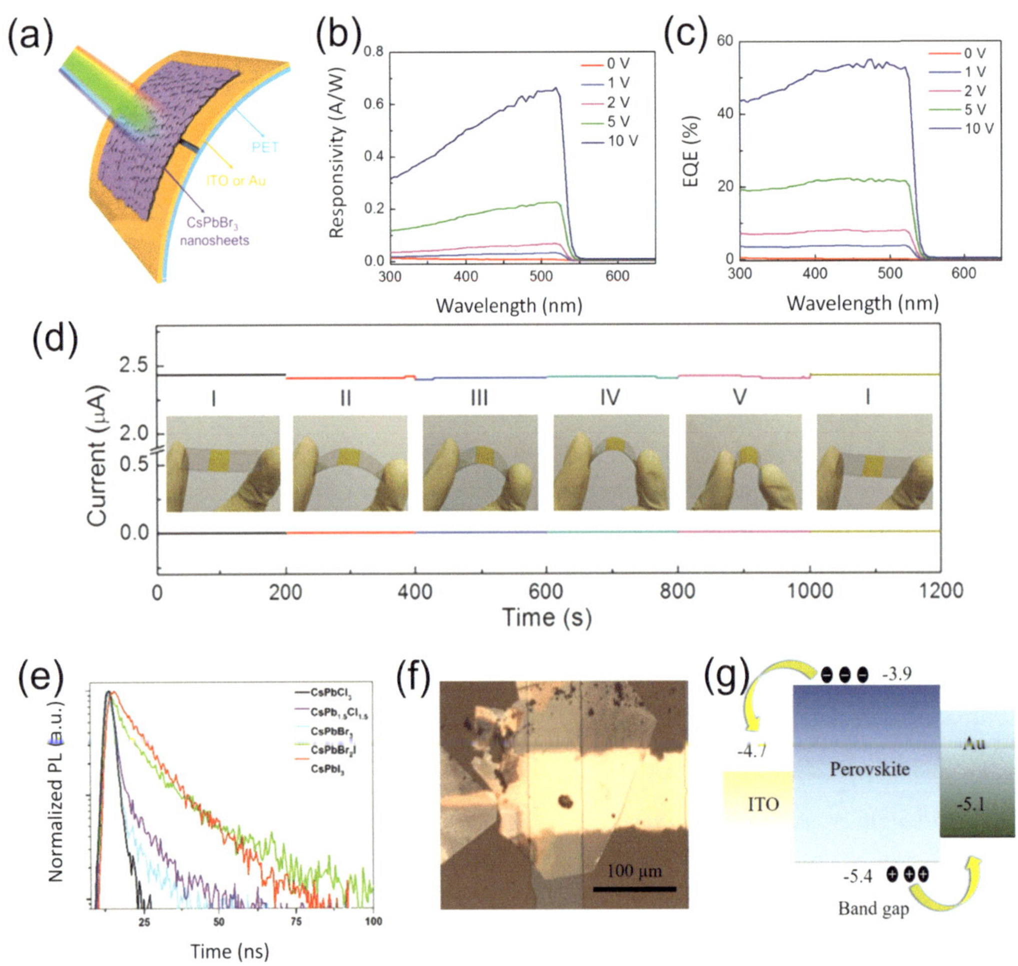

Figure 7. (a) Schematic diagram of the photodetector structure based on $CsPbBr_3$ nanosheets [76]. (b) Responsivity [76] and (c) EQE curves of $CsPbBr_3$ nanosheets based photodetectors under different forward biases [76]. (d) I–t curves of the flexible photodetector at different bending states with an applied bias voltage of 5.0 V. The insets show corresponding photographs of the device under the different bending states [76]. (e) Time-resolved PL decay transient spectra for $CsPbBr_3$ nanosheet thin films [76]. (f) Optical microscopic image of the device structure for the $CH_3NH_3PbI_3$ vertical-type photodetector [78]. (g) The band alignment for the ITO/perovskite/Au vertical heterostructure [79].

Device component and structure	Perovskite structure	Rise-decay time (ms)	On/off ratio	Responsivity (A W^{-1})	Ref.
Au/$(C_4H_9NH_3)_2PbBr_4$/graphene	Single crystal	–	10^3	2100	[58]
Glass/$(C_4H_9NH_3)_2PbI_4$/Au	Nanocrystal	28.4/27.5	1×10^2	3.00	[74]
Glass/$(C_4H_9NH_3)_2(CH_3NH_3)Pb_2I_7$/Au	Nanocrystal	8.4/7.5	1×10^2	7.31	[74]
Glass/$(C_4H_9NH_3)_2PbI_4$/Au	Nanocrystal	10.0/7.5	1×10^3	12.78	[74]
Pt/$(TMP)_{1.5}[Bi_2I_7Cl_2]$/Pt	Single crystal	–	3.4	–	[75]
ITO/$CsPbBr_3$/ITO	Nanosheet	0.019/0.024	$> 10^4$	0.64	[76]
Si/SiO_2/$CH_3NH_3PbI_3$/Ti/Au	Nanosheet	–	1×10^2	22	[77]
Au/$CsPbBr_3$/Au	Nanosheet	14.7/15.2	1×10^2	–	[78]
ITO/$CH_3NH_3PbI_3$/Au	Nanosheet	320/330	–	0.036	[79]
Au/$CH_3NH_3PbI_3$/Au	Nanosheet	230/190	–	–	[79]

Table 1. Comparison of the performances of 2D perovskites-based photodetectors.

photodetector were 19 and 25 μs, respectively. Meanwhile, the photodetectors were resistant to external bending stress as the photocurrents and dark currents remained unchanged at different bending states (**Figure 7d**). Also, after bending and recovering for >1000 times, the photodetector still showed high light on/off ratio, showing very high stability. In contrast to serious degradation of $MAPbBr_3$ photodetectors, $CsPbBr_3$ nanosheet photodetector almost remained unchanged under ambient environment. These results open up the opportunity for the practical application of 2D perovskite materials used as photodetectors due to their high stability [76]. In another example, Liu et al. reported a combined solution processing and vapor-phase conversion method to synthesize $CH_3NH_3PbI_3$ nanosheets as thin as a single unit cell (~ 1.3 nm), which were used in fabrication of a field-effect transistor [77]. This device displayed enhanced photocurrent upon illumination as well as high ratio of photocurrent to dark current, attributing to strong light-matter interaction and broad-band light-harvesting capability in a single-unit-cell nanosheet. Under illumination of a 405 nm laser, photoresponsivity can reach 22 A W^{-1}. Besides, the rise and decay time of the photodetector shorter than 20 and 40 ms exhibited a much faster response time than bulk films, showing potential application for optoelectronic switches and photodetectors.

Cyan-emitting $CsPbBr_3$ nanosheets based photodetector was reported by Lv et al. [78]. The source-drain current of $CsPbBr_3$ nanosheet film in the dark was detected to be on the order of 10^{-12} A due to the existence of long-chain organic ligands on the nanosheet surface. Under illumination of a laser diode at 450 nm, a significant increase in photocurrents was observed. Besides, excellent optical switching and stability of this photodetector were also proved under 1 Hz pulse laser (450 nm) at a fixed light density of 13 mW cm^{-2} with a bias voltage of 1 V. Through fitting the photocurrent response curve (**Figure 7e**), the rise time was estimated to be 17.8 ms and the decay times were determined to be 14.7 and 15.2 ms [78]. Li et al. used a combined solution-processed and vapor-phase conversion method for the synthesis of 2D $CH_3NH_3PbI_3$ nanosheets with a thickness of ~20 nm [79]. Vertical- and planar-type photodetectors based on $CH_3NH_3PbI_3$

nanosheets were tested to evaluate their optoelectronic properties (**Figure 7f**). With increasing the incident laser power, the photocurrent increased to 600 nA at the power of 98.56 μW. The rise and decay times of the photodetector were measured to be 320 and 330 ms, respectively. Comparable to planar-type photodetector, the vertical-type photodetector displayed higher photoresponsivity, about 36 mA W^{-1} under excitation power of 50.82 μW, resulting from short transmission distance of the photocarriers between the electrodes in the vertical photodetector and appropriate band energy distribution (**Figure 7g**). All these results might open a new avenue for the investigation of 2D perovskites including their optical, electronic and optoelectronic properties and for potential application as practical photodetectors.

Other photodetectors based on 2D halide perovskites have also been reported. **Table 1** compares the structure and key performance parameters of recently reported photodetectors.

5. Theoretical analysis and simulations

First-principles calculations based on density functional theory (DFT) have been used to simulate the electronic structures of many body systems from atomic level quantum mechanics. Recently, perovskite-based materials including lead-free perovskites, 3D perovskites, especially 2D perovskites have been studied using first-principles calculations, exploring their electron/hole effective masses, band gaps, carrier mobility, theoretical absorption spectra and other properties related to their potential applications. Besides DFT calculations, the local density approximation (LDA) and the generalized gradient approximation (GGA) functional have also been used in many systems due to their simplicity. However, these two methods usually underestimate band gaps of solids by 30–100%. Therefore, Green-function approximation (GWA) [80], time-dependent DFT (TDDFT), [81] HSE06, [82] and PBE0 hybrid functionals, [83] and more recent Delta self-consistent-field method [84] were mainly employed in theoretical calculations of perovskites. Spin-orbit coupling (SOC) is required to be included in calculations with the presence of heavy atoms to achieve a more accurate prediction of electronic structures.

For lead-free peroskites, first-principles quantum mechanical calculations were used to investigate their thermodynamic stability and photovoltaic-related properties. To this end, a series of design metrics including thermodynamic and crystallographic stability, light absorption, carrier effective mass, dopability, exciton binding and others should be considered [85]. For example, the latest calculations have identified 18 compounds out of ~100 possible candidates as most-likely targets (**Figure 8a**). Furthermore, the structure-property relationship was also studied and showed that changes in microscopic structures could significantly affect photovoltaic performance. Theoretical computations could serve as a roadmap for facilitating experimental discovery of alternative solar-energy hybrid perovskites. Sn- and Ge-based alternatives were reported to be more stable than Pb-based perovskites. However, they displayed poor performance in solar cells largely because of their intrinsic p-type conductivity and decrease in photocurrent density, and fill factor. Therefore, other factors beyond intrinsic bulk properties should be included in order to evaluate their performance more reliably.

Using DFT the electronic band structure and density of states of some 2D perovskites were computed, which confirmed that these 2D perovskites are indeed direct band gap semiconductors.

Their valence band mostly consists of I 5p states with a small amount of Pb 6 s, while their conduction band primarily consists of Pb 6p states. For example, the band gap is 1.99 eV for $(BA)_2PbI_4$, 1.78 eV for $(BA)_2(MA)_2Pb_3I_{10}$ and 0.96 eV for $(BA)_2(MA)_3Pb_4I_{13}$ (**Figure 8b** and **c**), respectively. It is found that the band gap decreases with increasing the number of perovskite layers. Besides, the structural stability of these materials was predicted by DFT, finding that non-centrosymmetric structures in 2D perovskites with a large number of layers are favorable. However, there is still a complex relationship between the perovskite layer thickness and lateral dimension [59]. Akkerman et al. reported the SOC-DFT calculations to investigate the electronic properties of $CsPbBr_3$ NPLs (**Figure 8d**), which is highly consistent with the experimentally observed bandgap evolution [70]. SOC effects can determine electronic and optical properties of perovskites as well as their quantum confinement behavior. However, deviations become obvious as the quantum wells decreased to ultrasmall dimension. This is mainly attributed to the tunneling of the electron/hole wave functions outside the well boundaries such as the lack of effectively

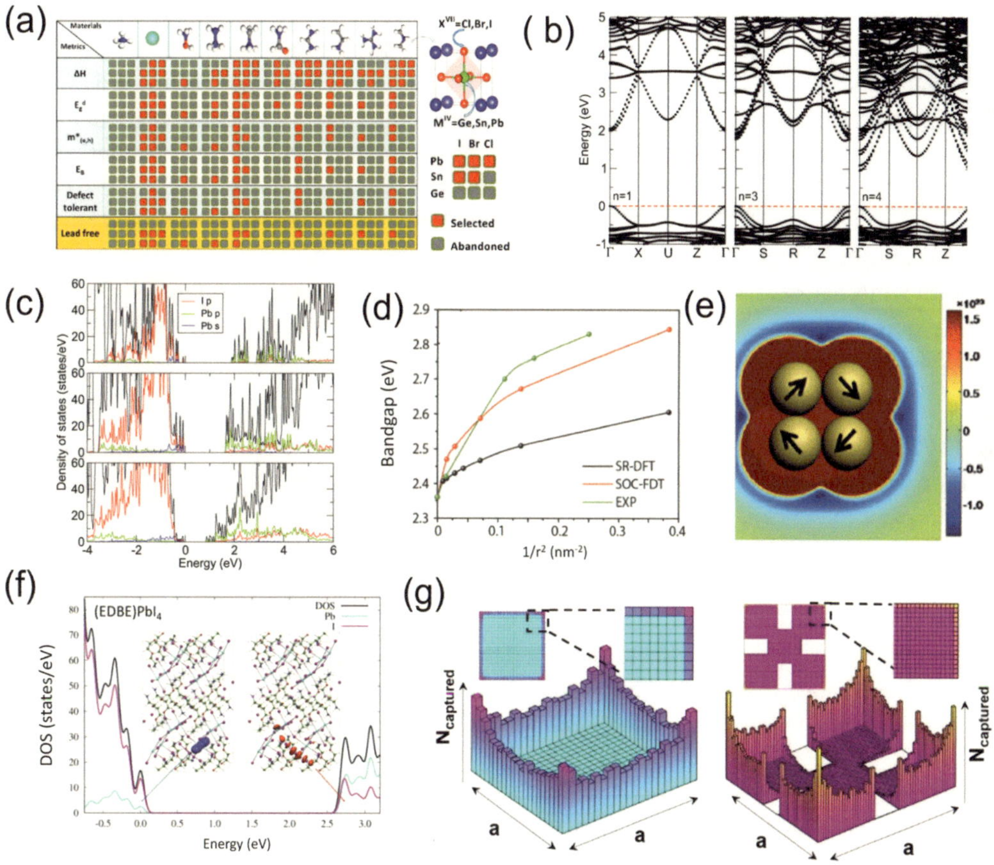

Figure 8. (a) Screening process for the potential superior bulk-material-intrinsic photovoltaic performance by first-principle calculations [85]. (b) Electronic band structure [59] and (c) DOS results of the polar configurations in $(BA)_2(MA)_{n-1}Pb_nI_{3n+1}$ [59]. (d) Band gaps from experiment and SR-, SOC-DFT calculations [70]. (e) Simulations of total potential field in a four-dot alignment [53]. (f) DOS curves of the neutral lead vacancy and isodensity plot of the defect states in $(EDBE)PbI_4$ [73]. (g) 3D bar charts showing Monte Carlo simulation results for a square (left) and fractal (right) seeds [50].

infinite potential barriers at the NPL border. In addition, the effective masses and the exciton both reduced with increasing quantum confinement, suggesting ambipolar transport occurred in $CsPbBr_3$ crystals.

To better understand the self-organization mechanisms of $CH_3NH_3PbBr_3$ QDs (or nanodots (NDs)) into NPLs, Liu et al. performed theoretical simulations based on the classical Derjaguin-Landau-Verwey-Overbeek (DLVO) theory [53]. They suggested that during the self-organization process, two or four coalescent QDs might be considered as stable units. Two coalescent QDs attracted each other from bottom and top to form a four-dot alignment under dipole-dipole attraction (**Figure 8e**). With a combination of VDW and electrostatic driving forces, the self-organization of $CH_3NH_3PbBr_3$ QDs preferred via side-side orientation to form nanoplatelets. Consequently, the axis size of NPLs is about four-fold that of building block QDs.

Cortecchia et al. have performed ab initio calculations to investigate the structural effects on the defectivity and energetic landscape for (EDBE)PbI_4. For the neutral Pb vacancy, formation of I_3^- trimer is preferred in (EDBE)PbI_4 by ~1 eV, which is driven by under-coordinated iodine atoms at the organic/inorganic interface with partial charge stabilized by $EDBE^{2+}$ [73]. However, iodine atoms per organic cation in $(NBT)_2PbI_4$ has a stronger electrostatic interaction, which prevents the vicinal iodine motion from forming I_3^-. Hybrid DFT calculations including spin-orbit coupling (PBE0-SOC) was also conducted to verify the formation of I_3^- trimer in (EDBE)PbI_4 (**Figure 8f**). An optical transition at 0.21 eV corresponds to the highest occupied molecular orbital (HOMO)-Lowest unoccupied molecular orbital (LUMO) transition of I_3^-, leading to the emergence of Stokes-shifted, broadened luminescence according to the relaxation mechanism for trapped charge carriers. With the templating cation assisting charge localization, the increased structural distortions in the PbX_6 octahedral coordination in (EDBE) PbI_4 lower the energy barrier for self-trapping and thus stabilize the trimer I_3^-.

In addition, Monte Carlo calculations were used to explain the growth mechanism of 2D sheets of $MAPbCl_3$ on mica [50]. The VDW nucleation and growth model was proposed (**Figure 8g**) [50]. The competition between the VDW diffusion and cohesive energy of perovskites can in principle increase the thickness of perovskite sheets. This means much thicker layered perovskites could be prepared through surface passivation. Furthermore, the development of two-photon-pumped laser combined theoretical simulations (e.g. two-photon absorption (TPA)) would enable to study certain perovskites towards understanding their nonlinear optical properties. For example, Chen et al. have recently revealed that the TPA process proceeds through a virtual level and then ended in an exciton band state. This observation would lay a solid physic basis for the development of high-performance nonlinear optical perovskite materials [86].

6. Concluding remarks and outlook

Metal halide perovskites have increasingly attracted research interests in recent years, largely because of their remarkable optical and electronic properties such as high carrier mobility, small exciton binding energy, high photoluminescence quantum yield, narrow bandwidth, long charge diffusion length and broadly tunable bandgap. They can further exhibit unique

http://dx.doi.org/10.5772/intechopen.71032

optical properties and enhanced stability in their 2D form due to quantum confinement effects, which has led to increasing interest in the design and assembly of 2D perovskites for their applications in photodetectors. This chapter mainly focused on the introduction of synthesis methods, structure characterizations for both 2D layered perovskites and 2D perovskites in shape, along with summarizing photodetection application of these two types of 2D perovskites.

Researchers have made tremendous progress in studying 2D perovskite-based photodetection techniques, but some challenges from both basic science and device engineering perspective remain to be overcome. Recent studies have shown the feasibility of preparation of perovskite nanosheets and nanoplatelets, but the control of their morphology and thickness needs to be improved. Due to the quantum confinement effects, the thickness of nanoplatelets or nanosheets is a critical factor for their optical properties. Therefore, the development of improved synthetic methods for better controlling morphology and thickness are indeed required, which has been well achieved in conventional semiconductors and metal nanocrystals. With the development of two-photon-pumped laser for studying $CsPbBr_3$ QDs, the nonlinear optical properties of perovskites have just been taken to the front of the research, such as two-photon absorption (TPA). Although 2D metal halide perovskites have displayed excellent optical properties, water and air stability is still a detrimental problem towards their practical applications. Some recent researches have attempted to solve this issue. For example, organic molecular capping, silica or Al_2O_3 coatings were used to improve the stability, which has been to some extent successful. However, further efforts still need to find more general ways with broad suitability for 2D perovskites. Meanwhile, another important issue, similar to stability, is the toxicity of Pb element, which is one of the most insidious factors for the environment and human. Albeit with the excellent properties of lead-containing 2D perovskites, it still exposes as a detriment factor for their applications. Many attempts have been made to substitute Pb with nontoxic metal elements, but the resulting perovskites normally do not have the same level performance. For example, Sn-based perovskites have appropriate bandgaps for optoelectronic applications, but they display much lower stability than Pb-containing perovskites due to the oxidation of Sn^{2+} to Sn^{4+} under ambient conditions. For Ge- and Bi-based perovskites, they indeed display high stability. However, these two types of perovskites possess a large bandgap, which limits the light absorption in long-wavelength range. Therefore, it is high desirable to design new 2D layered perovskites with appropriate bandgap and large absorption coefficient and high stability.

In terms of photodetection applications, being comparable to conventional colloidal quantum-dot-based photodetectors much improvement is needed for perovskites. For example, perovskite-based photoconductors suffer from slow photoresponse, low sensitivity and large electrical hysteresis. Some recent efforts have partially overcome these issues by using surface passivation, phototransistor gating and fabricating hybrid bilayer photodetectors. Interfacing perovskite layers with other functional nanomaterials is proposed to be an effective way for fabricating perovskite-based optoelectronic devices. In a broader perspective, perovskite-based materials could be tuned and engineered for fabrication of new-type devices that are endowed with new functions such as multi-spectral sensing.

In conclusion, in view of their novel structures and optical properties we would believe that 2D perovskites are on the way becoming promising materials for next-generation optoelectronic devices.

Acknowledgements

This work was supported by Independent Research Fund Denmark-Nature Sciences (DFF-FNU, to Q.C.) and the Villum Foundation (to Q.C. and Y.T.). Y.T. is grateful for the Villum Foundation supported postdoc fellowship. X.C. is grateful for the support from the Chinese Scholarship Council (PhD scholarship No. 201406170040).

Author details

Yingying Tang, Xianyi Cao and Qijin Chi*

*Address all correspondence to: cq@kemi.dtu.dk

Department of Chemistry, Technical University of Denmark, Kongens Lyngby, Denmark

References

[1] Büchele P, Richter M, Tedde SF, Matt GJ, Ankah GN, Fischer R, Biele M, Metzger W, Lilliu S, Bikondoa O, Macdonald JE, Brabec CJ, Kraus T, Lemmer U, Schmidt O. X-ray imaging with scintillator-sensitized hybrid organic photodetectors. Nature Photonics. 2015;**9**:843-848. DOI: 10.1038/nphoton.2015.216

[2] Huang SM, Huang SJ, Yan YJ, SH Y, Chou M, Yang HW, Chang YS, Chen RS. Extremely high-performance visible light photodetector in the Sb_2SeTe_2 nanoflake. Scientific Reports. 2017;**7**:45413. DOI: 10.1038/srep45413

[3] Sobhani A, Knight MW, Wang YM, Zheng B, King NS, Brown LV, Fang ZY, Nordlander P, Halas NJ. Narrowband photodetection in the near-infrared with a plasmon-induced hot electron device. Nature Communications. 2013;**4**:1643. DOI: 10.1038/ncomms2642

[4] Karim A, Andersson JY. Infrared detectors: Advances, challenges and new technologies. IOP Conference Series: Materials Science and Engineering. 2013;**51**:012001. DOI: 10.1088/1757-899X/51/1/012001

[5] Jansen-van Vuuren RD, Pivrikas A, Pandey AK, Burn PL. Colour selective organic photodetectors utilizing ketocyanine-cored dendrimers. Journal of Materials Chemistry C. 2013;**1**:3532-3543. DOI: 10.1039/C3TC30472H

[6] Ko WS, Bhattacharya I, Tran TTD, Ng KW, Gerke SA, Chang-Hasnain C. Ultrahigh responsivity-bandwidth product in a compact InP nanopillar phototransistor directly grown on silicon. Scientific Reports. 2016;**6**:33368. DOI: 10.1038/srep33368

[7] Mokkapati S, Jagadish C. III-V compound SC for optoelectronic devices. Materials Today. 2009;**12**:22-32. DOI: 10.1016/S1369-7021(09)70110-5

[8] de Arquer PGF, Armin A, Meredith P, Sargent EH. Solution-processed semiconductors for next-generation photodetectors. Nature Reviews Materials. 2017;**2**:16100. DOI: 10.1038/natrevmats.2016.100

[9] McNeill R, Siudak R, Wardlaw J, Weiss D. Electronic conduction in polymers. I. The chemical structure of polypyrrole. Australian Journal of Chemistry. 1963;**16**:1056-1075. DOI: 10.1071/CH9631056

[10] Coffey DC, Ferguson AJ, Kopidakis N, Rumbles G. Photovoltaic charge generation in organic semiconductors based on long-range energy transfer. ACS Nano. 2010;**4**:5437-5445. DOI: 10.1021/nn101106b

[11] Huang J, Miragliotta J, Becknell A, Katz HE. Hydroxy-terminated organic semiconductor-based field-effect transistors for phosphonate vapor detection. Journal of the American Chemical Society. 2007;**129**:9366-9376. DOI: 10.1021/ja068964z

[12] Nath D, Dey P, Deb D, Rakshit JK, Roy JN. Fabrication and characterization of organic semiconductor based photodetector for optical communication. CSI Transactions on ICT. 2017;**5**:149-160. DOI: 10.1007/s40012-016-0150-8

[13] Guo YL, Liu C, Tanaka H, Nakamura E. Air-stable and solution-processable perovskite photodetectors for solar-blind UV and visible light. The Journal of Physical Chemistry Letters. 2015;**6**:535-539. DOI: 10.1021/jz502717g

[14] Pei YL, Pei RH, Liang XC, Wang YH, Liu L, Chen HB, Liang J. CdS-nanowires flexible photo-detector with Ag-nanowires electrode based on non-transfer process. Scientific Reports. 2016;**6**:21551. DOI: 10.1038/srep21551

[15] Oertel DC, Bawendia MG. Photodetectors based on treated CdSe quantum-dot films. Applied Physics Letters. 2005;**87**:213505. DOI: 10.1063/1.2136227

[16] Mi LF, YQ Y, Wang H, Zhang Y, Yao XD, Chang YJ, Wang XM, Jiang Y. Ultrasensitive PbS-quantum-dot photodetectors for visible–near-infrared light through surface atomic-ligand exchange. Particle & Particle Systems Characterization. 2015;**32**:1102-1109. DOI: 10.1002/ppsc.201500138

[17] Zhou YY, Zhu K. Perovskite solar cells shine in the "Valley of the Sun". ACS Energy Letter. 2016;**1**:64-67. DOI: 10.1021/acsenergylett.6b00069

[18] Congreve DN, Weidman MC, Seitz M, Paritmongkol W, Dahod NS, Tisdale WA. Tunable light-emitting diodes utilizing quantum-confined layered perovskite emitters. ACS Photonics. 2017;**4**:476-481. DOI: 10.1021/acsphotonics.6b00963

[19] Sutherland BR, Johnston AK, Ip AH, JX X, Adinolfi V, Kanjanaboos P, Sargent EH. Sensitive, fast, and stable perovskite photodetectors exploiting interface engineering. ACS Photonics. 2015;**2**:1117-1123. DOI: 10.1021/acsphotonics.5b00164

[20] Wolf SD, Holovsky J, Moon SJ, Löper P, Niesen B, Ledinsky M, Haug FJ, Yum JH, Ballif C. Organometallic halide perovskites: Sharp optical absorption edge and its relation to photovoltaic performance. The Journal of Physical Chemistry Letters. 2014;**5**:1035-1039. DOI: 10.1021/jz500279b

[21] Du Y, Cai H, Wu Y, Xing Z, Li Z, Xu J, Huang L, Ni J, Li J, Zhang J. Enhanced planar perovskite solar cells with efficiency exceeding 16% via reducing the oxygen vacancy defect state in titanium oxide electrode. Physical Chemistry Chemical Physics. 2017;**19**:13679-13686. DOI: 10.1039/c7cp01936j

[22] Guo Z, Manser JS, Wan Y, Kamat PV, Huang LB. Spatial and temporal imaging of long-range charge transport in perovskite thin films by ultrafast microscopy. Nature Communications. 2015;**6**:7471. DOI: 10.1038/ncomms8471

[23] Wang YC, Zhang YM, Liu YT, Pang TQ, ZY H, Zhu YJ, Luan SZ, Jia RX. Temperature-dependence studies of organolead halide perovskite-based metal/semiconductor/metal photodetectors. RSC Advances. 2017;**7**:20206-20211. DOI: 10.1039/C7RA01496A

[24] Su L, Zhao ZX, Li HY, Yuan J, Wang ZL, Cao GZ, Zhu G. High-performance organolead halide perovskite-based self-powered triboelectric photodetector. ACS Nano. 2015;**9**:11310-11316. DOI: 10.1021/acsnano.5b04995

[25] Zhao FY, Xu K, Luo X, Lv WL, Peng YQ, Wang Y, Lu FP, Xu SN. Ultrasensitivity broadband photodetectors based on perovskite: Research on film crystallization and electrode optimization. Organic Electronics. 2017;**46**:35-43. DOI: 10.1016/j.orgel.2017.03.023

[26] Park Y, Lee S, Park HJ, Baac HW, Yoo G, Heo J. Hybrid metal-halide perovskite-MoS_2 phototransistor. Journal of Nanoscience and Nanotechnology. 2016;**16**:11722-11726. DOI: 10.1166/jnn.2016.13581

[27] Nam KT, Shelby SA, Choi PH, Marciel AB, Chen R, Tan L, Chu TK, Mesch RA, Lee BC, Connolly MD, Kisielowski C, Zuckermann RN. Free-floating ultrathin two-dimensional crystals from sequence-specific peptoid polymers. Nature Materials. 2010;**9**:454-460. DOI: 10.1038/nmat2742

[28] Zhang H. Ultrathin two-dimensional nanomaterials. ACS Nano. 2015;**9**:9451-9469. DOI: 10.1021/acsnano.5b05040

[29] Lin Z, Carvalho BR, Kahn E, Lv RT, Rao R, Terrones H, Pimenta MA, Terrones M. Defect engineering of two-dimensional transition metal dichalcogenides. 2D Mater. 2016;**3**:022002. DOI: 10.1088/2053-1583/3/2/022002

[30] Cai SL, Zhang WG, Zuckermann RN, Li ZT, Zhao X, Liu Y. The organic flatland-recent advances in synthetic 2D organic layers. Advanced Materials. 2015;**27**:5762-5770. DOI: 10.1002/adma.201500124

[31] Stine R, Lee WK, WhitenerJr KE, Robinson JT, Sheehan PE. Chemical stability of graphene fluoride produced by exposure to XeF_2. Nano Letter. 2013;**13**:4311-4316. DOI: 10.1021/nl4021039

[32] Luan VH, Tien HN, Hoa LT, Hien NTM, Oh ES, Chung JS, Kim EJ, Choi WM, Kong BS, Hur SH. Synthesis of a highly conductive and large surface area graphene oxide hydrogel and its use in a supercapacitor. Journal of Materials Chemistry A. 2013;**1**:208-211. DOI: 10.1039/C2TA00444E

[33] Samuels AJ, Carey JD. Engineering graphene conductivity for flexible and high-frequency applications. ACS Applied Materials & Interfaces. 2015;**7**:22246-22255. DOI: 10.1021/acsami.5b05140

[34] Jang MS, Brar VW, Sherrott MC, Lopez JJ, Kim L, Kim S, Choi M, Atwater HA. Tunable large resonant absorption in a midinfrared graphene Salisbury screen. Physical Review B. 2014;**90**:165409. DOI: 10.1103/PhysRevB.90.165409

[35] Cassabois G, Valvin P, Gil B. Hexagonal boron nitride is an indirect bandgap semiconductor. Nature Photonics. 2016;**10**:262-266. DOI: 10.1038/nphoton.2015.277

[36] Mak KF, Shan J. Photonics and optoelectronics of 2D semiconductor transition metal dichalcogenides. Nature Photonics. 2016;**10**:216-226. DOI: 10.1038/nphoton.2015.282

[37] XG Y, Marks TG, Facchetti A. Metal oxides for optoelectronic applications. Nature Materials. 2016;**15**:383-396. DOI: 10.1038/nmat4599

[38] Pizzi G, Gibertini M, Dib E, Marzari N, Iannaccone G, Fiori G. Performance of arsenene and antimonene double-gate MOSFETs from first principles. Nature Communications. 2016;**7**:12585. DOI: 10.1038/ncomms12585

[39] Khandelwal A, Mani K, Karigerasi MH, Lahiri I. Phosphorene–the two-dimensional black phosphorous: Properties, synthesis and applications. Materials Science and Engineering: B. 2017;**221**:17-34. DOI: 10.1016/j.mseb.2017.03.011

[40] Sichert JA, Tong Y, Mutz N, Vollmer M, Fischer S, Milowska KZ, Cortadella RG, Nickel B, Cardenas-Daw C, Stolarczyk JK, Urban AS, Feldmann J. Quantum size effect in organometal halide perovskite nanoplatelets. Nano Letter. 2015;**15**:6521-6527. DOI: 10.1021/acs.nanolett.5b02985

[41] Zhang YP, Liu JY, Wang ZY, Xue YZ, QD O, Polavarapu L, Zheng JL, Qia X, Bao QL. Synthesis, properties, and optical applications of low-dimensional perovskites. Chemical Communications. 2016;**52**:13637-13655. DOI: 10.1039/c6cc06425f

[42] Jeon T, Kim SJ, Yoon J, Byun J, Hong HR, Lee TW, Kim JS, Shin B, Kim SO. Hybrid perovskites: Effective crystal growth for optoelectronic applications. Advanced Energy Materials. 2017:1602596. DOI: 10.1002/aenm.201602596

[43] Sun SB, Yuan D, Xu Y, Wang AF, Deng ZT. Ligand-mediated synthesis of shape-controlled cesium lead halide perovskite nanocrystals via reprecipitation process at room temperature. ACS Nano. 2016;**10**:3648-3657. DOI: 10.1021/acsnano.5b08193

[44] Bekenstein Y, Koscher BA, Eaton SW, Yang PD, Alivisatos AP. Highly luminescent colloidal nanoplates of perovskite cesium lead halide and their oriented assemblies. Journal of American Chemical Society. 2015;**137**:16008-16011. DOI: 10.1021/jacs.5b11199

[45] Pan AZ, He B, Fan XY, Liu ZK, Urban JJ, Alivisatos AP, He L, Liu Y. Insight into the ligand-mediated synthesis of colloidal $CsPbBr_3$ perovskite nanocrystals: The role of organic acid, base, and cesium precursors. ACS Nano. 2016;**10**:7943-7954. DOI: 10.1021/acsnano.6b03863

[46] Shi YM, Li HN, Li LJ. Recent advances in controlled synthesis of two-dimensional transition metal dichalcogenides via vapour deposition techniques. Chemical Society Reviews. 2015;**44**:2744-2756. DOI: 10.1039/c4cs00256c

[47] Nayak PK, Horbatenko Y, Ahn S, Kim G, Lee JU, Ma KY, Jang AR, Lim H, Kim D, Ryu s CH, Park N, Shin HS. Probing evolution of twist-angle-dependent interlayer excitons in $MoSe_2$/WSe_2 van der Waals heterostructures. ACS Nano. 2017;**11**:4041-4050. DOI: 10.1021/acsnano.7b00640

[48] Ha ST, Liu XF, Zhang Q, Giovanni D, Sum TC, Xiong QH. Synthesis of organic–inorganic lead halide perovskite nanoplatelets: Towards high-performance perovskite solar cells and optoelectronic devices. Advanced Optical Materials. 2014;**2**:838-844. DOI: 10.1002/adom.201400106

[49] Xing GC, Mathews N, Sun SY, Lim SS, Lam YM, Grätzel M, Mhaisalkar S, Sum TC. Long-range balanced electron- and hole-transport lengths in organic-inorganic $CH_3NH_3PbI_3$. Science. 2013;**342**:344. DOI: 10.1126/science.1243167

[50] Wang YP, Shi YF, Xin GQ, Lian J, Shi J. Two-dimensional van der Waals expitaxy kinetics in a three-dimensional perovskite halide. Crystal Growth & Design. 2015;**15**:4741-4749. DOI: 10.1021/acs.cgd.5b00949

[51] Wang YL, Guan X, Li DH, Cheng HC, Duan XD, Lin ZY, Duan XF. Chemical vapor deposition growth of single-crystalline cesium lead halide microplatelets and heterostructures for optoelectronic applications. Nano Research. 2017;**10**:1223-1233. DOI: 10.1007/s12274-016-1317-1

[52] Wang Q, Li D, Cheng HC, Li Y, Chen CY, Yin A, Zhao Z, Lin Z, Wu H, He Q, Ding M, Liu Y, Huang Y, Duan X. Wafer-scale growth of large arrays of perovskite microplate crystals for functional electronics and optoelectronics. Science Advances. 2015;**1**:e1500613. DOI: 10.1126/sciadv.1500613

[53] Liu LG, Huang S, Pan LF, Shi LJ, Zou BS, Deng LG, Zhong HZ. Colloidal synthesis of $CH_3NH_3PbBr_3$ nanoplatelets with polarized emission through self-organization. Angewandte Chemie International Edition. 2017;**56**:1780-1783. DOI: 10.1002/anie.201610619

[54] Jodlowski AD, Yépez A, Luque R, Camacho L, de Miguel G. Benign-by-benign solventless mechanochemical synthesis of three-, two-, and one dimensional hybrid perovskites. Angewandte Chemie International Edition. 2016;**55**:14972-14977. DOI: 10.1002/anie.201607397

[55] Mitzi DB. Synthesis, crystal structure, and optical and thermal properties of $(C_4H_9NH_3)_2MI_4$ (M = Ge, Sn, Pb). Chemistry of Materials. 1996;**8**:791-800. DOI: 10.1021/cm9505097

[56] Calabrese J, Jones NL, Harlow RL, Herron N, Thorn DL, Wang Y. Preparation and characterization of layered lead halide compounds. Journal of American Chemical Society. 1991;**113**:2328-2330. DOI: 10.1021/ja00006a076

[57] Xu Z, Mitzi DB, Medeiros DR. $[(CH_3)_3NCH_2CH_2NH_3]SnI_4$: A layered perovskite with quaternary/primary ammonium dications and short interlayer iodine-iodine contacts. Inorganic Chemistry. 2003;**42**:1400-1402. DOI: 10.1021/ic0261981

[58] Tan ZJ, Wu Y, Hong H, Yin JB, Zhang JC, Lin L, Wang MZ, Sun X, Sun LZ, Huang YC, Liu KH, Liu ZF, Peng HL. Two-dimensional $(C_4H_9NH_3)_2PbBr_4$ perovskite crystals for high-performance photodetector. Journal of American Chemical Society. 2016;**138**:16612-16615. DOI: 10.1021/jacs.6b11683

[59] Stoumpos CC, Cao DH, Clark DJ, Young J, Rondinelli JM, Jang JI, Hupp JT, Kanatzidis MG. Ruddlesden-popper hybrid lead iodide perovskite 2D homologous semiconductors. Chemistry of Materials. 2016;**28**:2852-2867. DOI: 10.1021/acs.chemmater.6b00847

[60] Smith IC, Hoke ET, Solis-Ibarra D, McGehee MD, Karunadasa HI. A layered hybrid perovskite solar-cell absorber with enhanced moisture stability. Angewandte Chemie International Edition. 2014;**53**:11232-11235. DOI: 10.1002/anie.201406466

[61] Leng MY, Chen ZW, Yang Y, Li Z, Zeng K, Li KH, Niu GD, He YS, Zhou QC, Tang J. Lead-free, blue emitting bismuth halide perovskite quantum dots. Angewandte Chemie International Edition. 2016;**55**:15012-15016. DOI: 10.1002/anie.201608160

[62] Wang KH, Wu L, Li L, Yao HB, Qian HS, Yu SH. Large-scale synthesis of highly luminescent perovskite-related $CsPb_2Br_5$ nanoplatelets and their fast anion exchange. Angewandte Chemie International Edition. 2016;**55**:8328-8332. DOI: 10.1002/anie.201602787

[63] Vargas B, Ramos E, Pérez-Gutiérrez E, Alonso JC, Solis-Ibarra D. A direct bandgap copper-antimony halide perovskite. Journal of American Chemical Society. DOI: 10.1021/jacs.7b04119

[64] Levchuk I, Herre P, Brandl M, Osvet A, Hock R, Peukert W, Schweizer P, Spiecker E, Batentschuk M, Brabec CJ. Ligand-assisted thickness tailoring of highly luminescent colloidal $CH_3NH_3PbX_3$ (X = Br and I) perovskite nanoplatelets. Chemical Communications. 2017;**53**:244-247. DOI: 10.1039/c6cc09266g

[65] Zhang Q, Su R, Liu XF, Xing J, Sum TC, Xiong QH. High quality whispering-gallery-mode lasing from cesium lead halide perovskite nanoplatelets. Advanced Functional Materials. 2016;**26**:6238-6245. DOI: 10.1002/adfm.201601690

[66] Zhang Q, Ha ST, Liu XF, Sum TC, Xiong QH. Room-temperature near-infrared high-Q perovskite whispering-gallery planar nanolasers. Nano Letter. 2014;**14**:5995-6001. DOI: 10.1021/nl503057g

[67] Liao Q, Hu K, Zhang HH, Wang XD, Yao JN, Fu HB. Perovskite microdisk microlasers self-assembled from solution. Advanced Materials. 2015;**27**:3405-3410. DOI: 10.1002/adma.201500449

[68] Dou L, Wong AB, Yu Y, Lai M, Kornienko N, Eaton SW, Fu A, Bischak CG, Ma J, Ding T, Ginsberg NS, Wang LW, Alivisatos AP, Yang P. Atomically thin two-dimensional organic-inorganic hybrid perovskites. Science. 2015;**349**:1518-1521. DOI: 10.1126/science.aac7660

[69] Yang S, Niu WX, Wang AL, Fan ZX, Chen B, Tan CL, Lu QP, Zhang H. Ultrathin two-dimensional organic–inorganic hybrid perovskite nanosheets with bright, tunable photoluminescence and high stability. Angewandte Chemie International Edition. 2017;**56**:1-5. DOI: 10.1002/adma.201605448

[70] Akkerman QA, Motti SC, Kandada ARS, Mosconi E, D'Innocenzo V, Bertoni G, Marras S, Kamino BA, Miranda L, Angelis FD, Petrozza A, Prato M, Manna L. Solution synthesis approach to colloidal cesium lead halide perovskite nanoplatelets with monolayer-level thickness control. Journal of American Chemical Society. 2016;**138**:1010-1016. DOI: 10.1021/jacs.5b12124

[71] Billing DG, Lemmerer A. Synthesis, characterization and phase transitions in the inorganic-organic layered perovskite-type hybrids $[(C_nH_{2n+1}NH_3)_2PbI_4]$, n = 4, 5 and 6. Acta Crystallographic Section B: Structural Science. 2007;**63**:735-747. DOI: 10.1107/S0108768107031758

[72] Dohner ER, Jaffe A, Bradshaw LR, Karunadasa HI. Intrinsic white-light emission from layered hybrid perovskites. Journal of American Chemical Society. 2014;**136**:13154-13157. DOI: 10.1021/ja507086b

[73] Cortecchia D, Neutzner S, Kandada ARS, Mosconi E, Meggiolaro D, Angelis FD, Soci C, Petrozza A. Broadband emission in two-dimensional hybrid perovskites: The role of structural deformation. Journal of American Chemical Society. 2017;**139**:39-42. DOI: 10.1021/jacs.6b10390

[74] Zhou JC, Chu YL, Huang J. Photodetectors based on two-dimensional layer-structured hybrid lead iodide perovskite semiconductors. ACS Applied Materials & Interfaces. 2016;**8**:25660-25666. DOI: 10.1021/acsami.6b09489

[75] Li MQ, YQ H, Bi LY, Zhang HL, Wang YY, Zheng YZ. Structure tunable organic-inorganic bismuth halides for an enhanced two-dimensional lead-free light-harvesting material. Chemistry of Materials. DOI: 10.1021/acs.chemmater.7b01017

[76] Song JZ, LM X, Li JH, Xue J, Dong YH, Li XM, Zeng HB. Monolayer and few-layer all-inorganic perovskites as a new family of two-dimensional semiconductors for printable optoelectronic devices. Advanced Materials. 2016;**28**:4861-4869. DOI: 10.1002/adma.201600225

[77] Liu JY, Xue YZ, Wang ZY, Xu ZQ, Zheng CX, Weber B, Song JC, Wang YS, Lu YY, Zhang YP, Bao QL. Two-dimensional $CH_3NH_3PbI_3$ perovskite: Synthesis and optoelectronic application. ACS Nano. 2016;**10**:3536-3542. DOI: 10.1021/acsnano.5b07791

[78] Lv LF, YB X, Fang HH, Luo WJ, FJ X, Liu LM, Wang BW, Zhang XF, Yang D, WD H, Dong AG. Generalized colloidal synthesis of high-quality, two-dimensional cesium lead halide perovskite nanosheets and their applications in photodetectors. Nanoscale. 2016;**8**:13589-13596. DOI: 10.1039/c6nr03428d

[79] Li PF, Shivananju BN, Zhang YP, Li SJ, Bao QL. High performance photodetector based on 2D $CH_3NH_3PbI_3$ perovskite nanosheets. Journal of Physics D: Applied Physics. 2017;**50**:094002. DOI: 10.1088/1361-6463/aa5623

[80] Stan A, Dahlen NE, Leeuwen RV. Levels of self-consistency in the GW approximation. The Journal of Chemical Physics. 2009;**130**:114105. DOI: 10.1063/1.3089567

[81] Imamura Y, Nakai H. Time-dependent density functional theory (TDDFT) calculations for core-excited states: Assessment of an exchange functional combining the Becke88 and van Leeuwen–Baerends-type functionals. Chemical Physics Letters. 2006;**419**:297-303. DOI: 10.1016/j.cplett.2005.11.084

[82] Wen XD, Martin RL, Scuseria GE, Rudin SP, Batista ER, Burrell AK. Screened hybrid and DFT + U studies of the structural, electronic, and optical properties of U_3O_8. Journal of Physics: Condensed Matter. 2013;**25**:025501. DOI: 10.1088/0953-8984/25/2/025501

[83] Betzinger M, Friedrich C, Blügel S. Hybrid functionals within the all-electron FLAPW method: implementation and applications of PBE0. Physical Review B. 2010;**81**:195117. DOI: 10.1103/PhysRevB.81.195117

[84] Gavnholt J, Olsen T, Engelund M, Schiøtz J. Delta self-consistent field as a method to obtain potential energy surfaces of excited molecules on surfaces. Physical Review B. 2008;**78**. DOI: 075441, 10.1103/PhysRevB.78.075441

[85] Yang DW, Lv J, Zhao XG, QL X, YH F, Zhan YQ, Zunger A, Zhang LJ. Functionality-directed screening of Pb-free hybrid organic-inorganic perovskites with desired intrinsic photovoltaic functionalities. Chemistry of Materials. 2017;**29**:524-538. DOI: 10.1021/acs.chemmater.6b03221

[86] Chen JS, Žídek K, Chábera P, Liu DZ, Cheng PF, Nuuttila L, Al-Marri MJ, Lehtivuori H, Messing ME, Han KL, Zheng KB, Pullerits T. Size- and wavelength-dependent two-photon absorption cross-section of $CsPbBr_3$ perovskite quantum dots. The Journal of Physical Chemistry Letters. 2017;**8**:2316-2321. DOI: 10.1021/acs.chemmater.6b03221

Chapter 5

Design and Development of Two-Dimensional Strained Layer Superlattice (SLS) Detector Arrays for IR Applications

Ashok K. Sood, John W. Zeller, Roger E. Welser,
Yash R. Puri, Nibir K. Dhar,
Priyalal S. Wijewarnasuriya and Sanjay Krishna

Additional information is available at the end of the chapter

http://dx.doi.org/10.5772/intechopen.71328

Abstract

The implementation of strained layer superlattices (SLS) for detection of infrared (IR) radiation has enabled compact, high performance IR detectors and two-dimensional focal plane arrays (FPAs). Since initially proposed three decades ago, SLS detectors exploiting type II band structures existing in the InAs/GaSb material system have become integral components in high resolution thermal detection and imaging systems. The extensive technological progress occurring in this area is attributed in part to the band structure flexibility offered by the nearly lattice-matched InAs/AlSb/Ga(In)Sb material system, enabling the operating IR wavelength range to be tailored through adjustment of the constituent strained layer compositions and/or thicknesses. This has led to the development of many advanced type II SLS device concepts and architectures for low-noise detectors and FPAs operating from the short-wavelength infrared (SWIR) to very long-wavelength infrared (VLWIR) bands. These include double heterostructures and unipolar-barrier structures such as graded-gap M-, W-, and N-structures, nBn, pMp, and pBn detectors, and complementary barrier infrared detector (CBIRD) and pBiBn designs. These diverse type II SLS detector architectures have provided researchers with expanded capabilities to optimize detector and FPA performance to further benefit a broad range of electro-optical/IR applications.

Keywords: strained layer superlattice, infrared sensors, focal plane array, thermal imaging, type II

1. Introduction

Infrared (IR) sensors have been and continue to be developed for a variety of commercial and military systems applications, which include short-wavelength infrared (SWIR), mid-wavelength infrared (MWIR), long-wavelength infrared (LWIR), and very long-wavelength infrared (VLWIR) detectors [1]. Conventional SWIR detectors operate over the ~0.9–2.5 μm region above the visible band [2]. MWIR sensors sensitive in the 3–5 μm region have typically been based on InSb or HgCdTe (MCT) [3]. In addition, LWIR and VLWIR sensors for 8–12 μm applications commonly utilize MCT or microbolometer technology [4]. Quantum well infrared photodetectors (QWIPs) are another relatively recent IR sensing technology, which can be designed to operate over the MWIR to VLWIR range.

While large format III–V InSb focal plane arrays (FPAs) have been developed, the fixed bandgap of InSb detectors restricts their operation to the MWIR, and such devices are limited to cryogenic temperatures [5, 6]. MCT detectors are known for high quantum efficiencies (above 90%) and detectivities, but require expensive and relatively scarce CdZnTe substrates for optimal performance, and nonuniform growth defects are a substantial limiting factor for MCT-based FPAs [7]. In addition, performance limitations of inexpensive microbolometers, particularly in relation to sensitivity and speed, are well-known [8].

Since their initial development, IR photodetectors based on strained layer superlattices have drawn much interest from research and commercial sectors in recent decades. SLS detectors typically exploit type II band structures existing in InAs/GaSb, which constitutes an attractive material system for developing MWIR and LWIR detectors with advanced properties. As illustrated in **Figure 1(a)** [9], type II superlattices are characterized by a staggered band alignment in which the conduction band of the InAs layer is lower than the valence band of the InGaSb layer [10]. The implementation of barrier layers in the photoconductor structure in type II SLS detectors prevents current flow of minority and/or majority carriers, leading to higher performance antimonide-based FPAs [11]. In such structures, varying the thickness of the InAs layers can have a substantial impact on the spectral response (**Figure 1(b)**) [9].

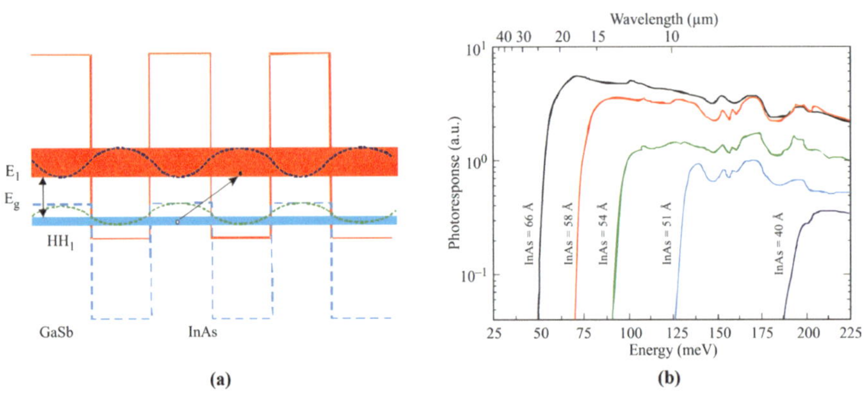

Figure 1. Band edge diagram of InAs/GaSb SLS, illustrating the confined electron and hole minibands which form the energy bandgap; and (b) charted change in cutoff wavelength with InAs thickness for fixed GaSb thickness of 40 Å in type II SLS detector [9].

In the nearly lattice-matched InAs/AlSb/Ga(In)Sb 6.1 Å family material system, type I (nested, or straddling), type II staggered, and type II broken-gap (misaligned, or type III) energy-band alignments are all realizable [8]. The 6.1 Å family material includes wide, medium, and narrow gap components, with GaSb, insulating AlSb, and high mobility InAs (see **Figure 2**) [12, 13]. Type II InAs/GaSb SLS detectors offer broad design band structure flexibility: it is possible in these materials to tailor the IR operating wavelength from 3 μm to around 30 μm, covering most of the practical IR wavelength spectrum [14]. For example, the effective bandgap (and corresponding detection wavelength) of an InAs/Ga(In)Sb type II SLS can be varied continuously (theoretically in the range of 0 to about 400 meV) by changing the thicknesses of constituent layers and/or ternary compound composition [11]. Since the electron-hole overlap in such detectors is controlled by the thicknesses of the constituent layers, it is possible to fabricate small bandgap materials using "mid-gap" semiconductors.

The strain within the superlattice (SL) layers in type II SLS detectors create a large splitting between the heavy-hole and light-hole bands in the ternary superlattices [15]. This reduces the hole-hole Auger recombination process and increases the minority carrier lifetime, thereby improving the device detectivity (D^*) and lowering the required operating temperature [16]. By optimizing the oscillator strength in this material system, a large quantum efficiency and responsivity can likewise be obtained [17]. In addition, type II SLS detectors based on the 6.1 Å family of materials can be passively cooled, thus reducing the cryocooler burden, and these take advantage of the relatively large installed III–V material manufacturing base [18]. These properties have enabled the fabrication of large format IR FPAs based on type II SLS suitable for high-resolution thermal imaging applications including space-borne surveillance systems, low-background night vision, and missile detection [19–26].

InAs/GaSb in type II SLS is the only known type of IR detector material having theoretically higher performance than precedential HgCdTe (MCT), due primarily to longer Auger lifetimes [28]. Also, the larger effective mass in InAs/Ga(In)Sb leads to a reduction of tunneling currents in type II SLS detectors compared with MCT detectors of the same bandgap, while high electron mobility and diffusivity is maintained [29, 30]. In addition, III–V materials offer

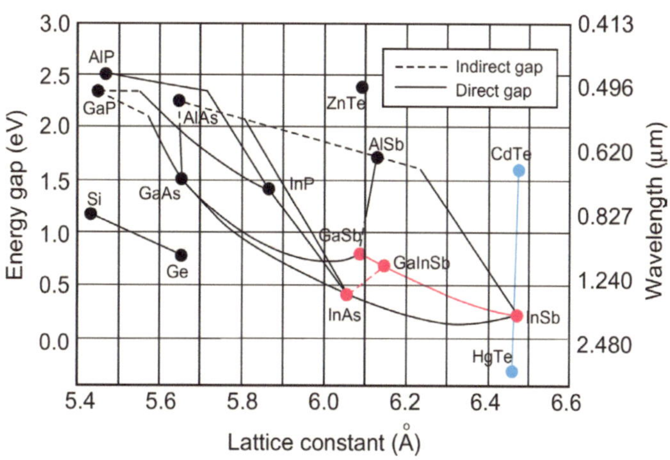

Figure 2. Compositional and wavelength/energy gap dependence diagram of antimonide-based III–V material systems [12].

much stronger chemical bonds compared to MCT, potentially providing higher chemical stability and thus better producibility [8]. Furthermore, SLS detectors can detect normal incident IR radiation, in contrast to *n*-type QWIPs that are thus prevented due to polarization selection rules of intersubband transitions that limit quantum efficiencies to 10–20% [31]. These features, combined with the suppression of Auger recombination due in part to the decoupled spin-orbital splitting with SL bandgap, make InAs/Ga(In)Sb SLS a very competitive material system for next generation IR detectors and FPA technologies. **Figure 3** presents a comparison of thermal detectivities versus temperature calculated for various detector technologies, including SLS, HCT, and QWIP [27].

The existence of type II staggered energy band alignments in the 6.1 Å material family allow the development of many advanced heterojunction device concepts and architectures for reduced noise IR SLS detectors, one of which is shown in **Figure 4** [11]. These include graded-gap W- [19], M- [20], and N-structures [32]; buried junction nBn [21], pBp [22], and pBn [33] designs; and complementary barrier infrared detector (CBIRD) [23] and pBiBn [24] architectures. These device implementations are the result of exploitation of the material, electrical, and optical properties in type II SL materials for optimization of detector performance, which can benefit and advance a diverse array of applications.

This chapter will cover the development of type II superlattice materials and detector device architectures for next generation IR FPA technologies, primarily for MWIR and longer wavelength applications. Section 2 discusses the progress and achievements of type II SL-based technological development up to the present. This includes the historical development of detector architectures for low-noise IR detection and their noteworthy performance features, as well as a discussion of the nature and relevance of the different IR bands to SLS detectors and FPAs. Section 3 presents a theoretical analysis of noise components common to SLS detectors and their respective impact on design considerations and performance. Section 4 focuses on the compositions, features, and benefits of the various optimized type II SLS detector architectures developed. In particular, relevant aspects among individual dark current

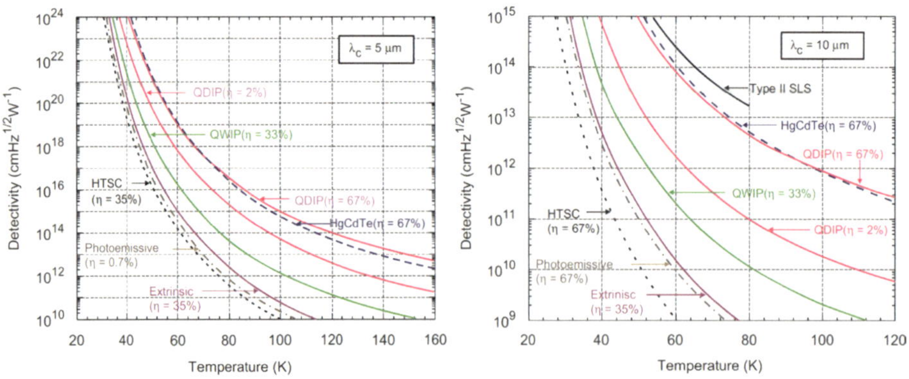

(a) **(b)**

Figure 3. Calculated thermal detectivity as function of temperature for various MWIR (λ_c = 5 μm) (a) and LWIR (λ_c = 10 μm) (b) photodetectors, where predicted quantum efficiencies are indicated [27].

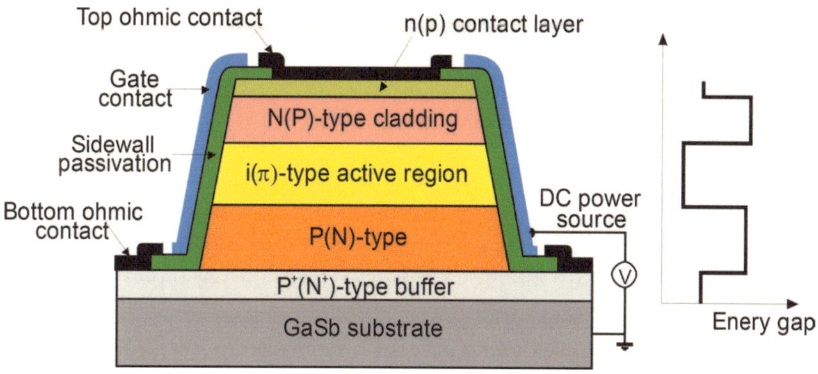

Figure 4. Schematic band diagram of *n-i-p* double heterostructure antimonide-based III–V photodiode [11].

mechanisms with respect to the material quality, operating conditions, and performance of the type II SLS detector structures will be addressed. Finally, in Section 5, notable trends and practical progress made in the development of type II SLS detectors and FPAs are discussed that have had and/or are expected to have a significant impact on the status of IR sensing and imaging technology.

2. Technological development of type II SLS detectors

2.1. Historical development of SLS detectors

2.1.1. Development of type II SLS detectors

The developmental timeline of type II SLS detector technology commenced in the 1970s when investigations were carried out, most notably by Esaki and Tsu, of near lattice-matched InAs/GaSb superlattices systems as alternatives to GaAs/AlGaAs superlattices [34]. The first fundamental superlattices structure, based on InAsSb/InAsSb and capable of achieving small infrared energy gaps with strong interactions between them, was reported in 1977 by Sai-Halasz et al. [35]. A decade later, in 1987, Smith and Mailhiot proposed an InAsSb/InSb type II SLS for IR detection applications, which was predicted theoretically to provide favorable optical properties including small diode tunneling currents and good mobilities and diffusion lengths [36]. By 1997, the device material quality had been greatly improved and band-to-band (BTB) tunneling currents suppressed, enabling fabrication of devices using the molecular beam epitaxy (MBE) process. One early MWIR SLS detector device developed around this time demonstrated a responsivity of 2 A/W, as well as Johnson noise-limited detectivity in excess of 10^{12} cm$\sqrt{\text{Hz}}$/W, or Jones [37].

In 2007, Rodriguez et al. demonstrated the first type II SLS nBn detector that was capable of performing up to 300 K due to reduction of both Shockley-Read-Hall (SRH) generation currents and surface recombination currents [7]. The pBiBn design (LWIR) was first introduced in 2010 by Gautam et al., which due to its unipolar blocking layers exhibited comparatively low (at the time) dark current density of 1.2 mA/cm^2 [24], and in 2012 this group demonstrated MWIR pBiBn detectors capable room temperature operation [38]. That same year DeCuir et al. reported a type II SLS pBiBn structure with 50% cutoff wavelength of 8.7 μm that demonstrated low dark current density of ~10 μA/cm^2 at −50 mV [39].

By 2014, LWIR nBn photodetectors had been developed demonstrating peak responsivity of ~4.5 A/W at 7.9 μm, corresponding to a quantum efficiency of 54%, though the dark current density (e.g., 440 μA/cm^2 at 77 K) was considerably higher than that for pBiBn structure devices [43]. **Figure 5(a)** and **(b)** compare the dark current performance of type II SLS detectors from various groups with a "Rule 07" model providing a theoretical prediction of MCT detector performance as a function of wavelength and over time, respectively [7, 110, 40–42]. Likewise, **Figure 6(a)** and **(b)** chart the carrier lifetimes at 77 K for HgCdTe and type II SLS detectors operating at MWIR and LWIR wavelengths, respectively, as a function of doping concentration [11].

2.1.2. Development of type II SLS FPAs

The development of type II SLS FPAs brought a degree of practical utility that had not previously been realized to SLS device research and development. The first FPA based on type II SLS detector pixels, reported in 2002 by Cabanski et al., was a MBE-grown, 256 × 256 pixel, 40 μm pitch MWIR device that demonstrated a noise equivalent temperature difference (NETD) of under 12 mK [44]. In 2006, the first LWIR FPA based on type II SLS material was reported by Razeghi et al., which incorporated an M-structure SL design to enhance its wavelength tunability [45]. That same year a dual-band FPA was demonstrated, combining spectral selective detection in the 3.0–4.1 μm and 4.1–5.0 μm wavelength ranges at each pixel [46].

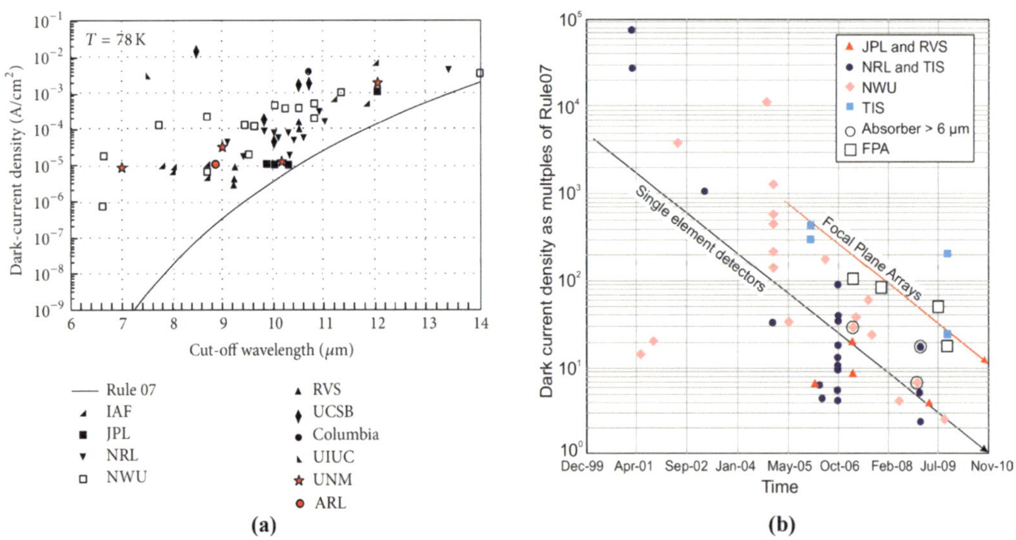

Figure 5. (a) Dark current densities plotted against cutoff wavelength for type II SLS non-barrier and barrier detectors at 78 K [7, 11]. The solid line indicates the dark current density calculated using the empirical "rule 07" model, and the circle indicates results from DeCuir et al. [39] (b) Detector dark current density as multiple of rule 07 [40–42]. The black line represents the trend line of dark current reduction over time for single element detectors, while the red line shows this trend for FPAs.

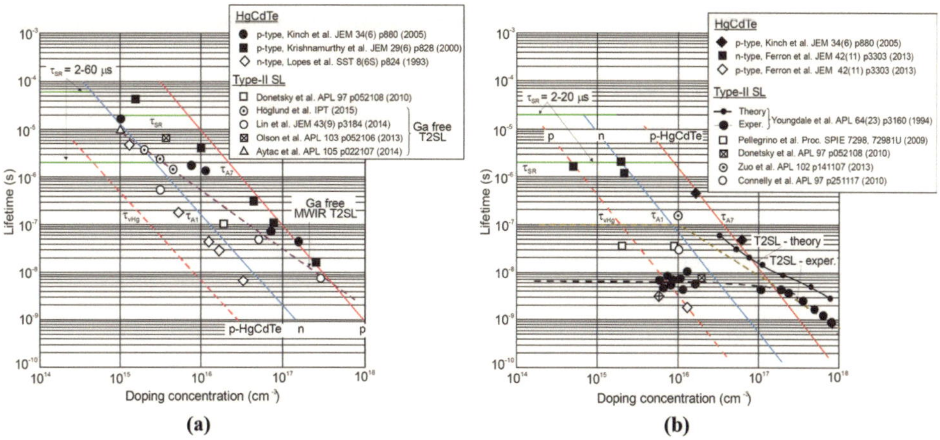

Figure 6. Carrier lifetimes for (a) HgCdTe and type II SLS MWIR detectors, and (b) HgCdTe and type II SLS LWIR detectors, at 77 K as a function of doping concentration. The dashed lines follow experimental data [11].

The first MWIR SLS FPA with megapixel (1024 × 1024) resolution was reported in 2009 by Hill et al., from which detectivities as high as 8 × 10^{13} Jones were measured [47]. The following year, Gunapala et al. demonstrated the first type II SLS megapixel FPA for LWIR detection, which yielded a NETD of 53 mK at 80 K [48]. From 2010 to 2012, further large format LWIR type II SLS FPAs were demonstrated; these were characterized by pixel pitches of around 20 μm and peak quantum efficiencies as high as 96% [49, 50]. In 2016, Klipstein et al. reported a 640 × 512 type II SLS LWIR FPA with pixel pitch as small as 15 μm [51]. **Figure 7(a)–(c)** show the detector pixel architecture for a recently reported type II SLS MWIR FPA featuring a *p-i-n* design, a top-view image of the detector pixel array, and a captured image from the FPA, respectively [40].

2.2. Impact of IR bands on SLS performance

2.2.1. Short-wavelength infrared (SWIR)

In contrast to the longer wavelength IR bands, the SWIR range (~0.9–2.5 μm) involves reflected light rather than thermal radiation, thus producing more visible-like images [52]. The shorter

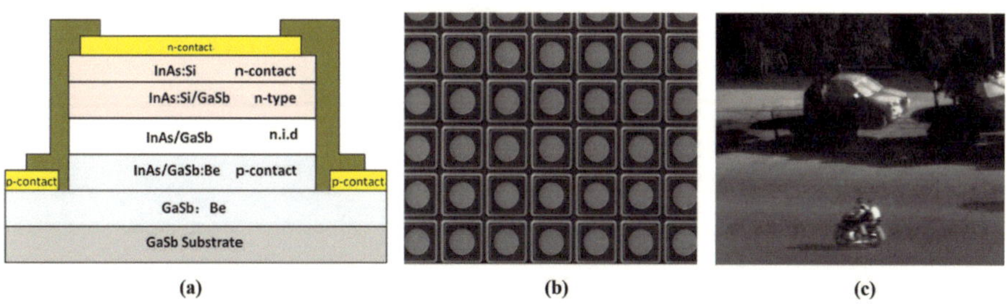

Figure 7. (a) Cross-section schematic of type II InAs/GaSb SLS MWIR FPA detector pixel. (b) Top-view SEM image of type II SLS FPA chip; and (c) outdoor image from FPA at 80 K [40].

wavelengths over this range also enable smaller SWIR FPA pixel sizes. Development of SWIR detectors based on type II SLS is comparatively challenging but has been accomplished, for example, by adding a third AlAs layer to the superlattice allowing the effective bandgap to be raised (**Figure 8**) [53], though this necessitates some degree of lattice mismatch [54]. SWIR detection has been combined on single SLS devices with both MWIR and LWIR detection capability to offer both active and passive multiband imaging [52]. SWIR detection applications include telecommunications, astronomical observation, remote sensing, spectroradiometry, spectrophotometry, and hostile mortar fire detection [53, 55].

2.2.2. Mid-wavelength infrared (MWIR)

MWIR detection involves passive imaging of electromagnetic wavelengths in the of 3–5 μm atmospheric transparent window. Rather than sensing reflected radiation as occurring at shorter wavelengths, detection of MWIR and longer wavelengths involves capturing thermal signatures, or temperature contrasts. A primary goal for MWIR detection has been to elevate the operating temperature; however, in this regard care must be taken in temperature regimes where carriers are close to the intrinsic, since at this point the thermally generated carrier concentration becomes comparable to the doping concentration [56].

The MWIR and LWIR bands are both commonly utilized for thermal IR detection, but there are significant differences between them that impact their respective detection and imaging performance. For example, the scene contrast at 300 K is 3.5–4% in the MWIR band compared to 1.7% for the LWIR band, resulting in higher attainable NETDs for the latter [57]. In addition, undesirable atmospheric water vapor effects are often present for MWIR wavelengths up to ~7.5 μm that can greatly decrease contrast at significant distances and/or in humid environments [58]. MWIR detection is utilized for many applications (a number of which are also common to LWIR), which include aerial and satellite reconnaissance, target tracking, navigation, and object identification [54].

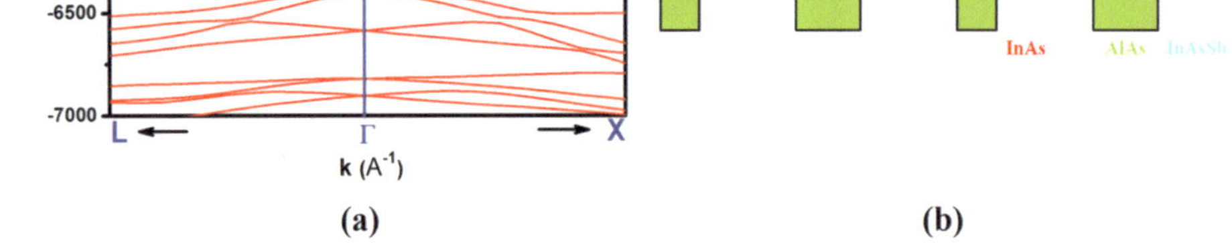

Figure 8. (a) Band structure of SWIR SLS design around C-point, where energy levels are referenced to the vacuum level. (b) Superlattice structure and formation of effective bandgap, where colored rectangles represent forbidden bandgaps of the different layers in the SL [53].

2.2.3. *Long-wavelength infrared (LWIR)*

The LWIR spectral band typically refers to wavelengths over the 8–14 μm range, though it can likewise be meant to describe the 8–12 μm wavelength range. LWIR detectors have traditionally incorporated relatively inexpensive microbolometer technology with low power requirements, but associated drawbacks include higher noise and lower sensitivities and speeds [8]. Based on blackbody physics, LWIR type II SL-based imaging systems offer higher temperature dynamic range and greater in-band photon radiance for a given target temperature, as well as improvements in speed, temperature range, uniformity, and stability over conventional MWIR detectors such as those based on InSb [59].

In LWIR SL-based detectors and FPAs, the decreasing energy gaps at extended IR wavelengths cause the absorbing detector materials to become more sensitive not only to bulk material properties but also to the surface states of the exposed sidewalls, so a challenge in such detectors has been to decrease the leakage current at the sidewalls of the devices in addition to the bulk dark current [54].

Since humans emit IR radiation with a peak intensity corresponding to approximately 9.2 μm, the LWIR band is best suited for applications involving observing or monitoring people in some capacity [60]. LWIR detection applications include industrial inspection, high-speed imaging, capturing high temperature targets, atmospheric absorption monitoring, and enhanced night vision [61].

2.2.4. *Very long-wavelength infrared (VLWIR)*

Various type II SLS detectors and FPAs have been developed capable of detection and imaging of VLWIR wavelengths (~12–30 μm) involving narrower bandgaps than for MWIR and LWIR. At present, VLWIR detector technology is dominated by MCT and extrinsic silicon blocked impurity band (BIB) detectors. Type II SLS detectors have been shown theoretically to offer better VLWIR performance above the 14 μm cutoff (beyond this issues with nonuniformity and excessive dark current arise) than MCT and BIB detectors, the latter of which require very low operating temperatures due to high thermal ionization [62, 63].

There have been considerable improvements in the optical and electrical qualities of VLWIR ternary SL materials in recent years, particularly lower Ga content, which has been achieved largely through optimization of molecular beam epitaxy (MBE) growth parameters. These have resulted in better device performance characteristics, including longer minority carrier lifetimes (e.g., ~140 ns) for VLWIR type II SLS detectors [8].

Relatively strong dark currents and low differential resistance area products (R_0A) at zero bias have limited the ability of type II SLS VLWIR FPAs to match readout integrated circuits (ROICs) for practical imaging devices [64]. Nevertheless, much progress has been made in recent years in the development of VLWIR SLS FPAs. For example, as shown **Figure 9**, a type II InAs/InAsSb VLWIR FPA grown on a GaSb substrate was reported in 2017 that produced sharp thermal images [64], and another recently that exhibited peak responsivity of 4.8 A/W and detectivity of 1.4×10^{10} Jones [65]. Potential applications of VLWIR detection and imaging include long-range ballistic missile defense, space-based astronomy, space-borne remote sensing, remote pollution monitoring, and meteorological monitoring [66, 67].

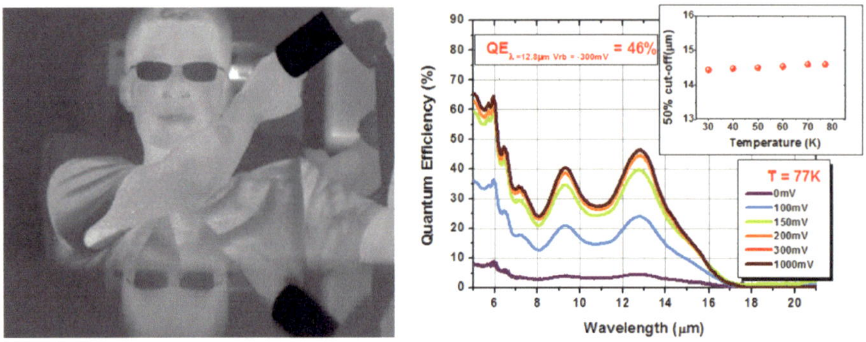

(a) (b)

Figure 9. (a) Thermal image acquired from 320 × 256 VLWIR FPA at 65 K, in which mirror image can be seen reflected on the surface of a desk [64]. (b) Quantum efficiency spectrum with varying bias at 77 K; inset shows 50% cutoff wavelength versus temperature [65].

3. Noise contributions in SLS detectors

3.1. Impact of noise on performance

3.1.1. SLS detectors

Although InAs/Ga(In)Sb type II SLS been investigated for the past three decades, many of the fundamental material and physical properties are still not well understood. This can pose challenges in identifying the major noise mechanisms for various device architectures, which is an important step towards facilitating low-noise detector designs.

For photodetectors, dark current is basically the accumulation of current contributions from diffusion current in bulk *p* and *n* regions, generation-recombination (GR) current in the depletion region, band-to-band tunneling, trap-assisted tunneling (TAT), and surface leakage current [68]. It was discovered that the InAs/GaSb 6.1 Å material family in type II SLS detectors reduces adverse BTB/TAT currents and GR Auger contribution to the total dark current [8]. In such SLS detectors, the placement of barrier layers provides a means to effectively filter the various dark current components.

3.1.2. SLS focal plane arrays (FPAs)

Among the major issues limiting the performance of IR FPAs are charge handling capacity of the readout, residual nonuniformity of the imager, and dark current of the sensitive material [41]. The effects of dark current and its spatial variation are more prominent in LWIR imagers operating at the longer wavelength end of their operational range, and therefore even more pronounced for VLWIR. The ratio of photocurrent to dark current is another major metric, as dark current (which is very sensitive to any small detector temperature fluctuation) when high fills the wells and forces a shorter integration time, resulting in a higher temporal (background limited) NETD. Therefore, high quantum efficiency (QE) and/or faster optics for FPAs are required if dark currents are not kept well below the photocurrent level. In **Figure 10(a)** and **(b)**, predicted dark current density and noise equivalent temperature difference (NEDT), respectively, versus temperature are compared for type II SLS and MCT detectors [57].

3.2. Major dark current contributions in SLS devices

In this section the major dark current components are discussed. We have sought to address the more relevant aspects among the individual dark current mechanisms with respect to material quality and device operation conditions.

3.2.1. Diffusion current

Diffusion current is present at any semiconductor junction, and it contributes to detector dark current under both forward and reverse bias. The diffusion current can be represented by the following equation [69]:

$$J_{diff} = n_i^2(T)\sqrt{qk_B T}\left(\frac{1}{N_A}\sqrt{\frac{\mu_e}{\tau_e}} + \frac{1}{N_D}\sqrt{\frac{\mu_h}{\tau_h}}\right)(e^{qV/k_BT} - 1) \tag{1}$$

where $n_i(T)$ is the intrinsic carrier concentration, q is the electron charge, T is temperature, k_B is the Boltzmann constant, N_A and N_D are acceptor and donor densities, respectively, μ_e, τ_e, and μ_h, τ_h, are mobility and lifetime for electrons and holes, respectively, and V is the applied bias voltage.

The shorter lifetimes in the absorption regions of InAs/Ga(In)Sb SLS detectors compared to those in IR detectors based on alternative material systems generally result in higher diffusion dark currents. However, the diffusion current is also dependent on the doping concentration in the junction, and the doping profile may be manipulated (e.g., through *in situ* doping profile engineering during material growth) with the objective of minimizing the diffusion current.

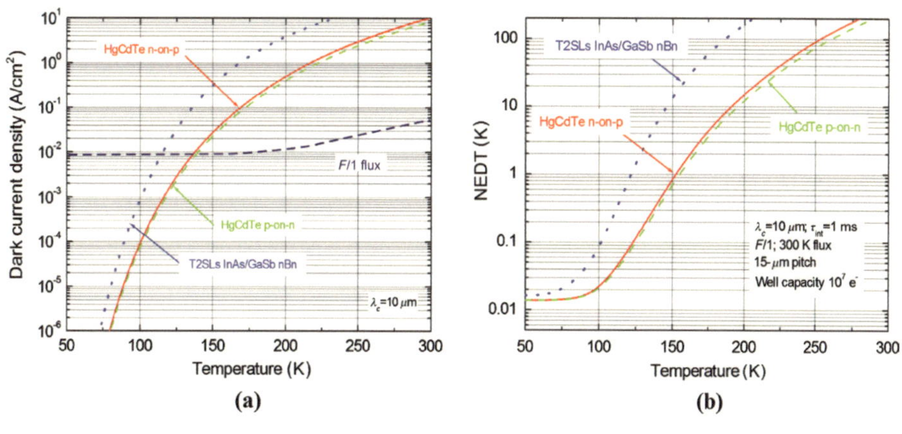

Figure 10. Calculated dark current density (a) and NETD (b) vs. temperature for HgCdTe and type II SLS InAs/GaSb nBn detectors having a cutoff wavelength (λ_c) of 10 μm [57].

3.2.2. Shockley-Read-Hall (SRH) generation-recombination (GR) current

Shockley-Read-Hall (SRH) GR dark current is mostly contributed from mid-gap defects and trap states inside semiconductor materials. These states can act as generation or recombination sites, particularly in narrow bandgap materials. In most experimental investigations performed the SRH GR current remained the major dark current component, particularly at lower operating temperatures and smaller reverse biases [38]. One means to suppress GR current is by reducing the depletion region width.

The SRH current, which is proportional to the intrinsic carrier concentration, contributes to the device noise in both bias regimes. The SRH dark current is expressed by Eq. (2) [70, 71]:

$$J_{SRH} = \frac{qn_i W}{\tau_{SRH}} \frac{2k_B T}{q(V_{bi} - V)} \sinh\left(\frac{qV}{2k_B T}\right) f(b) \tag{2}$$

$$f(b) = \int_0^{\infty} \frac{1}{u^2 + 2bu + 1} \qquad b = e^{-qV/2k_BT} \cosh\left(\frac{E_t - E_i}{k_B T}\right)$$

where W is the depletion region width, τ_{SHR} is the generation-recombination (SRH) lifetime, V_{bi} is the built-in voltage, E_i is the intrinsic Fermi level, and E_t is the trap energy level.

It is generally believed that SRH processes are the dominating factor contributing to shorter carrier lifetimes (<100 ns) in both MWIR and LWIR type II SLS detectors [72]. For LWIR binary SLS absorbers, the rapid increase in the Auger coefficient with increasing cutoff wavelength results in short electron lifetimes that are typically in the range of 15–30 ns at 77 K, and even shorter for VLWIR [15]. SRH process statistical theory suggests the occurrence of longer carrier lifetimes in bulk InAs than in bulk GaSb material based on the respective locations of the stabilized Fermi level, from which it has been theorized that native Ga-mediated defects generated during the binary SL growth are responsible for the SRH-limited minority carrier lifetimes observed in InAs/GaSb type II SLS detectors [11].

3.2.3. Trap-assisted tunneling (TAT)

TAT current becomes the dominate current mechanism under higher electrical fields with larger values of reverse bias [38]. It occurs at or near the depletion region, where minority carriers enter a trap state from the valence band and then tunnel into the conduction band. TAT is modeled by the following Eq. [73]:

$$J_{TAT} = \frac{q^2 m_T V M^2 N_t}{8\pi \hbar^3 (E_g - E_t)} \exp\left(\frac{4\sqrt{2 m_T}\,(E_g - E_t)^2}{3q\hbar F(V)}\right) \tag{3}$$

where m_T is the reduced tunneling effective mass, N_t is the trap density, M is the transition matrix element associated with the trap, $\hbar = h/2\pi$; h being Planck's constant, and $F(V)$ is the electric field.

In **Figure 11(a)** and **(b)**, the contributions of diffusion, GR, and TAT currents to dark current and dark current density, respectively, are plotted versus applied bias voltage [39, 74].

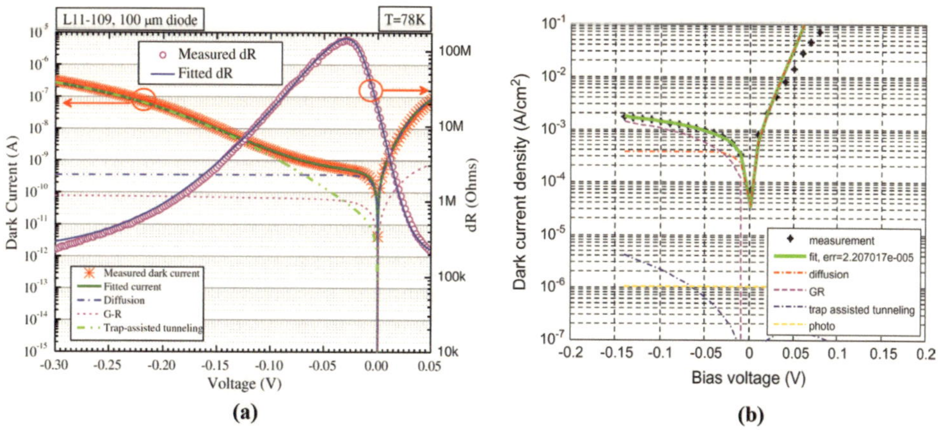

Figure 11. (a) Dark current density versus voltage for 100 μm diameter type II SLS (pBiBn) detector diode, taking into account diffusion, GR, and TAT currents. The associated differential resistance (dR) is also shown and fitted with the extrapolated dR [39]. (b) Decomposition of SLS structure device dark current at 130 K into different dark current components plotted versus bias voltage [74].

3.2.4. Band-to-band (BTB) tunneling

BTB tunneling current results from direct tunneling of carriers from the valence band into the conduction band. Like TAT, BTB is more pronounced in higher electrical field regimes and under greater reverse bias. The BTB current is also larger for higher doping levels in the active region [54].

An expression for BTB current is given by the following equation [75, 76]:

$$J_{BTB} = \frac{q^3 F(V) V}{4\pi^2 \hbar^2} \sqrt{\frac{2m_T}{E_g}} \exp\left(\frac{4\sqrt{2m_T E_g^3}}{3q\hbar F(V)}\right) \tag{4}$$

where *F(V)* is the electrical field under different bias voltages. Eq. (4) indicates that the tunneling dark current is strongly dependent on the bandgap, and consequently plays a more significant role in LWIR and VLWIR detectors.

3.2.5. Surface leakage current

Excessive surface leakage currents are known to limit the performance of mesa etched InAs/GaSb SL detectors, especially for LWIR detection [68]. During the FPA pixel isolation process, the periodic crystal structure can terminate abruptly, resulting in the formation of unsatisfied (dangling) chemical bonds responsible for generation of surface states within the bandgap [7]. These states cause pinning of the surface Fermi level near the mid-gap, consequently enhancing the surface leakage current [77]. Passivation, which for this purpose usually involves deposition of a relatively thick layer of dielectric material (usually SiO_2), reduces the number of surface states and is thus critical for minimizing surface leakage currents, especially in low bandgap materials [38, 78]. In SLS detectors, and in particular nBn structures, the placement and position of the barrier(s) can serve as a means to effectively reduce the surface leakage current [8].

4. Type II SLS detector design technologies

4.1. Overview

A majority of type II SLS detector structures are variations of the general double heterostructure (DH) design, such as the *p-i-n* detector device depicted in **Figure 12** with associated band structure [41, 79]. This type of structure comprises an unintentionally doped intrinsic region sandwiched between heavily doped *p* and *n* layers of larger bandgap material [11]. The DH architecture was initially intended to optimize quantum efficiency while minimizing transport barriers at the heterointerfaces [80]. It provides practical benefits over basic homojunction devices for type II SLS detectors, most notably the incorporation of current blocking layers either in the conduction or valence band that reduce multiple dark current components including GR and BTB tunneling currents.

In this section, primary classes of type II SLS detector designs are described, and in certain cases illustrated. The present classifications for the different types of SLS detectors are somewhat ambiguous in that multiple and overlapping categories exist; however, in the following discussion, attempts will be made to clarify potential conflicts. In consideration of space constraints, other less common type II SLS detectors are not covered here in detail, which include, but are not limited to, the shallow etch mesa isolation (SEMI), a type of *n*-on-*p* graded-gap "W" photodiode structure [8]; hybrid superlattice (HSL) structure [81]; and strain-compensated InAs/AlInSb superlattice barrier (ALSL-B) [82].

4.2. Buried junction

Devices generally classified as *buried junction* include nBn, pBp, and pBn architectures [6]. The pBp architecture may be considered the inverse of the nBn structure, the main distinctions being that the minority carriers for nBn are holes while those for pBp are electrons, and the barriers each differ in height relative to the valence and conduction bands [83]. These structures likewise belong to the *unipolar barrier* category, which basically describes the inclusion of a single barrier that blocks one carrier type (electron or hole) but allows for the other to flow unimpededly [79]. A unipolar barrier can be implemented either outside the depletion region in the *p*-type layer of a SLS device, or at the edge of the *n*-type absorbing layer near the

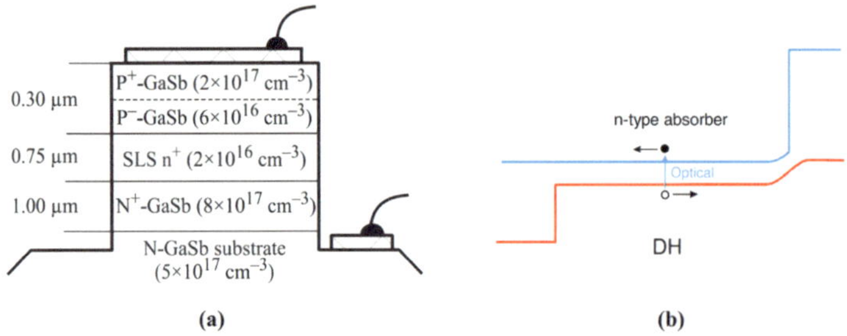

Figure 12. (a) Cross-section schematic of InAs/GaInSb SLS DH *p-i-n* detector [79]. (b) Corresponding bandgap diagram typical of DH photodiode [41].

junction [8]. To be properly implemented into a low-noise photodetector design, the unipolar carrier blocking layers must be sufficiently thick, and the conduction band offset to valence band offset ratio (CBO/VBO) large enough, to effectively block majority carriers and thereby suppress thermionic emission and direct tunneling dark current.

4.2.1. *nBn design*

Due to the absence of majority carrier flow, the nBn detector (**Figure 13**) [84, 85] is essentially a photoconductor with unity gain, in which the junction (space charge region) is replaced by an electron blocking unipolar barrier (B) and the *p*-contact by an *n*-contact [8]. The wide bandgap undoped barrier layer is chosen to have minimal valence band offset [86]. Compared to the standard *p-i-n* design, detectors based on the nBn architecture have shown promise in suppressing currents associated with SRH centers, the GR component of dark current, and mesa lateral surface imperfections, thereby reducing the required temperature of operation [87, 88].

4.2.2. *pBn design*

In pBn structure devices, the *p-n* junction can either be located at the interface between the heavily doped *p*-type material and the lower doped barrier, or within the lower barrier itself [33]. This type of architecture is preferable when a large barrier in the conduction band offset is not achievable, allowing for zero bias operation [89].

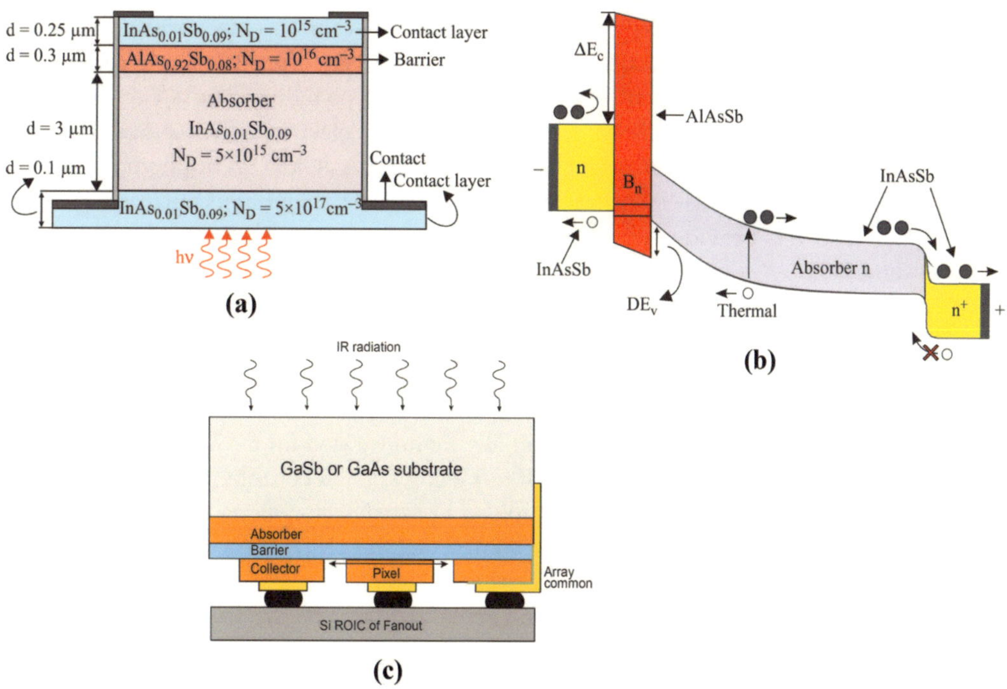

Figure 13. MWIR InAsSb/AlAsSb nBn SLS detector: (a) device structure, and (b) simulated energy band diagram under reverse bias conditions [84]; and (c) potential detector array (FPA) architecture [85].

4.2.3. pBp design

The pMp design consists of two *p*-doped superlattice active regions and a thin valence band barrier, producing electrical transport due to minority carriers (electrons) with low current density [22]. A large energy offset in the valence band prevents the movement of minority electrons but allows photogenerated majority carrier current to flow freely [29]. The higher mobility of electrons in comparison with holes in SLS pBp detectors results in greater photoresponse. The biasing requirements for SLS pBp detectors were demonstrated to be lower compared to those reported for nBn detectors [68, 79]. *D** and QE performance of MWIR and LWIR InAs/GaSb pBp detectors has been shown to be superior to that of comparable QWIP detectors, but somewhat inferior to the performance of MCT detectors operating over the same bands [90].

4.3. Supercell superlattices

Structures selectively designated as *supercell superlattices* are further variations of the unipolar barrier structure concept that constitute somewhat more complex designs than buried junction devices. Three primary varieties of supercell SLS architectures exist: M-, W-, and N-structures. As might be expected, these were named according to the similarities between the shapes of the respective letters and the band alignment of the constituent detector layers.

4.3.1. M-structure

M-structure detectors, originally developed for controlling BTB tunneling and consequently diffusion current for LWIR wavelengths, were subsequently found to accomplish this as well for the MWIR to allow increased active region doping levels [91]. The M-structure comprises a thin AlSb barrier inserted in the center of a GaSb layer in a type II binary InAs/GaSb superlattice [13]. The high bandgap AlSb layer effectively blocks the electron and hole dark currents, thereby improving the R_0A and *D** performance in type II InAs/GaSb LWIR photodiodes [92]. The large carrier effective mass associated with the InAs/GaSb material in M-structure detectors has been shown to provide more freedom in tuning the bandgap compared to other type II SLS designs [93].

Two variations of M-structure SLS devices are the pMp and p-π-M-n designs. These differ primarily in their usage of the M-structure barrier, which is employed to block hole majority carriers in the pMp architecture [94], while in p-π-M-n devices to suppress tunneling current [95].

The unipolar barrier pMp device concept, initially demonstrated for the VLWIR regime, was developed to reduce dark current due to BTB tunneling and GR contributions [54]. **Figure 14** shows the band structure and working principle of a pMp detector, along with key associated performance characteristics [22, 95]. Devices comprising the pMp architecture consist of two *p*-doped superlattice active regions and a thin valence band M-barrier having zero conduction band discontinuity with respect to the *p*-type active regions [79]. By using the M-structure SL as a barrier region, the band alignments can be experimentally controlled, allowing for efficient extraction of photocurrent under applied biases of less than 50 meV [22]. It has been demonstrated that the dark current in a pMp detector can be decreased by an order of magnitude by reducing the doping concentration in the contact layer [96].

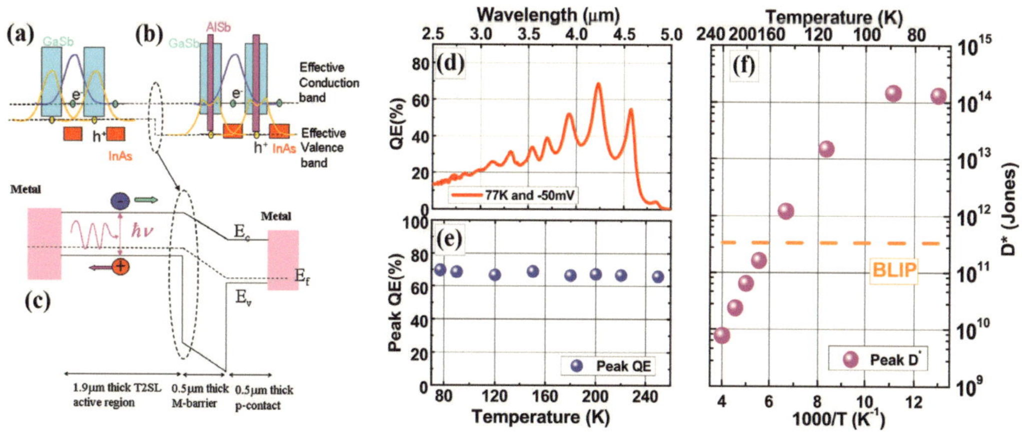

Figure 14. Band alignment and the creation of effective bandgap in (a) InAa/GaSb SL and (b) M-structure SL. (c) Schematic diagram and working principle of pMp design: the M-barrier blocks transport of majority holes, while allowing diffusion of minority electrons and photogenerated carriers from the active region [22]. (d) Quantum efficiency of photodetector at 77 K and −150 mV bias; (e) saturated QE at peak responsivity from 77 to 250 K; and (f) evolution of peak detectivity with temperature, where peak detectivity crosses calculated background-limited line at 165 K [95].

The p-π-M-n design consists of an M-barrier lightly *n*-doped to prevent excessive quantum efficiency bias dependence, inserted between the *p* and *n* regions of a standard p-π-n structure [8, 50]. In comparison with a basic p-π-n type design, the electric field in the depletion region of a p-π-M-n detector is reduced, and the tunneling barrier between the *p* and the *n* regions spatially broadened [95]. The first reported p-π-M-n detector which comprised a 500 nm M-barrier exhibited R_0A of 200 Ω cm^2 at 77 K, approximately one order of magnitude higher than that of a comparable p-π-n device without an M-layer barrier [94]. The dark current density J and R_0A versus bias voltage, and QE vs. wavelength, of a p-π-M-n detector are plotted in **Figure 15** [54]. Large format LWIR FPAs have been realized using for this architecture, one version of which demonstrated a QE of 78% and NETD of 23.6 mK at 81 K using 0.13 ms integration time [50].

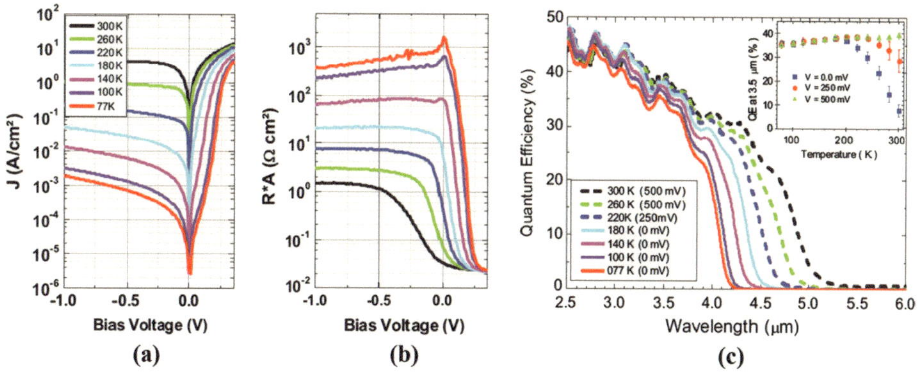

Figure 15. (a) Dark current density and (b) R_0A characteristics of a p-π-M-n SLS detector as a function of temperature. (c) Quantum efficiency spectra of the device at different operating temperatures, where inset shows temperature-dependent QE at a wavelength of 3.5 μm for different reverse biases [54].

4.3.2. W-structure

The W-structured type-II SLS detector architecture reduces dark current due to strong suppression of BTB tunneling and GR processes [19]. Initially developed to increase the gain of MWIR lasers [13], it was reported to offer an order of magnitude improvement in dark current performance [12] and R_0A comparable to MCT. As illustrated in **Figure 16**, this type of design consists of InAs electron wells surrounding a GaInSb hole well and positioned between two quaternary AlGaInSb barriers [10]. The barriers enhance the electron–hole overlap while nearly localizing the wavefunctions, thereby increasing absorption near the band edge [80]. A subsequent generation W-structured LWIR detector was reported that employed a graded bandgap *p-i-n* structure and featured design improvements such as optimization to the barrier height that significantly increased minority carrier mobility and improved the quantum efficiency [97].

4.3.3. N-structure

The N-structure detector design consists of a *p-i-n* structure comprising two monolayer AlSb electron barriers inserted symmetrically between InAs and GaSb layers along the growth direction [8]. In this type of structure, the AlSb barrier pushes the electron and hole wavefunctions towards the layer edges under bias [98]. By improving the overlap of the spatially separated electron and hole wavefunctions, the absorption is increased while the dark current is reduced, enhancing the directivity [32].

4.4. Complementary barrier

A more sophisticated class of type II SLS structures contain complementary barriers that block both electron and holes, and are considered further variations of the double heterostructure design [99]. Two comparably designed representatives of this class were designated

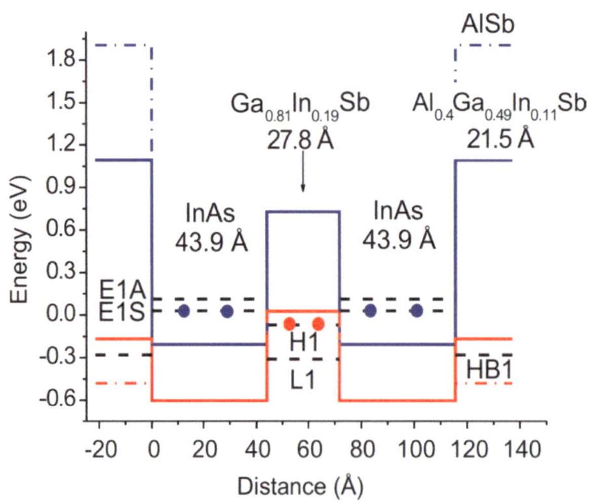

Figure 16. Band profile of enhanced W-structure type II SLS photodiode [10].

complementary barrier infrared detector (*CBIRD*) by Ting et al. [23], and *pBiBn* by Gautam et al. [19]. These types of detectors have advantages in suppressing dark currents through blocking both majority and minority carriers, as well as circumventing technological problems arising from the need to make ohmic contacts with the widegap layers [100].

4.4.1. Complementary barrier infrared detector (CBIRD)

A CBIRD structure consists of a lightly doped *p*-type InAs/GaSb SL absorber surrounded by unipolar hole and electron blocking barriers [23]. As can be seen from **Figure 17(a)**, the electron barrier (eB) exists in the conduction band and the hole barrier (hB) in the valence band, where the two barriers complement one another to impede the flow of dark current [77]. A heavily doped *n*-type InAsSb layer adjacent to the eB SL acts as the bottom contact layer [9]. The barrier layers are designed to have approximately zero conduction band and valence band offsets with respect to the SL absorber. The absorber superlattice in CBIRD detectors is doped lightly *p*-type to improve the minority carrier (electron) mobility, and thus provide more favorable electron transport properties leading to enhanced performance including higher directivity [99]. **Figure 17(b)** plots the dark current density vs. applied bias for a CBIRD device [23].

4.4.2. pBiBn detector

The quantum-engineered pBiBn SLS detector structure [24] may be considered a hybrid between a conventional *p-i-n* structure and unipolar barrier concepts [101]. Like the CBIRD design, it was designed to mitigate the higher dark current in *p-n* junction and *p-i-n* photodetectors through incorporation of two unipolar carrier blocking barriers [18]. The first generation pBiBn detector demonstrated a significant improvement in performance over conventional *p-i-n* designs, with a fourfold increase in detectivity in the LWIR [24].

The band profile for a pBiBn design is shown in **Figure 18(a)**, and the detector layer structure given in **Figure 18(b)** [24, 101, 102]. It consists of an *n*-contact layer, followed by a hB layer, an

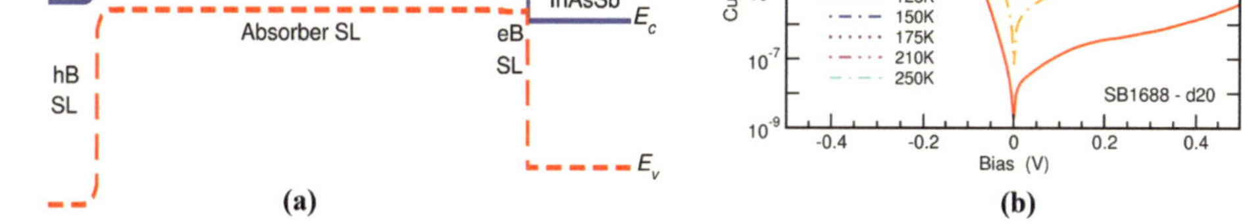

Figure 17. (a) Calculated zero-bias energy band diagram of a CBIRD structure, where a LWIR InAs/GaSb SL absorber is surrounded by an InAs/AlSb SL unipolar hB and shorter period eB [99]. (b) Dark current density as a function of applied bias for CBIRD detector, measured at various temperatures ranging from 65-250 K [23].

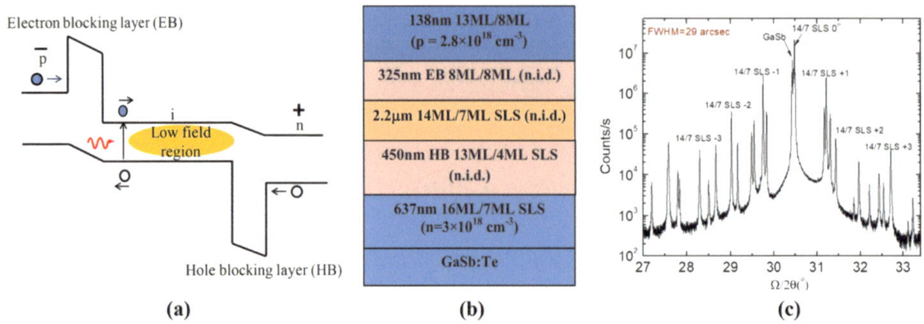

Figure 18. (a) Schematic drawing of band profile and (b) heterostructure schematic of pBiBn SLS detector [101, 102]. (c) XRD pattern of pBiBn material structure [24].

intrinsic absorber *i*-region, followed by an eB layer, and finally a *p*-contact layer. The unipolar carrier blocking layers (eB and hB) surrounding the type II InAs/GaSb SL absorber in the pBiBn structure comprises of heterojunctions with unique energy band alignments, for which the CBO ΔE_c for the eB layer (or VBO ΔE_v for the hB layer) is equivalent to the bandgap difference ΔE_g. The pBiBn band alignment and carrier transport differ from that in DH designs: instead of the electrons and holes being confined in the narrow bandgap material, a wider bandgap material with only a conduction band offset or a valence band offset is adopted. **Figure 18(c)** presents the X-ray diffraction (XRD) pattern from characterization of a pBiBn device [24].

The inclusion of hole and electron barriers flanking both sides of the absorber serves to lower diffusion currents by blocking the minority carriers from either side of the junction, while allowing unimpeded flow of photogenerated carriers. It also limits the voltage and substantially reduces the built-in electrical field in the absorber region [39]. This drop in electrical potential across the pBiBn active region is small compared to that in a conventional *pin* design because of

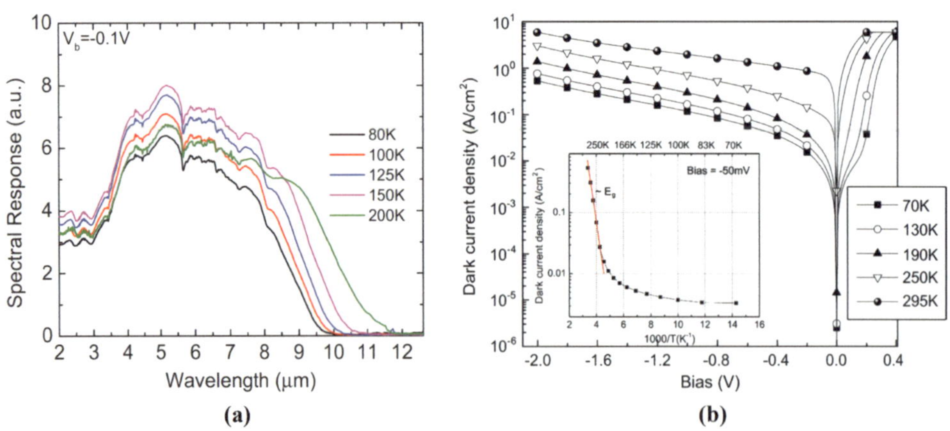

Figure 19. (a) Current–voltage (I-V) characteristics of type II SLS pBiBn detector; inset shows temperature dependent dark current density measured under applied bias of −50 mV [104]. (b) Measured spectral response for pBiBn detector as a function of temperature [102].

the eB and hB layers [103], which have wider bandgaps than the absorber region. The reduction of electric field within the absorber in the pBiBn structure effects a smaller depletion region, which further reduces the SRH dark current and suppresses TAT and BTB tunneling dark current components [24, 38]. **Figure 19(a)** and **(b)** plot I-V characteristics versus applied bias and spectral response vs. wavelength, respectively, for SLS pBiBn detectors [102, 104].

5. Type II SLS detectors: practical progress and trends

5.1. High operating temperature (HOT)

For advanced IR photodetectors and FPAs, high temperature operability is of great interest and in high demand for many applications. High operating temperature (HOT: ~150–300 K and above) capability enables significant advantages in cost, size, weight, and power (CSWaP) reduction resulting from the lessening of cooling load [105]. Initial efforts for developing HOT detectors concentrated on photoconductors and photoelectromagnetic (PEM) devices, material improvement to lower generation-recombination leakage mechanisms, alternate materials such as cascade devices, and SLS barrier structures such as nBn [106]. Type II MWIR SLS detectors and FPAs have certain advantages for HOT operation in relation to InSb, MCT, and QWIP devices, in many instances offering comparable performance at significantly higher operating temperatures [107].

Type II SLS detectors based on the InAs/GaSb/AlSb system have demonstrated HOT operation in both MWIR and LWIR regimes [38]. Achieving this typically requires minimizing the residual doping level in the intrinsic region, which is highly sensitive to crystal growth conditions [31]. **Figure 20(a)** compares images captured with a MWIR SLS FPA at different operating temperatures, and **Figure 20(b)** plots corresponding NETD versus temperature at different integration times [108, 109]. A MWIR pBiBn InAs/GaSb type II SLS was reported

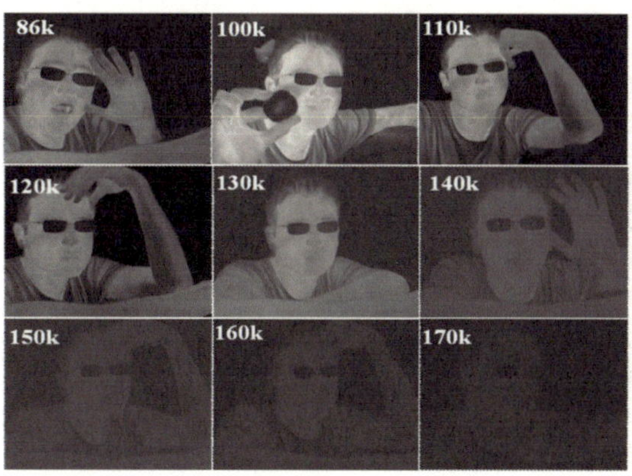

(a)

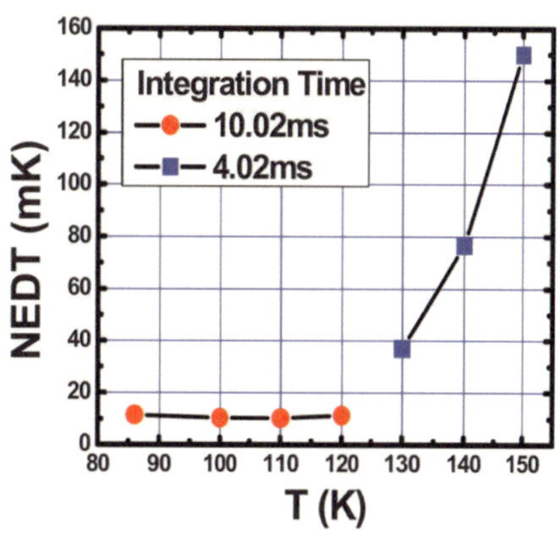

(b)

Figure 20. (a) MWIR images captured with type II SLS FPA over a range of temperatures [108]. (b) Corresponding NEDT measured at 86–150 K using *f*/2.3 optics [109].

that demonstrated operation up to 295 K, but the dark current density of the device was high and the spectral response relatively low, ruling out its practical use in detector applications at room temperature [38]. However, HOT nBn detectors were shown at 300 K to reach higher detectivities in comparison with standard InAsSb photodiodes [84, 110].

5.2. Large format FPAs

Recent improvements in the maturation of growth and fabrication process technologies, design of process tools and optics, reproducibility and yield, and spatial uniformity of final FPA devices have enabled and facilitated the transition to larger format FPAs [51, 111]. FPAs having resolutions up to 1024 × 1024 with pixel pitches 18 μm and smaller have been developed [48–51]. Large format, small pixel pitch FPAs clearly possess a number of benefits for IR imaging applications including larger fields of view and higher angular resolutions [77]. This is particularly advantageous for target tracking and surveillance applications, for example, by enabling longer detection ranges for a target tracking system with a fixed optical aperture [8].

A potential means of making the production of large format arrays more practical and commercially viable is to employ larger diameter substrate wafers. For example, only a single LWIR array with 18 μm pitch can be processed on a 2 in. diameter wafer, while one 3 in. wafer can allocate up to four such 1 k × 1 k arrays or one 2 k × 2 k array [113]. Consequently, in recent years there has been heightened demand for high performance material on non-native substrates that are cheaper and larger than GaSb despite significant (~7%) lattice mismatch [56]. **Figure 21** presents physical features [50] and a captured image [112] from a megapixel MWIR SLS FPA.

5.3. Multiband IR detection

Multiband (or multicolor) detectors are useful for various military and civil IR imaging applications involving identification of temperature differences and determination of thermal characteristics of objects, including remote sensing and target identification [37]. Multiband

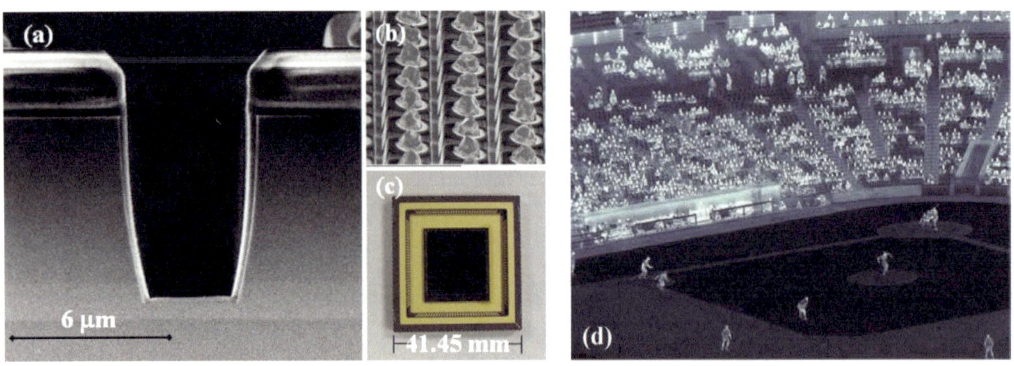

Figure 21. (a) Cross-sectional scanning electron micrograph of etched trench between two pixels in large format (1 k × 1 k) LWIR type II n-M-π-p SLS FPA; (b) FPA pixels with uniform indium bumps; and (c) FPA mounted on a chip carrier [50]. (d) Captured image of scene from baseball game using MWIR nBn FPA, also with 1 k × 1 k pixel resolution [112].

SLS photodetectors have been developed that combine wavelength bands from the SWIR to VLWIR. In addition, type II SLS multiband technology is compatible with commercially available ROICs, enabling development of many types of multiband SLS FPAs.

More conventional designs combining detection of MWIR and LWIR wavelength bands commonly consist of separate LWIR and MWIR superlattices separated by an AlGaSb unipolar barrier. Such MWIR/LWIR dual-band detectors typically possess pBp architectures comprising vertical designs based on two back-to-back InAs/GaSb SLS photodiodes separated by a common ground contact layer [90]. In such devices, properly biased unipolar barriers and thicker absorber regions block the migration of photogenerated carriers to reduce spectral crosstalk between the different absorber regions (channels) [100]. In multiband SLS detectors and FPAs, the selection of the desired band is typically achieved through alternating the polarity of the applied bias.

Successful fabrication of multiband SLS photodetectors requires the development of separate materials sensitive to each band, which necessitates prioritizing the optimization of the optical and electrical performance for each channel [114]. An advantage of type II SLS multiband detectors and FPAs compared to other technologies is relative design simplicity, while low hole mobility and limited lateral diffusion remain significant challenges [15].

Most multiband detectors and FPAs developed thus far are dual-band (i.e., dual-color). InAs/GaSb type II SLS FPAs with separate LWIR and MWIR absorbers have been developed with measured dark current densities in the low $\mu A/cm^2$ range and specific detectivities of up to 5×10^{11} Jones [30]. **Figure 22(a)** shows band diagrams for a dual-band MWIR/LWIR SLS pBp detector [115], and **Figure 22(b)** and **(c)** feature corresponding MWIR and LWIR images, respectively, from a dual-band detector demonstrating the contrast in appearance when imaging an optical filter [91].

Recent years have seen the development of various three-color SLS detector designs [52, 100, 116]. A MWIR/LWIR/VLWIR SLS detector has been reported within the past year in which the spectral crosstalk was reduced by controlling minority carrier transport through doping of the two active regions [116]. As seen in **Figure 23**, an innovative triple-band SWIR/

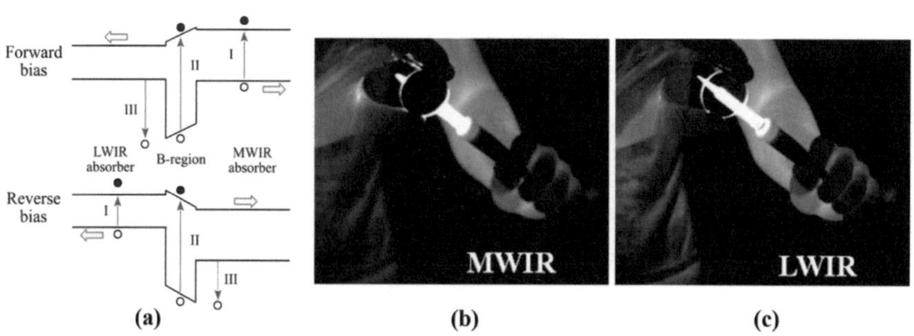

Figure 22. (a) Band alignment under forward and reverse biases for dual-band type II InAs/GaSb SLS photodetector based on a pBp architecture, showing effective optical transitions contributing to photocurrents (having dependence on the bias polarity) [115]. Images from (b) MWIR and (c) LWIR channels of dual-band FPA, capturing 11.3 μm narrow bandpass optical filter at 81 K [91].

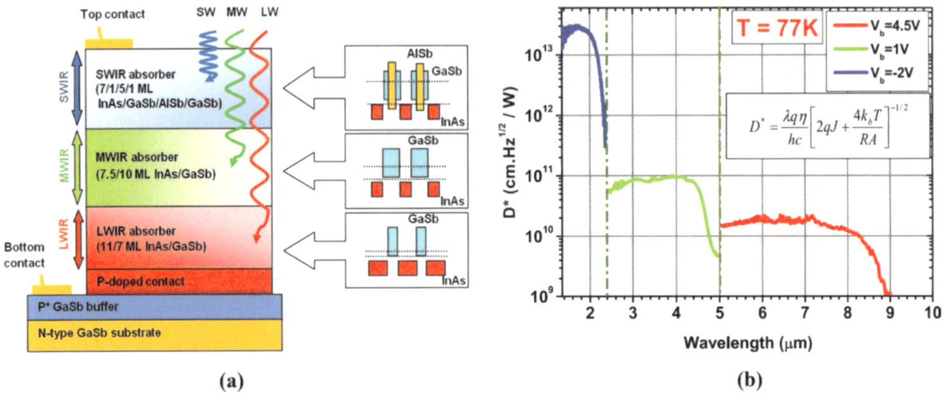

Figure 23. (a) Schematic diagram of triple-band SWIR/MWIR/LWIR photodiode structure with two terminal contacts and schematic band alignment of superlattices in three absorption layers. (b) Plotted directivities for the three channels at the wavelengths of interest at 77 K [52].

MWIR/LWIR device design was likewise reported recently, capable of sequentially performing as three individual single color photodetectors using only two terminals [52]. This device employed conduction band offsets and different doping levels between absorption layers to control the turn-on voltages for the individual channels.

5.4. High-speed imaging

The higher quantum efficiencies and reduced cooling requirements offered by type II SLS detectors and FPAs make them well suited for high frame rate applications with low integration times. InAs/GaSb SLS devices incorporate a potentially high-speed detector material in comparison to uncooled microbolometers, for which time constant constraints limit frame rates to under ~30 fps [110]. These properties make certain types of SLS detectors and FPAs

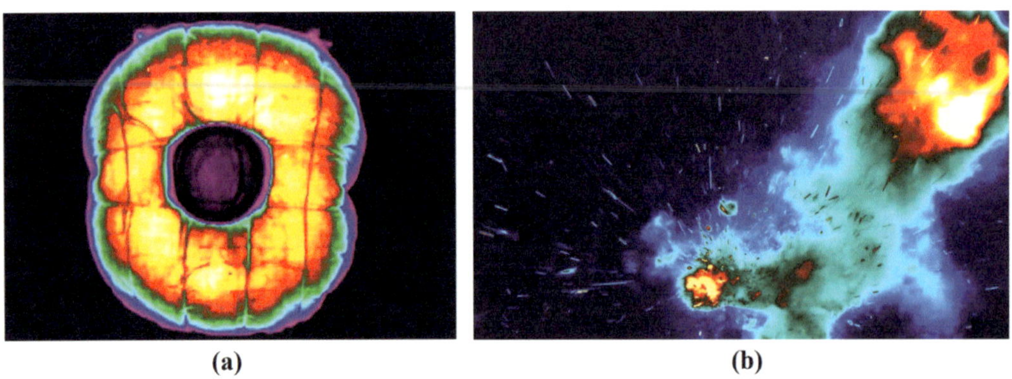

Figure 24. (a) Airbag deployment imaged by SLS camera with 160 μs integration time [58]. (b) Thermal image of munitions testing captured at 1000 fps and 640 × 512 resolution [117].

practical for detecting rapid temperature rises in concert with rapid scene motion, particularly at temperatures ~300 K and below [58].

While frame rates of 60 Hz are common for type II SLS detectors reported in recent years, for certain applications significantly higher capture speeds are required. Combined with new higher speed ROIC designs that are more linear across their entire dynamic range, SLS detectors can now accommodate integration times below 270 ns, providing frame rates of over 180 fps at 1028 × 1024 resolution, or over 1000 fps at 640 × 512 resolution [59, 117]. Such higher speed, faster frame rate detection capability has significant advantages in applications such as safety analysis and evaluation (**Figure 24(a)**) [58], ballistic and munitions testing (**Figure 24(b)**) [117], and missile seekers and missile warning systems [44].

6. Summary and conclusions

Type II superlattices possess a robustness and practical utility originating from the flexibility to control and manipulate the material system for practically any desired target of operability, while inheriting the benefits of mature III–V semiconductor growth and fabrication process technology [56]. The broad array of SLS detector/FPA architectures in existence today provides researchers with wide-ranging ability to control the various noise and dark current mechanisms to optimize SLS detector performance for electro-optical/IR applications. Progress and development in areas such as high temperature operability, multiband detection, and high-speed imaging have showcased some of the diverse and unique features of type II InAs/GaSb SLS devices. These have led to practical, real-world improvements in the field, many of which would not be possible using alternative IR detector technologies and material systems.

Type II SLS detectors and two dimensional FPAs based on the InAs/GaSb material system can theoretically provide higher performance than precedential IR detection technologies, including QWIP and HgCdTe detectors [28]. While in general type II SLS devices have yet to achieve such a noted benchmark, substantial progress in this area has been made and continues to be undertaken, as discussed in this Chapter. Further future progress and technical development in practical devices are still needed, potentially involving new discoveries and technological breakthroughs in material and structural engineering, for type II SLS to emerge as a leading technology. Nevertheless, these devices have had a definite and lasting impact the area of IR detection and imaging, and expectations are that type II SLS will remain a key enabling technology in the field for years and possibly decades to come.

Acknowledgements

The authors thank Dr. Jay Lewis of DARPA/MTO for technical discussions and guidance, as well as Ms. Susan Nicholas and Mr. Oscar Cerna of DARPA for their ongoing support. This research was developed with funding from the Defense Advanced Research Projects Agency

(DARPA). The views, opinions and/or findings expressed are those of the author and should not be interpreted as representing the official views or policies of the Department of Defense or the U.S. Government.

Author details

Ashok K. Sood[1*], John W. Zeller[1], Roger E. Welser[1], Yash R. Puri[1], Nibir K. Dhar[2], Priyalal S. Wijewarnasuriya[3] and Sanjay Krishna[4]

*Address all correspondence to: aksood@magnoliaoptical.com

1 Magnolia Optical Technologies, Woburn, MA, USA

2 U.S. Army Night Vision & Electronic Sensors Directorate, Fort Belvoir, VA, USA

3 U.S. Army Research Laboratory, Adelphi, MD, USA

4 Department of Electrical and Computer Engineering, Ohio State University, Columbus, OH, USA

References

[1] Pollehn H, Choi KK, Svensson S, Dhar N. IR material research at the Army Research Laboratory. In: Proceedings of SPIE, Infrared Technology and Applications XXXIII; 14 May 2007; Orlando, Florida. SPIE; 2007;**6542**:65420C. DOI: 10.1117/12.723659

[2] Martin TJ, Cohen MJ, Dries JC, Lange MJ. InGaAs/InP focal plane arrays for visible light imaging. In: Proceeding of SPIE, Infrared Technology and Applications XXX; 30 August 2004; Orlando, Florida. SPIE; 2004;**5406**: p. 39. DOI: 10.1117/12.542523

[3] Davis M, Devitt J, Greiner M, Rawe R, Timlin A, Wade D. Advanced FPA technology development at CMC electronics. In: Proceedings of SPIE, Infrared Systems and Photoelectronic Technology; 2-6 August 2004; Denver, Colorado. SPIE; 2004;**5563**: p. 63. DOI: 10.1117/12.565665

[4] Norton PW, Kohin M, Dovidio M, Becker B. Commercialization of uncooled infrared technology. In: Proceedings of SPIE, Infrared Systems and Photoelectronic Technology; 21 October 2004; Denver, Colorado. SPIE; 2004. p. 55-61. DOI: 10.1117/12.565663

[5] Myers S, Plis E, Khoshakhlagh A, Kim HS, Sharma Y, Dawson R, Krishna S, Gin A. The effect of absorber doping on electrical and optical properties of nBn based type-II InAs/GaSb strained layer superlattice infrared detectors. Applied Physics Letters. 2009;**95**:121110. DOI: 10.1063/1.3230069

[6] Tidrow MZ. Type II strained layer superlattice: A potential future IR solution. Infrared Physics & Technology. 2009;**52**:322-325. DOI: 10.1016/j.infrared.2009.05.028

[7] Plis EA. InAs/GaSb type-II superlattice detectors. Advances in Electronics. 2014;**2014**:1-12. DOI: 10.1155/2014/246769

[8] Martyniuk P, Antoszewski J, Martyniuk M, Faraone L, Rogalski A. New concepts in infrared photodetector designs. Applied Physics Reviews. 2014;**1**:041102. DOI: 10.1063/1.4896193

[9] Rogalski A. History of infrared detectors. Opto-Electronics Review. 2012;**20**:279-308. DOI: 10.2478/s11772-012-0037-7

[10] Rogalski A, Antoszewski J, Faraone L. Third-generation infrared photodetector arrays. Journal of Applied Physics. 2009;**105**:091101. DOI: 10.1063/1.3099572

[11] Rogalski A, Kopytko M, Martyniuk P. InAs/GaSb type-II superlattice infrared detectors: three decades of development. In: Proceedings of SPIE, Infrared Technology and Applications XLIII; 3 May 2017; Anaheim, California. SPIE; 2017;**10177**: p. 1017715. DOI: 10.1117/12.2272817

[12] Rogalski A. Material considerations for third generation infrared photon detectors. Infrared Physics & Technology. 2007;**50**:240-252. DOI: 10.1016/j.infrared.2006.10.015

[13] Aifer EH, Tischler JG, Warner JH, Vurgaftman I, Kim JC, Meyer JR, Bennett BR, Whitman LJ, Jackson EM, Lorentzen JR. W-structured type-II superlattice based long and very-long wavelength infrared photodiodes. In: Proceedings of SPIE, Quantum Sensing and Nanophotonic Devices II; 25 March 2005; San Jose, California. SPIE; 2005;**5732**:259-272. DOI: 10.1117/12.597134

[14] Plis EA, Kutty MN, Krishna S. Passivation techniques for InAs/GaSb strained layer superlattice detectors. Laser & Photonics Reviews. 2013;**7**:45-59. DOI: 10.1002/lpor.201100029

[15] Haugan HJ, Olson BV, Brown GJ, Kadlec EA, Kim JK, Shaner EA. Significantly enhanced carrier lifetimes of very long-wave infrared absorbers based on strained-layer InAs/GaInSb superlattices. Optical Engineering. 2017;**56**:091604. DOI: 10.1117/1.OE.56.9.091604

[16] Rogalski A. Competitive technologies of third generation infrared photon detectors. Opto-electronics Review. 2006;**14**:84-98. DOI: 10.2478/s11772-006-0012-2

[17] Rodriguez JB, Plis E, Bishop G, Sharma YD, Kim H, Dawson LR, Krishna S. nBn structure based on InAs/GaSb type-II strained layer superlattices. Applied Physics Letters. 2007;**91**:043514. DOI: 10.1063/1.2760153

[18] Treider LA, Morath CP, Cowan VM, Tian ZB, Krishna S. Radiometric characterization of an LWIR, type-II strained layer superlattice pBiBn photodetector. Infrared Physics & Technology. 2015;**70**:70-75. DOI: 10.1016/j.infrared.2014.09.043

[19] Vurgaftman I, Aifer EH, Canedy CL, Tischler JG, Meyer JR, Warner JH, Jackson EM, Hildebrandt G, Sullivan GJ. Graded band gap for dark-current suppression in long-wave infrared W-structured type-II superlattice. Applied Physics Letters. 2006;**89**:121114. DOI: 10.1063/1.2356697

[20] Maimon S, Wicks GW. nBn detector, an infrared detector with reduced dark current and higher operating temperature. Applied Physics Letters. 2006;**89**:151109. DOI: 10.1063/1.2360235

[21] Nguyen BM, Razeghi M, Nathan V, Brown GJ. Type-II M structure photodiodes: an alternative material design for mid-wave to long wavelength infrared regimes. In: Proceedings of SPIE, Quantum Sensing and Nanophotonic Devices IV; 12 February 2007; San Jose, California. SPIE; 2007;**6479**: p. 64790S. DOI: 10.1117/12.711588

[22] Nguyen BM, Bogdanov S, Pour SA, Razeghi M. Minority electron unipolar photodetectors based on type II InAs/GaSb/AlSb superlattices for very long wavelength infrared detection. Applied Physics Letters. 2009;**95**:183502. DOI: 10.1063/1.3258489

[23] Ting DZY, Hill CJ, Soibel A, Keo SA, Mumolo JM, Nguyen J, Gunapala SD. A high-performance long wavelength superlattice complementary barrier infrared detector. Applied Physics Letters. 2009;**95**:023508. DOI: 10.1063/1.3177333

[24] Gautam N, Kim HS, Kutty MN, Plis E, Dawson LR, Krishna S. Performance improvement of longwave infrared photodetector based on type-II InAs/GaSb superlattices using unipolar current blocking layers. Applied Physics Letters. 2010;**96**:231107. DOI: 10.1063/1.3446967

[25] Tian Z, Hinkey RT, Yang RQ, Qiu Y, Lubyshev D, Fastenau JM, Liu AWK, Johnson MB. Interband cascade infrared photodetectors with enhanced electron barriers and p-type superlattice absorbers. Journal of Applied Physics. 2012;**111**:024510. DOI: 10.1063/1.3678003

[26] Yang RQ, Tian Z, Cai Z, Klem JF, Johnson MB, Liu HC. Interband cascade infrared photodetectors with superlattice absorbers. Journal of Applied Physics. 2010;**107**:054514. DOI: 10.1063/1.3327415

[27] Martyniuk P, Rogalski A. Quantum-dot infrared photodetectors: Status and outlook. Progress in Quantum Electronics. 2008;**32**:89-120. DOI: 10.1016/j.pquantelec.2008.07.001

[28] Rhiger DR. Performance comparison of long-wavelength infrared type II superlattice devices with HgCdTe. Journal of Electronic Materials. 2011;**40**:1815-1822. DOI: 10.1007/s11664-011-1653-6

[29] Plis E, Myers S, Ramirez D, Smith EP, Rhiger D, Chen C, Phillips JD, Krishna S. Dual color longwave InAs/GaSb type-II strained layer superlattice detectors. Infrared Physics & Technology. 2015;**70**:93-98. DOI: 10.1016/j.infrared.2014.09.027

[30] Plis E, Naydenkov M, Myers S, Klein B, Gautam N, Krishna SS, Smith EP, Johnson S, Krishna S. Dual-band pBp detectors based on InAs/GaSb strained layer superlattices. Infrared Physics & Technology. 2013;**59**:28-31. DOI: 10.1016/j.infrared.2012.12.005

[31] Li JV, Hill CJ, Mumolo J, Gunapala S, Mou S, Chuang SL. Midinfrared type-II InAs/GaSb superlattice photodiodes toward room temperature operation. Applied Physics Letters. 2008;**93**:163505. DOI: 10.1063/1.2949744

[32] Salihoglu O, Muti A, Kutluer K, Tansel T, Turan R, Ergun Y, Aydinli A. "N" structure for type-II superlattice photodetectors. Applied Physics Letters. 2012;**101**:073505. DOI: 10.1063/1.4745841

[33] Hood AD, Evans AJ, Ikhlassi A, Lee DL, Tennant WE. LWIR strained-layer superlattice materials and devices at Teledyne imaging sensors. Journal of Electronic Materials. 2010;**39**:1001-1006. DOI: 10.1007/s11664-010-1091-x

[34] Esaki L, Tsu R. Superlattice and negative differential conductivity in semiconductors. IBM Journal of Research and Development. 1970;**14**:61-65. DOI: 10.1147/rd.141.0061

[35] Sai-Halasz GA, Tsu R, Esaki L. A new semiconductor superlattice. Applied Physics Letters. 1977;**30**:651-653. DOI: 10.1063/1.89273

[36] Smith DL, Mailhiot C. Proposal for strained type II superlattice infrared detectors. Journal of Applied Physics. 1987;**62**:2545-2548. DOI: 10.1063/1.339468

[37] Fuchs F, Weimer U, Pletschen W, Schmitz J, Ahlswede E, Walther M, Wagner J, Koidl P. High performance InAs/$Ga_{1-x}In_xSb$ superlattice infrared photodiodes. Applied Physics Letters. 1997;**71**:3251-3253. DOI: 10.1063/1.120551

[38] Gautam N, Myers S, Barve AV, Klein B, Smith EP, Rhiger D, Plis E, Kutty MN, Henry N, Schuler-Sandy T, Krishna S. Band engineered HOT midwave infrared detectors based on type-II InAs/GaSb strained layer superlattices. Infrared Physics & Technology. 2013; **59**:72-77. DOI: 10.1016/j.infrared.2012.12.017

[39] DeCuir EA, Meissner GP, Wijewarnasuriya PS, Gautam N, Krishna S, Dhar NK, Welser RE, Sood AK. Long-wave type-II superlattice detectors with unipolar electron and hole barriers. Optical Engineering. 2012;**51**:124001. DOI: 10.1117/1.OE.51.12.124001

[40] Zhou X, Li D, Huang J, Zhang Y, Mu Y, Ma W, Tie X, Zuo D. Mid-wavelength type II InAs/GaSb superlattice infrared focal plane arrays. Infrared Physics & Technology. 2016;**78**:263-267. DOI: 10.1016/j.infrared.2016.08.014

[41] Rogalski A. Recent progress in infrared detector technologies. Infrared Physics & Technology. 2011;**54**:136-154. DOI: 10.1016/j.infrared.2010.12.003

[42] Zheng L, Tidrow M, Aitcheson L, O'Connor J, Brown S. Developing high-performance III–V superlattice IRFPAs for defense: challenges and solutions. Proceedings of SPIE, Infrared Technology and Applications XXXVI; 23 April 2010; Orlando, Florida. SPIE; 2010;**7660**: p. 76601E. DOI: 10.1117/12.852239

[43] Haddadi A, Chen G, Chevallier R, Hoang AM, Razeghi M. InAs/$InAs_{1-x}Sb_x$ type-II superlattices for high performance long wavelength infrared detection. Applied Physics Letters. 2014;**105**:121104. DOI: 10.1063/1.4896271

[44] Cabanski W, Münzberg M, Rode W, Wendler J, Ziegler J, Fleißner J, Fuchs F, Rehm R, Schmitz J, Schneider H, Walther M. Third-generation focal plane array IR detection modules and applications. In: Proceedings of SPIE, Infrared Technology and Applications XXX; 30 August 2004. Orlando, Florida. SPIE; 2005;**5406**:184-192. DOI: 10.1117/12.542186

[45] Razeghi M, Delaunay PY, Nguyen BM, Hood A, Hoffman D, McClintock R, Wei Y, Michel E, Nathan V, Tidrow MZ. First demonstration of ~10 microns FPAs in InAs/GaSb SLS. In: Digest of the LEOS Summer Topical Meetings; 7-19 July 2006; Quebec City, Que., Canada. New York: IEEE; 2006. p. 20. DOI: 10.1109/LEOSST.2006.1694049

[46] Münzberg M, Breiter R, Cabanski W, Lutz H, Wendler J, Ziegler J, Rehm R, Walther M. Multi spectral IR detection modules and applications. In: Proceedings of SPIE, Infrared Technology and Applications XXXII; 5 May 2006; Orlando (Kissimmee), Florida. SPIE; 2006;**6206**: p. 620627. DOI: 10.1117/12.665467

[47] Hill CJ, Soibel A, Keo SA, Mumolo JM, Ting DZ, Gunapala SD, Rhiger DR, Kvaas RE, Harris SF. Demonstration of mid-and long-wavelength infrared antimonide-based focal plane arrays. In: Proceedings of SPIE, Infrared Technology and Applications XXXV; 7 April 2009; Orlando, Florida. SPIE; 2009;**7298**: p. 729804. DOI: 10.1117/12.818692

[48] Gunapala SD, Ting DZ, Hill CJ, Nguyen J, Soibel A, Rafol SB, Keo SA, Mumolo JM, Lee MC, Liu JK, Yang B. Demonstration of a 1024 × 1024 pixel InAs-GaSb superlattice focal plane array. IEEE Photonics Technology Letters. 2010;**22**:1856-1858. DOI: 10.1109/LPT.2010.2089677

[49] Haddadi A, Ramezani-Darvish S, Chen G, Hoang AM, Nguyen BM, Razeghi M. High operability 1024 × 1024 long wavelength type-II superlattice focal plane array. IEEE Journal of Quantum Electronics. 2012;**48**:221-228. DOI: 10.1109/JQE.2011.2175903

[50] Manurkar P, Ramezani-Darvish S, Nguyen BM, Razeghi M, Hubbs J. High performance long wavelength infrared mega-pixel focal plane array based on type-II superlattices. Applied Physics Letters. 2010;**97**:193505. DOI: 10.1063/1.3514244

[51] Klipstein PC, Avnon E, Azulai D, Benny Y, Fraenkel R, Glozman A, Hojman E, Klin O, Krasovitsky L, Langof L, Lukomsky I. Type II superlattice technology for LWIR detectors. In: Proceedings of SPIE, Infrared Technology and Applications XLII; 20 May 2016; Baltimore, Maryland. SPIE; 2016;**9819**: p. 98190T. DOI: 10.1117/12.2222776

[52] Hoang AM, Dehzangi A, Adhikary S, Razeghi M. High performance bias-selectable three-color short-wave/mid-wave/long-wave infrared photodetectors based on type-II InAs/GaSb/AlSb superlattices. Scientific Reports. 2016;**6**:24144. DOI: 10.1038/srep24144

[53] Haddadi A, Suo XV, Adhikary S, Dianat P, Chevallier R, Hoang AM, Razeghi M. High-performance short-wavelength infrared photodetectors based on type-II InAs/$InAs_{1-x}Sb_x/_{AlAs1-x}Sb_x$ superlattices. Applied Physics Letters. 2015;**107**:141104. DOI: 10.1063/1.4932518

[54] Razeghi M, Haddadi A, Hoang AM, Huang EK, Chen G, Bogdanov S, Darvish SR, Callewaert F, McClintock R. Advances in antimonide-based type-II superlattices for infrared detection and imaging at center for quantum devices. Infrared Physics & Technology. 2013;**59**:41-52. DOI: 10.1016/j.infrared.2012.12.008

[55] Sood AK, Zeller JW, Richwine RA, Puri YR, Efstathiadis H, Haldar P, Dhar NK, Polla DL. SiGe based visible-NIR photodetector technology for optoelectronic applications. In: Yasin M, Arof H, Harum SW, editors. Advances in Optical Fiber Technology: Fundamental Optical Phenomena and Applications. Rijeka: InTech; 2015. p. 315-351. DOI: 10.5772/59065

[56] Razeghi M, Nguyen BM. Advances in mid-infrared detection and imaging: A key issues review. Reports on Progress in Physics. 2014;**77**:082401. DOI: 10.1088/0034-4885/77/8/082401

[57] Martyniuk P, Rogalski A. Performance comparison of barrier detectors and HgCdTe photodiodes. Optical Engineering. 2014;**53**:106105. DOI: 10.1117/1.OE.53.10.106105

[58] Richards A. Advantages of strained-layer superlattice detectors for high-speed thermal events. In: Proceedings of SPIE, Quantum Sensing and Nano Electronics and Photonics XIV; 27 January 2017; San Francisco, California. SPIE; 2017;**10111**: p. 101111B. DOI: 10.1117/12.2250039

[59] Bainter C, High-speed IR. Detectors aid ballistic testing. Photonics Spectra. 2017;**51**:63-67

[60] Klein B. Defect investigations in InAs/GaSb type-II strained layer superlattice [thesis]. Albuquerque: The University of New Mexico; 2014

[61] Salvetti O, Ronchi LA, Corsi C, Rogalski A, Strojnik M. Advanced infrared technology and applications. Advances in Optical Technologies. 2013;**2013**:1-2. DOI: 10.1155/2013/459074

[62] Rogalski A. Progress in focal plane array technologies. Progress in Quantum Electronics. 2012;**36**:342-473. DOI: 10.1016/j.pquantelec.2012.07.001

[63] Hood A, Razeghi M, Aifer EH, Brown GJ. On the performance and surface passivation of type II InAs/ GaSb superlattice photodiodes for the very-long-wavelength infrared. Applied Physics Letters. 2005;**87**:151113. DOI: 10.1063/1.2089170

[64] Han X, Xiang W, Hao HY, Jiang DW, Sun YY, Wang GW, Xu YQ, Niu ZC. Very long wavelength infrared focal plane arrays with 50% cutoff wavelength based on type-II InAs/GaSb superlattice. Chinese Physics B. 2017;**26**:018505. DOI: 10.1088/1674-1056/26/1/018505

[65] Hoang AM, Chen G, Chevallier R, Haddadi A, Razeghi M. High performance photodiodes based on InAs/InAsSb type-II superlattices for very long wavelength infrared detection. Applied Physics Letters. 2014;**104**:251105. DOI: 10.1063/1.4884947

[66] Wei Y, Gin A, Razeghi M, Brown GJ. Advanced InAs/GaSb superlattice photovoltaic detectors for very long wavelength infrared applications. Applied Physics Letters. 2002;**80**:3262-3264. DOI: 10.1063/1.1476395

[67] Li X, Jiang D, Zhang Y, Zhao L. Interface optimization and fabrication of InAs/GaSb type II superlattice for very long wavelength infrared photodetectors. Superlattices and Microstructures. 2016;**91**:238-243. DOI: 10.1016/j.spmi.2016.01.013

[68] Das S, Das U, Gautam N, Krishna S. Type-II InAs/GaSb photodiode array pixel isolation by femto-second laser anneal. Infrared Physics & Technology. 2016;**78**:162-166. DOI: 10.1016/j.infrared.2016.07.023

[69] Chakrabarti P, Gawarikar A, Mehta V, Garg D. Effect of trap-assisted tunneling (TAT) on the performance of homojunction mid-infrared photodetectors based on InAsSb. Journal of Microwaves, Optoelectronics and Electromagnetic Applications. 2006;**5**:1-4

[70] Rhiger DR, Kvaas RE, Harris SF, Hill CJ. Characterization of LWIR diodes on InAs/GaSb type-II superlattice material. Infrared Physics & Technology. 2009;**52**:304-309. DOI: 10.1016/j.infrared.2009.05.009

[71] Grein CH, Garland J, Flatte ME. Strained and unstrained layer superlattices for infrared detection. Journal of Electronic Materials. 2009;**38**:1800-1804. DOI: 10.1007/s11664-009-0757-8

[72] Cervera C, Jaworowicz K, Aït-Kaci H, Chaghi R, Rodriguez JB, Ribet-Mohamed I, Christol P. Temperature dependence performances of InAs/GaSb superlattice photodiode. Infrared Physics & Technology. 2011;**54**:258-262. DOI: 10.1016/j.infrared.2010.12.025

[73] Yang QK, Fuchs F, Schmitz J, Pletschen W. Investigation of trap-assisted tunneling current in InAs/(GaIn) Sb superlattice long-wavelength photodiodes. Applied Physics Letters. 2002;**81**:4757-4759. DOI: 10.1063/1.1529306

[74] Asplund C, von Würtemberg RM, Lantz D, Malm H, Martijn H, Plis E, Gautam N, Krishna S. Performance of mid-wave T2SL detectors with heterojunction barriers. Infrared Physics & Technology. 2013;**59**:22-27. DOI: 10.1016/j.infrared.2012.12.004

[75] Sah CT, Noyce RN, Shockley W. Carrier generation and recombination in pn junctions and pn junction characteristics. In: Proceedings of the IRE, September 1957. New York: IEEE; 1957;**45**:1228-1243. DOI: 10.1109/JRPROC.1957.278528

[76] Nguyen J, Ting DZ, Hill CJ, Soibel A, Keo SA, Gunapala SD. Dark current analysis of InAs/GaSb superlattices at low temperatures. Infrared Physics & Technology. 2009;**52**:317-321. DOI: 10.1016/j.infrared.2009.05.022

[77] Plis E, Kutty MN, Myers S, Kim HS, Gautam N, Dawson LR, Krishna S. Passivation of long-wave infrared InAs/GaSb strained layer superlattice detectors. Infrared Physics & Technology. 2011;**54**:252-257. DOI: 10.1016/j.infrared.2010.12.024

[78] Kim HS, Plis E, Gautam N, Myers S, Sharma Y, Dawson LR, Krishna S. Reduction of surface leakage current in InAs/GaSb strained layer long wavelength superlattice detectors using SU-8 passivation. Applied Physics Letters. 2010;**97**:143512. DOI: 10.1063/1.3499290

[79] Martyniuk P, Kopytko M, Rogalski A. Barrier infrared detectors. Opto-Electronics Review. 2014;**22**:127-146. DOI: 10.2478/s11772-014-0187-x

[80] Johnson JL, Samoska LA, Gossard AC, Merz JL, Jack MD, Chapman GR, Baumgratz BA, Kosai K, Johnson SM. Electrical and optical properties of infrared photodiodes using the $InAs/Ga_{1-x}In_x$ Sb superlattice in heterojunctions with GaSb. Journal of Applied Physics. 1996;**80**:1116-1127. DOI: 10.1063/1.362849

[81] Canedy CL, Aifer EH, Warner JH, Vurgaftman I, Jackson EM, Tischler JG, Powell SP, Olver K, Meyer JR, Tennant WE. Controlling dark current in type-II superlattice photodiodes. Infrared Physics & Technology. 2009;**52**:326-334. DOI: 10.1016/j.infrared.2009.09.004

[82] Aifer EH, Maximenko SI, Yakes MK, Yi C, Canedy CL, Vurgaftman I, Jackson EM, Nolde JA, Affouda CA, Gonzalez M, Meyer JR, Clark KP, Pinsukanjana PR. Recent developments

in type-II superlattice-based infrared detectors. In: Proceedings of SPIE, Infrared Technology and Applications XXXVI; 4 May 2010; Orlando, Florida. SPIE; 2010;**7660**: p. 76601Q. DOI: 10.1117/12.855028

[83] Myers S, Plis E, Morath C, Cowan V, Gautam N, Klein B, Kutty MN, Naydenkov M, Sandy TS, Krishna S. Comparison of superlattice based dual color nBn and pBp infrared detectors. In: Proceedings of SPIE, Infrared Sensors, Devices, and Applications, and Single Photon Imaging II; 17 September 2011; San Diego, California. SPIE; 2011;**8155**: p. 815507. DOI: 10.1117/12.894986

[84] Martyniuk P, Rogalski A. Theoretical modeling of InAsSb/AlAsSb barrier detectors for higher-operation-temperature conditions. Optical Engineering. 2014;**53**:017106. DOI: 10.1117/1.OE.53.1.01710

[85] Rogalski A, Martyniuk P, Kopytko M. Challenges of small-pixel infrared detectors: A review. Reports on Progress in Physics. 2016;**79**:046501. DOI: 10.1088/0034-4885/79/4/046501

[86] Kim HS, Plis E, Rodriguez JB, Bishop GD, Sharma YD, Dawson LR, Krishna S, Bundas J, Cook R, Burrows D, Dennis R. Mid-IR focal plane array based on type-II InAs/GaSb strain layer superlattice detector with nBn design. Applied Physics Letters. 2008;**92**:183502. DOI: 10.1063/1.2920764

[87] Khoshakhlagh A, Rodriguez JB, Plis E, Bishop GD, Sharma YD, Kim HS, Dawson LR, Krishna S. Bias dependent dual band response from InAs/Ga(in)Sb type II strain layer superlattice detectors. Applied Physics Letters. 2007;**91**:263504. DOI: 10.1063/1.2824819

[88] Klein B, Plis E, Kutty MN, Gautam N, Albrecht A, Myers S, Krishna S. Varshni parameters for InAs/GaSb strained layer superlattice infrared photodetectors. Journal of Physics D: Applied Physics. 2011;**44**:075102. DOI: 10.1088/0022-3727/44/7/075102

[89] Martyniuk P, Gawron W, Rogalski A. Theoretical modeling of HOT HgCdTe barrier detectors for the mid-wave infrared range. Journal of Electronic Materials. 2013;**42**:3309-3319. DOI: 10.1007/s11664-013-2737-2

[90] Plis EA, Krishna SS, Gautam N, Myers S, Krishna S. Bias switchable dual-band InAs/GaSb superlattice detector with pBp architecture. IEEE Photonics Journal. 2011;**3**:234-240. DOI: 10.1109/JPHOT.2011.2125949

[91] Razeghi M, Haddadi A, Hoang AM, Chen G, Ramezani-Darvish S, Bijjam P. High performance bias-selectable dual-band mid-/long-wavelength infrared photodetectors and focal plane arrays based on InAs/GaSb type-II superlattices. In: Proceedings of SPIE, Infrared Technology and Applications XXXIX; 11 June 2013; Baltimore, Maryland. SPIE; 2013;**8704**: p. 87040S. DOI: 10.1117/12.2019147

[92] Delaunay PY, Nguyen BM, Hoffman D, Huang EK, Razeghi M. Background limited performance of long wavelength infrared focal plane arrays fabricated from M-structure InAs-GaSb superlattices. IEEE Journal of Quantum Electronics. 2009;**45**:157-162. DOI: 10.1109/JQE.2008.2002667

[93] Hoang AM, Chen G, Haddadi A, Abdollahi Pour S, Razeghi M. Demonstration of shortwavelength infrared photodiodes based on type-II InAs/GaSb/AlSb superlattices. Applied Physics Letters. 2012;**100**:211101. DOI: 10.1063/1.4720094

[94] Nguyen BM, Hoffman D, Delaunay PY, Razeghi M. Dark current suppression in type II InAs/GaSb superlattice long wavelength infrared photodiodes with M-structure barrier. Applied Physics Letters. 2007;**91**:163511. DOI: 10.1063/1.2800808

[95] Chen G, Haddadi A, Hoang AM, Chevallier R, Razeghi M. Demonstration of type-II superlattice MWIR minority carrier unipolar imager for high operation temperature application. Optics Letters. 2015;**40**:45-47. DOI: 10.1364/OL.40.000045

[96] Nguyen BM, Chen G, Hoang AM, Abdollahi Pour S, Bogdanov S, Razeghi M. Effect of contact doping in superlattice-based minority carrier unipolar detectors. Applied Physics Letters. 2011;**99**:033501. DOI: 10.1063/1.3613927

[97] Aifer EH, Tischler JG, Warner JH, Vurgaftman I, Bewley WW, Meyer JR, Kim JC, Whitman LJ, Canedy CL, Jackson EM. W-structured type-II superlattice long-wave infrared photodiodes with high quantum efficiency. Applied Physics Letters. 2006;**89**:053519. DOI: 10.1063/1.2335509

[98] Salihoglu O, Hostut M, Tansel T, Kutluer K, Kilic A, Alyoruk M, Sevik C, Turan R, Ergun Y, Aydinli A. Electronic and optical properties of 4.2 μm "N" structured superlattice MWIR photodetectors. Infrared Physics & Technology. 2013;**59**:36-40. DOI: 10.1016/j.infrared.2012.12.007

[99] Ting DZ, Soibel A, Nguyen J, Höglund L, Khoshakhlagh A, Rafol B, Keo SA, Liao A, Mumolo JM, Liu JK, Gunapala SD. Type II superlattice barrier infrared detector. In: Proceedings of SPIE, Infrared Remote Sensing and Instrumentation XIX; 17 September 2011; San Diego, California. SPIE; 2011;**8154**: p. 81540L. DOI: 10.1117/12.896240

[100] Gautam N, Naydenkov M, Myers S, Barve AV, Plis E, Rotter T, Dawson LR, Krishna S. Three color infrared detector using InAs/GaSb superlattices with unipolar barriers. Applied Physics Letters. 2011;**98**:121106. DOI: 10.1063/1.3570687

[101] Tian Z, DeCuir EA, Wijewarnasuriya PS, Pattison JW, Gautam N, Krishna S, Dhar N, Welser RE, Sood AK. Low-dark current structures for long-wavelength type-II strained layer superlattice photodiodes. In: Proceedings of SPIE, Infrared Technology and Applications XXXIX; 11 June 2013; Baltimore, Maryland. SPIE; 2013;**8704**: p. 870415. DOI: 10.1117/12.2015489

[102] Gautam N, Myers S, Barve AV, Klein B, Smith EP, Rhiger DR, Kim HS, Tian ZB, Krishna S. Barrier engineered infrared photodetectors based on type-II InAs/GaSb strained layer superlattices. IEEE Journal of Quantum Electronics. 2013;**49**:211-217. DOI: 10.1109/JQE.2012.2236643

[103] Klipstein P. "XBn" barrier photodetectors for high sensitivity and high operating temperature infrared sensors. In: Proceedings of SPIE, Infrared Technology and

Applications XXXIV; 23 April 2008; Orlando, Florida. SPIE; 2008;**6940**: p. 69402U. DOI: 10.1117/12.778848

[104] Plis E, Klein B, Myers S, Gautam N, Smith EP, Krishna S. High operating temperature midwave infrared InAs/GaSb superlattice photodetectors on (111) GaSb substrates. IEEE Electron Device Letters. 2013;**34**:426-428. DOI: 10.1109/LED.2012.2236534

[105] Ariyawansa G, Reyner CJ, Steenbergen EH, Duran JM, Reding JD, Scheihing JE, Bourassa HR, Liang BL, Huffaker DL. InGaAs/InAsSb strained layer superlattices for mid-wave infrared detectors. Applied Physics Letters. 2016;**108**:022106. DOI: 10.1063/1.4939904

[106] Pusz W, Kowalewski A, Martyniuk P, Gawron W, Plis E, Krishna S, Rogalski A. Mid-wavelength infrared type-II InAs/GaSb superlattice interband cascade photodetectors. Optical Engineering. 2014;**53**:043107. DOI: 10.1117/1.OE.53.4.043107

[107] Delaunay PY, Razeghi M. Noise analysis in type-II InAs/GaSb focal plane arrays. Journal of Applied Physics. 2009;**106**:063110. DOI: 10.1063/1.3224958

[108] Razeghi M, Abdollahi Pour S, Huang E, Chen G, Haddadi A, Nguyen B. Type-II InAs/GaSb photodiodes and focal plane arrays aimed at high operating temperatures. Opto-Electronics Review. 2011;**19**:261-269. DOI: 10.2478/s11772-011-0028-0

[109] Pour SA, Huang EW, Chen G, Haddadi A, Nguyen BM, Razeghi M. High operating temperature midwave infrared photodiodes and focal plane arrays based on type-II InAs/GaSb superlattices. Applied Physics Letters. 2011;**98**:143501. DOI: 10.1063/1.3573867

[110] Wei Y, Hood A, Yau H, Gin A, Razeghi M, Tidrow MZ, Nathan V. Uncooled operation of type-II InAs/GaSb superlattice photodiodes in the midwavelength infrared range. Applied Physics Letters. 2005;**86**:233106. DOI: 10.1063/1.1947908

[111] Gautam N, Myers S, Barve AV, Klein B, Smith EP, Rhiger DR, Dawson LR, Krishna S. High operating temperature interband cascade midwave infrared detector based on type-II InAs/GaSb strained layer superlattice. Applied Physics Letters. 2012;**101**:021106. DOI: 10.1063/1.4733660

[112] Adams A, Rittenberg E. HOT IR sensors improve IR camera size, weight, and power. Laser Focus World. 2014;**50**:83-87

[113] Nguyen BM, Chen G, Hoang MA, Razeghi M. Growth and characterization of long-wavelength infrared type-II superlattice photodiodes on a 3-in GaSb wafer. IEEE Journal of Quantum Electronics. 2011;**47**:686-690. DOI: 10.1109/JQE.2010.2103049

[114] Hoang AM, Chen G, Haddadi A, Razeghi M. Demonstration of high performance bias-selectable dual-band short-/mid-wavelength infrared photodetectors based on type-II InAs/GaSb/AlSb superlattices. Applied Physics Letters. 2013;**102**:011108. DOI: 10.1063/1.4720094

[115] Lao YF, Pitigala PK, Unil Perera AG, Plis E, Krishna SS, Wijewarnasuriya PS. Band offsets and carrier dynamics of type-II InAs/GaSb superlattice photodetectors studied by

internal photoemission spectroscopy. Applied Physics Letters. 2013;**103**:181110. DOI: 10.1063/1.4827881

[116] Jiang DW, Xiang W, Guo FY, Hao HY, Han X, Li XC, Wang GW, Xu YQ, Yu QJ, Niu ZC. Low crosstalk three-color infrared detector by controlling the minority carriers type of InAs/GaSb superlattices for middle-long and very-long wavelength. Chinese Physics Letters. 2016;**33**:048502. DOI: doi.org/10.1088/0256-307X/33/4/048502

[117] Smith M. FLIR X6900sc Camera Captures Temperature Images at 1,000 Frames per Second [Internet]. 2016. Available from: http://www.industrytap.com/flir-x6900sc-camera-captures-temperature-images-at-1000-frames-per-second/35359 [Accessed: Jul 15, 2017]

Chapter 6

AlGaAs/GaAs Quantum Well Infrared Photodetectors

Michael A. Dem'yanenko, Dmitry G. Esaev,
Aleksandr I. Toropov, Natalia A. Valisheva,
Sergey A. Dvoretsky, Dmitry V. Dmitriev,
Dmitry V. Gulyaev, Vladimir A. Fateev,
Igor V. Marchishin, Dmitry Yu Protasov,
Anatoly P. Savchenko, Victor N. Ovsyuk and
Konstantin Zhuravlev

Additional information is available at the end of the chapter

http://dx.doi.org/10.5772/intechopen.71266

Abstract

In this article, we present an overview of a focal plane array (FPA) with 640 × 512 pixels based on the AlGaAs quantum well infrared photodetector (QWIP). The physical principles of the QWIP operation and their parameters for the spectral range of 8–10 μm have been discussed. The technology of the manufacturing FPA based on the QWIP structures with the pixels 384 × 288 and 640 × 512 has been demonstrated. The parameters of the manufactured 640 × 512 FPA with a step of 20 μm have been given. At the operating temperature of 72 K, the temperature resolution of QWIP focal plane arrays is less than 35 mK. The number of defective elements in the matrix does not exceed 0.5%. The stability and uniformity of the FPA have been demonstrated.

Keywords: GaAs/AlGaAs quantum well infrared photodetector, focal plane array, multiplexer, dark current, noise equivalent temperature difference

1. Introduction

The solution to a well-known problem of a finite rectangular quantum well (QW) indicates that the energy position of the quantum subbands is determined by several parameters, namely, the width of the quantum well, the height of the barriers, and the carrier effective masses. The possibility of a rather simple way of monitoring the wavelength of intersubband transitions in a quantum well to create photodetectors has attracted the attention of researchers for a long time [1]. Improving the technology of the molecular beam epitaxy (MBE) has allowed converting these dreams into reality. The quantum well infrared photodetector (QWIP) was first demonstrated on the basis of the AlGaAs/GaAs heterostructure in 1987 [2]. Permanent improvement of the QWIP design, epitaxial growth technologies of the AlGaAs layers, and techniques for device manufacturing allow using the QWIP based on AlGaAs/GaAs heterostructures not only for military purposes but also for various civilian tasks [3–7].

In this paper, we have described the current state of the manufacturing technology in the Rzhanov Institute of Semiconductor Physics of SB RAS for the focal plane array (FPA) based on the AlGaAs/GaAs QWIP structure.

2. The physical model of the QWIP

A typical heterostructure intended to detect the infrared radiation (IR) is shown in **Figure 1**. Such heterostructures are commonly formed by a thin (4–6 nm) GaAs QW located between the wide band gap AlGaAs barrier layers. The thickness of these barrier layers (L_b) is much larger than the width (L_w) of the quantum well and selected in a range of 40–55 nm to suppress a dark current caused by the electron tunneling between the neighboring QWs. The barrier height ΔE_c is determined by the aluminum mole fraction (x) in the barriers. The energy position of the

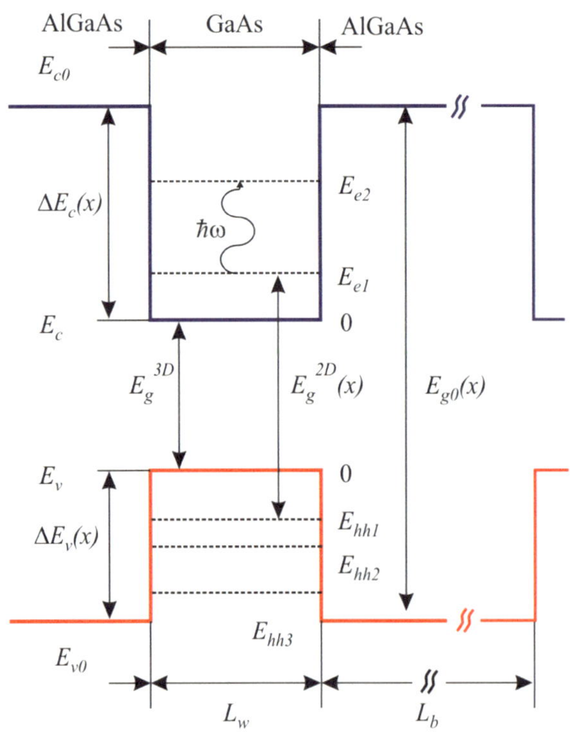

Figure 1. The energy structure of the GaAs/AlGaAs QW.

quantum levels depends not only on the quantum well thickness but also on x, as the barrier height is not infinite.

In such a structure, the electron moves from a ground quantum level $Ee1$ to the first excited level $Ee2$ at the absorption of the photon with the energy $\hbar\omega$. In contrast to the bulk case, the probability of intersubband transitions depends on the direction (polarization) of the electric field in the incident electromagnetic wave. In this case, the matrix element of the intersubband transition from the i-th level to the j-th level obtained in the first order of the perturbation theory is proportional $\langle jk'|e{\cdot}\widehat{p}|ik\rangle$ [8], where e is the polarization vector of a electromagnetic wave and $\widehat{p}$ is the momentum operator. In case of the polarization of the electromagnetic wave in the QW plane (in x or y directions), the polarization vector is equal to $e=(1,0,0)$ or $e=(0,1,0)$, whereas the operator is equal to $-i\hbar\,{\partial}/{\partial x}$ or $-i\hbar\,{\partial}/{\partial y}$, respectively. Thus, under the influence of this operator upon the total wave function

$$\Psi_{ik}(R) = A^{-1/2}\phi_i(z)e^{ik\cdot r} \tag{1}$$

such expressions as $\hbar k_{x,y}\langle jk' \mid ik\rangle$ being zero at the $i\neq j$ resulted from the orthogonality of the wave functions are to be obtained. The constant A in the expression (Eq.(1)) is the area of the quantum well.

Conversely, when the electromagnetic wave is polarized along the z axis, the operator $e\widehat{p}$ is equal to $-i\hbar\,{\partial}/{\partial z}$ and the expression $\langle jk'|e{\cdot}\widehat{p}|ik\rangle$ changes as:

$$\langle jk'|e{\cdot}\widehat{p}|ik\rangle = -i\hbar A^{-1/2}\int \phi_j^*(z)\frac{\partial}{\partial z}\phi_i(z)dz\int e^{i(k-k')\cdot r}d^2r \tag{2}$$

The first integral in expression (Eq.(2)) is nonzero for i and j with different parities, and the last integral is nonzero for $k=k'$. Therefore, optical transitions in the QW occur only in the presence of polarization in an electromagnetic wave along the coordinate z, i.e., in the direction of the heterostructure growth.

The concentration of the two-dimensional electron gas (2DEG) in the photodetector structures should be large enough to locate the ground quantum level below the Fermi level. The required position of the Fermi level is achieved at the 2DEG concentration (N_{2D}) approximately equal to (4–5) $\times$ 10^{11} cm^{-2}. This electron density is distributed over the QW width in proportion to the square of the ground state wave function. According to the Poisson equation, such a large number of negative charges result in the appearance of a negative curvature of the conduction band bottom in the quantum well. It is worth mentioning that the middle part of the QW is maximally distorted, where the value of the square of the cosine is maximum. However, it is the middle part of the QW that is commonly doped to fill the QWs with electrons. A positive charge of the ionized donors bends the bottom of the QW back and downward. Consequently, the quantum wells remain almost rectangular shape under such a method of doping. The potential diagrams, the energy of the quantum levels, and wave functions in such rectangular AlGaAs/GaAs/AlGaAs quantum wells calculated by the *nextnano* program are shown **Figure 2A** [9].

The doping of the structure outside the GaAs QW (so-called the modulation-doped QWs) leads to significant changes in the QW shape (**Figure 2B**). Actually, the separation of electrons

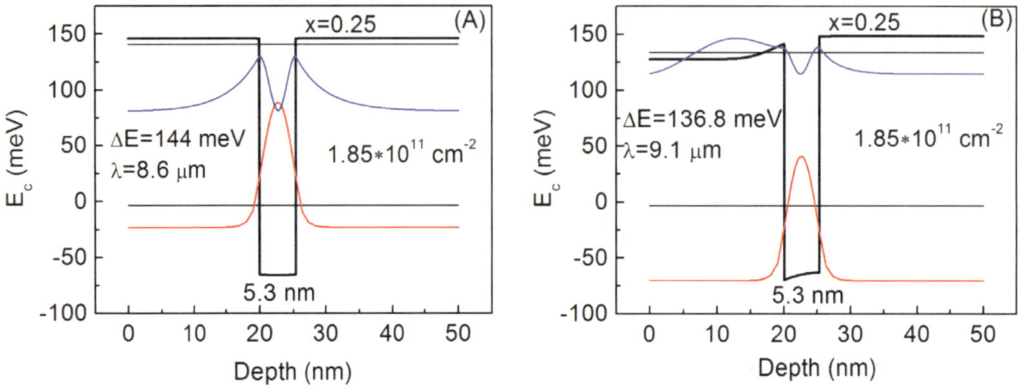

Figure 2. The calculated potential diagrams and wave functions for AlGaAs/GaAs/ AlGaAs QWs: (A) doped in the middle part of the quantum well; (B) with the one-side δ-doping with the 5-nm spacer.

and ionized donors causes the appearance of an electric field and, therefore, a noticeable bending of the band next to the quantum well region. Therefore, the energy of the quantum levels changes. The calculated energies of the ground and first-excited quantum levels resulted from a self-consistent calculation proved to be equal to 3.2/140.9 meV and −3.1/133.7 meV for the QW doped in middle part and the modulation-doped QW, respectively. These values are given with respect to the Fermi level that lies at the energy $\varepsilon = 0$. As one can easily estimate, the shift in the quantum levels resulted from the changes in the doping location leads to a shift in the intersubband absorption band by 0.5 μm.

The nomograms of the dependence of the intersubband transition line upon the QW width and the barrier height were calculated by the self-consistent solution of the Schrödinger and Poisson equations to compare the rectangular and modulation-doped QWs. **Figure 3** shows the calculated

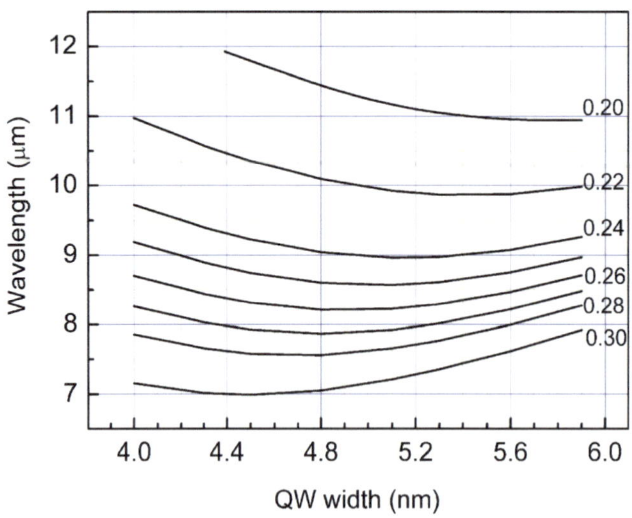

Figure 3. The nomogram showing the dependence of the wavelength upon the quantum well width and the barrier height (molar aluminum mole fraction) for a quantum well with the middle doped part.

nomogram for the QW doped in the middle part. As one can see from **Figure 3**, λmax is weakly dependent on the QW width in the wavelength region (8–10 μm) worth examining, whereas the required wavelength of the IR absorption is more difficult to obtain in the modulation-doped QWs. One more parameter is added, and the energy of the intersubband transition from the ground to the first excited level depends not only on the well width and the barrier height (or the Al mole fraction in the barrier) but also on the doping level. Therefore, the calculation was made for the modulated-doped QW at a fixed QW width equal to 5.6 nm (**Figure 4**).

The dark current flowing in the thermal equilibrium through the photodetector under a bias is an important characteristic of a QWIP. Its value is frequently used as a QWIP quality criterion. In QWIPs, the physical reason for the appearance of a dark current is the thermionic electron emission from the ground-filled quantum level to continuum states both above and below the energy barrier E_{c0} (**Figure 1**), due to its tunneling through the triangle barrier in the electric field.

The dark current density (I_d) in the structure shown in **Figure 1** consists of the thermoactivation and tunneling components, and it has the following general form [8]:

$$J(E) = q{\cdot}v(E)\left(\frac{m^*}{\pi\hbar^2(L_W + L_B)}\right)\int_{\varepsilon_1}^{\infty} f(\varepsilon, T)T(\varepsilon, E)dE \tag{3}$$

where $v(E) = \mu E\sqrt{1 + \left(\mu E/v_{sat}\right)^2}$ is the average electron drift velocity in the AlGaAs barrier layers, v_{sat} = 2.9×10^6 cm/s is the saturation velocity in the multiple quantum well (MQW), μ_n = 10^3 cm^2/V is the electron mobility in the AlGaAs with $x \approx 0.3$, ε_1 is the energy of the ground level in the quantum well. The fourth multiplier in (Eq. (3)) determines the concentration of the carriers participating in the conductivity, $f(\varepsilon)$ is the Fermi distribution function

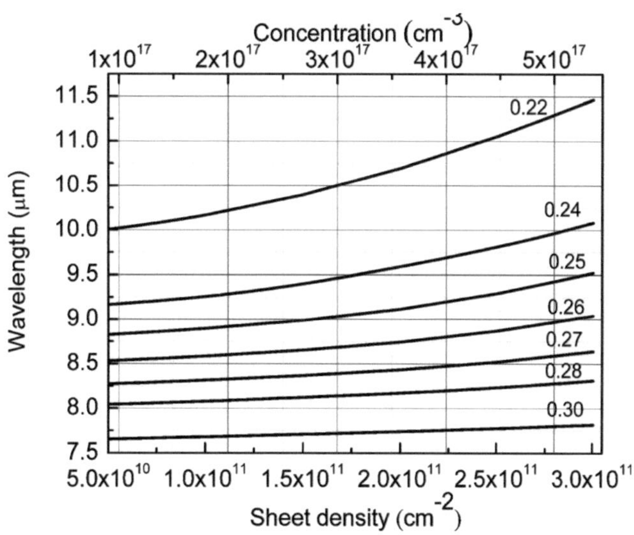

Figure 4. The nomogram showing the dependence of the wavelength upon the 2DEG concentration and the barrier height (aluminum mole fraction) in the modulation-doped QWs.

of the 2DEG, and $T(\varepsilon,E)$ is the probability of tunneling of the electrons from the GaAs layer to the states above the barrier. The electric field dependence of the $T(\varepsilon,E)$ reflects the effective barrier lowering for the hot electrons with a total energy $\varepsilon \cong V_b$. We can assume with good accuracy that for $\varepsilon > V_b$, the coefficient is $T(\varepsilon,E) \approx 1$, whereas for $\varepsilon < V_b$, the coefficient is $T(\varepsilon,E) = 0$.

The modeling with the help of *nextnano* has shown that the barrier lowering in an electric field can be well described by the expression:

$$V_b(E) = V_b^0 - qE\frac{L_W}{2} \quad (4)$$

Thus, the current density through the heterostructure is finally obtained as:

$$j(E) = q \cdot v(E)\left(\frac{m^*}{\pi\hbar^2(L_W + L_B)}\right)\exp\left(\frac{V_b^0 - qE\frac{L_W}{2} - (\varepsilon_1 + \varepsilon_F)}{kT}\right) \quad (5)$$

where k is the Boltzmann constant and T is the temperature. To calculate the dependence of the current density upon the donor concentration, the dependence of the activation energy $V_b^0 - (\varepsilon_1 + \varepsilon_F)$ upon the donor concentration for the QW of 5.7 nm width and the aluminum mole fraction in the barriers $x = 0.25$ was calculated by the nextnano program at T = 77 K. The doped middle part of the quantum well was 3.5-nm thick. The calculation results are shown in **Figure 5**.

As one can see from **Figure 5**, the barrier height decreases at the enhanced doping, which is caused by an increase in the Fermi level accompanied by an increase in the DEG concentration from 1.4×10^{11} to 3.4×10^{11} cm^{-2}.

Figure 6 shows the calculated dependences of the dark current density upon the applied electric field for different donor concentrations.

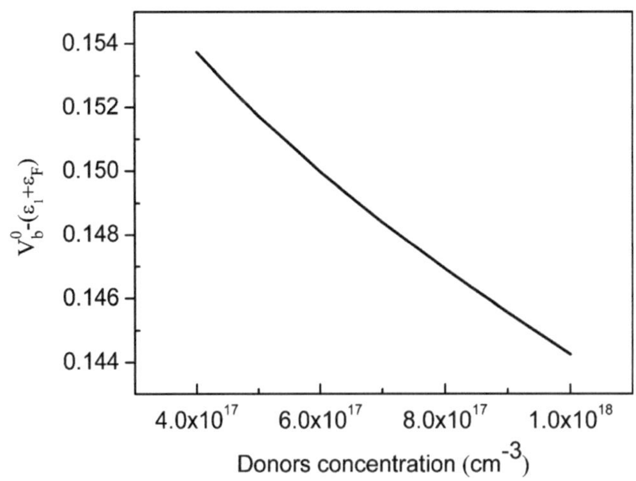

Figure 5. Dependence of the activation energy of the QWIP dark current upon the donor concentration.

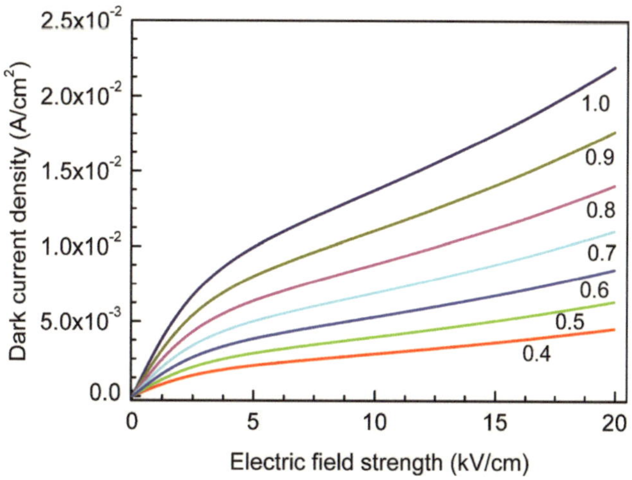

Figure 6. Dependence of the QWIP dark current upon the applied electric field for different donor concentrations. The donor concentration is indicated in units of 10^{18} cm^{-3} in the figure.

The optimum 2DEG concentration corresponding to the minimum value of the threshold power of the radiation (NEP) is estimated as follows. The value of NEP is given by:

$$NEP = \frac{I_n}{R} \tag{6}$$

where I_n is the noise current, R is the responsivity. The noise current depends on the dark current I_d as follows [10]:

$$I_n^2 = 4qg_n I_d \Delta f \tag{7}$$

where g_n is the coefficient of the noise gain and Δf is the width of the noise band. As one can see from the expression (Eq.(5)), the dark current increases exponentially at the increasing N_{2D}. Thus, for the degenerate 2DEG, the current depends on the 2DEG concentration N_{2D} (which is directly proportional to the donor concentration) as:

$$I_d \sim \exp\left(N_{2D}\pi\hbar^2/m^*kT\right) - 1 \tag{8}$$

In its turn, the responsivity is proportional to the absorption coefficient of a single QW α, which, in turn, is proportional to the DEG concentration [10]:

$$R \sim \alpha \sim N_{2D} \tag{9}$$

Thus, combining Eqs. (6)–(8), and Eq. (9), we obtain [11]:

$$NEP \sim \sqrt{e^{\rho}-1}/\rho \tag{10}$$

where $\rho = \pi^2 N_{2D}/m^*kT$. The minimum value of the expression given is achieved under the condition $\varrho = 1.6$ that is $N_{2D} = 2.7 \times 10^{11}$ or $N = 5.1 \times 10^{17}$ cm^{-3} for the quantum well of 5.3 nm width at the QWIP operating temperature of T = 70 K.

Thus, the maximum sensitivity of the GaAs/AlGaAs QWIP is expected to be around 8.6 μm to ensure the spectral range of the detector (8–10 μm). The calculations have shown that this condition is satisfied by such parameters of the doped QW as the well width $L_W = 5.3$ nm and aluminum mole fraction $x = 0.25$. When the ground state energy level is $\varepsilon_1 = 66.8$ meV, the energy of the first excited level is $\varepsilon_2 = 206.0$ meV, and the barrier height is $V_b = 211.0$ meV that corresponds to a fairly accurate coincidence of the first-exited quantum level and the edge of the potential barrier.

3. Properties of the heteroepitaxial GaAs/AlGaAs MQW structures

Multiple quantum wells GaAs/AlGaAs were obtained by the MBE on the GaAs (100) substrate with a buffer consisting of a 0.4-μm GaAs layer. The doped GaAs layers of 1.5 μm thickness were used as a base contact, whereas the upper ohmic contact was provided by the doped GaAs of 1.2 μm thickness.

It has been found that the best structural perfection of epitaxial GaAs layers and the greatest value of the carrier mobility can be obtained at a growth under conditions ensuring a (3 × 6) surface structure having a stoichiometric composition on the growth surface and being intermediate between the "As-stabilized" (2 × 4) and "Ga-stabilized" (4 × 2) surfaces. At the growth under such conditions, the concentrations of the gallium and arsenic vacancies on the surface are minimal, which provides a minimum concentration of the intrinsic point defects in the crystal. The deviation of the growth conditions toward the gallium or arsenic stabilization leads to the enrichment of the epitaxial layers by the defects of anion or cation sublattices, respectively. Since the (3 × 6) surface is observed in the narrow ranges of the substrate temperature and V/III flow ratio, the growth conditions around the transition from (2 × 4) to (3 × 6) surface can be recommended in order to obtain good process reproducibility.

Such growth conditions being applied, the necessary homogeneity and concentration of the growth defects across the wafer can be achieved. The distribution of the growth defect ("oval" defect) density is shown in **Figure 7**. One can see that the average density of the defects of 100–200 pcs/cm^2 can potentially lead to defectiveness of the photosensitive element matrix by no more than 0.2% (The number of pixels in the array of the photodetectors is 100–300 thousands). A small size of the growth defect in 4–6 microns suggests that even if one pixel is damaged, the neighboring pixels will remain unaffected.

The structure of the samples was examined by the method of studying the cross section of structures by the transmission electron microscopy (TEM). The studies were performed by the JEM-4000EX electron microscope by JEOL (Japan). The samples for the TEM were made in the geometry of the cross section by grinding them with their subsequent thinning during etching by Ar+ ions with an energy of 3–4 keV. The survey was conducted at an accelerating voltage of 400 kV. **Figure 8** shows the results of the TEM AlGaAs/GaAs heterostructure. The image represents no structural defects, which indicate a high quality of the heterostructures under study. The measurements carried out by the TEM methods demonstrate a good correspondence of the heterostructure layer thicknesses planned and obtained. Sharp changes in the

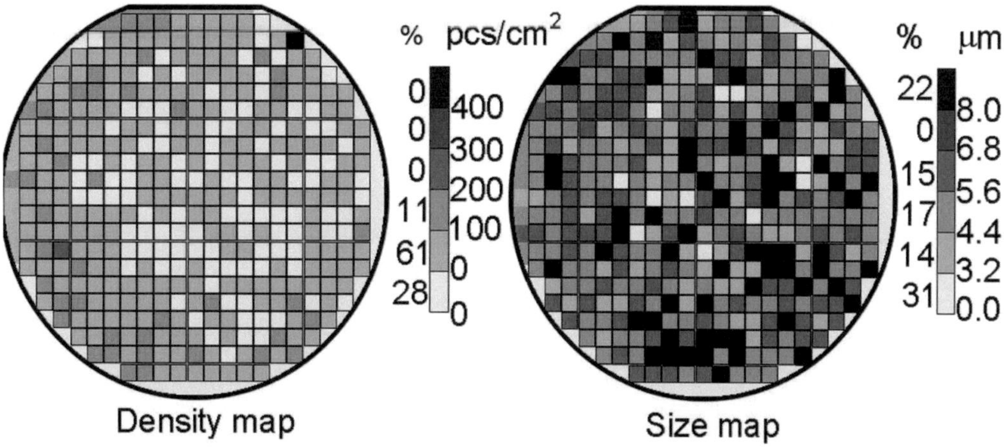

Figure 7. Distribution of surface defects of the epitaxial QWIP structure. The distribution of the density of defects (left) and defect sizes (right).

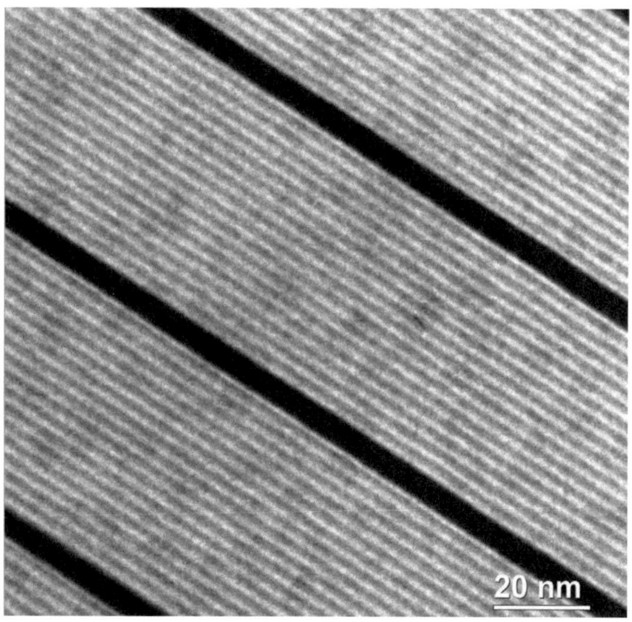

Figure 8. The cross section of the QWIP heterostructure.

intensity at the quantum well heterogeneities testify to fairly sharp and smooth transitions from one material to another.

The spectra of the piezomodulated reflection of the MQW structure in a visible spectral range at a liquid nitrogen temperature were measured to control the aluminum mole fraction in the AlGaAs barrier layers. Mechanical vibrations of the ceramic plate caused the modulation of the mechanical stresses in the structure, the modulations of the real $\Delta\varepsilon_r$ and the imaginary $\Delta\varepsilon_i$

parts of the dielectric constant, and as a consequence, the modulation of the reflection coefficient. The reflection variances ΔR are related to the variation of the dielectric constant by the Seraphin ratio [12]:

$$\frac{\Delta R}{R} = \alpha\Delta\varepsilon_r + \beta\Delta\varepsilon_i \tag{11}$$

where α and β are Seraphin coefficients. The energy position of the peaks in the piezoreflection spectrum corresponds to the electronic transitions in the structure under study. Typical piezoreflection spectra of the MQW structure are shown in **Figure 9A**. The aluminum mole fraction in the barrier layer is calculated from the peak energy. The peak at the 1.484 eV corresponds to the transitions from the levels of residual acceptors (neutral carbon atoms) to the conduction band in the GaAs substrate. The band gap of the GaAs at the 77 K is $E_g(GaAs) = 1.508$ eV [13], and the binding energy of the carbon atom is $\Delta = 25$ meV. Therefore, the calculated transition energy is expected to be $E_g(GaAs)\text{-}\Delta$ = 1.483 eV, which is in a good agreement with the experimental results. The peak at the 1.839 eV (or 674 nm) corresponds to the bound excitons in the $Al_XGa_{1\text{-}X}As$ barrier layers. The intermediate peaks between the 1.484 and 1.839 eV energies correspond to the transitions between the hole (light and heavy) and electron quantum levels in the MQWs [14]. Thus, the aluminum mole fraction in the barriers is determined from the energy position of the 1.839 eV line by a well-known relation between the bandgap width of the AlGaAs layer and the aluminum mole fraction (x) as $E_g(x) = E_g(0) + 1.427x + 0.041x^2$ [15].

The measurement of the QW width was carried out by means of the photoluminescence (PL) spectroscopy at a liquid nitrogen temperature along a line corresponding to the transition from the ground electron level to the ground level of the heavy holes *e1-h1*. According to calculations in the approximation of a rectangular quantum well of a finite depth for a quantum well with a width of 5 nm and an aluminum mole fraction of 0.26, the energies of the electron and

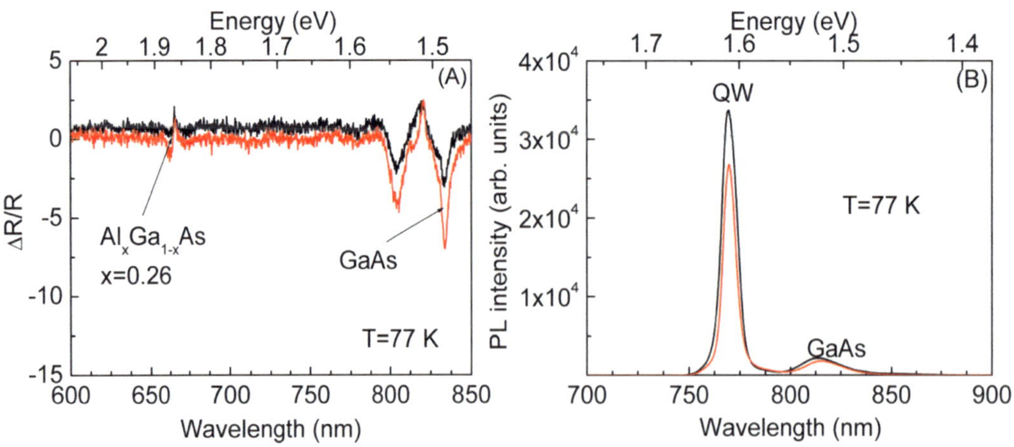

Figure 9. (A) The spectra of the piezomodulated reflection of the MQW structure. (B) The photoluminescence spectra of the MQW structure. The YAG laser was used to excite the PL.

hole ground quantum levels are 70 and 15 meV, respectively. Consequently, the energy of the *e1-h1* transition at a room temperature is expected about 1.6 eV. As one can see from **Figure 9B**, the measured maximum of the PL band *e1-h1* is located at 1.61 eV, which indicates the grown thickness of the layers being sufficiently accurate.

To control the electron concentration in the grown structures with MQWs, the C-V characteristics were measured. Those measurements were carried out with the specially fabricated Schottky barriers of the TiAu with a diameter of 200 μm and a C-V bridge operating at frequencies of 1 to 100 kHz. The upper contact layer had been previously etched and then measured. According to the measured C-V dependences, the concentration dependences upon the depletion region depth $N_{2D}(W)$ are determined from the relation:

$$N_{2D}(W) = {}^{2}\!/_{q\varepsilon}\varepsilon_0 S^2 \left(\frac{d(C^2)}{dV}\right)^{-1} \tag{12}$$

where q is the electron charge, C is the measured capacitance, V is the applied voltage, S is the area of the Schottky barrier, ε and ε_0 are the permittivity of the semiconductor and vacuum, respectively. The depth W, where the concentration of free carriers is determined, is $W = \varepsilon\varepsilon_0 S/C$, whereupon, the dependence of the sheet concentration with precision to constant upon the depletion region depth $\Delta\Gamma(W)$ was obtained by integrating $N_{2D}(W)$ (**Figure 10**). As one can see from the figure, all the experimental data obtained from different sample points agree with a high accuracy, which indicates a high uniformity of the electron density in the quantum wells over the wafer area. It should be noted that this method can be duplicated by a capacitive method where the mercury probe and profilometer operating at a frequency of 1 MHz are applied to measure the derivative.

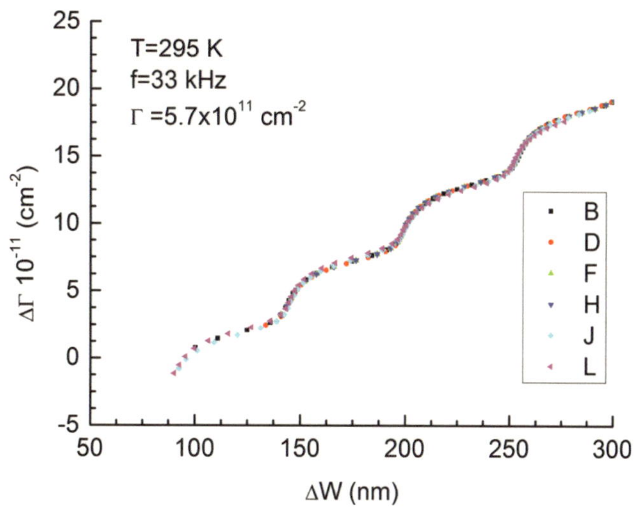

Figure 10. Distribution of the sheet concentration of the charge carriers ΔΓ in the quantum wells for six different points. These points are located along the wafer radius.

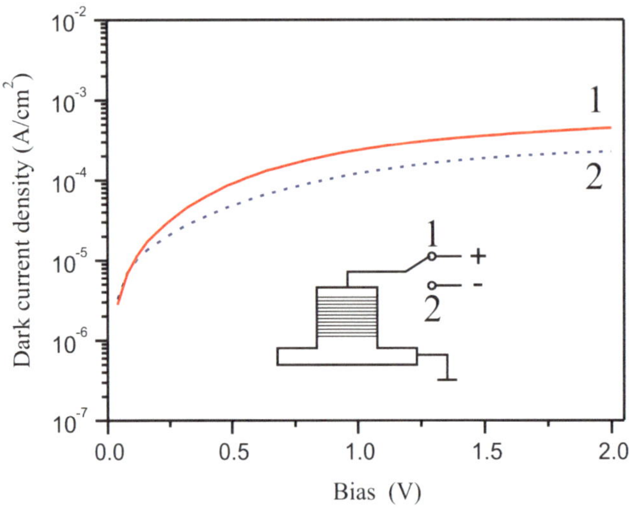

Figure 11. Field dependences of the dark current in the QWIP structure (the donor concentration is 2.5×10^{11} cm^{-2}) for various current directions at the 77 K temperature.

To calculate parameters of the multiplexer (capacitance and integration time), it is necessary to know the field dependences of the dark current. The measured dependencies of the dark current upon the bias for the QWIP structure made up of 30 periods of QW and barriers (L_b = 40 nm) with the pixel area 20 × 20 μm are presented in the **Figure 11**. As one can see from the figure, the current asymmetry is observed, which results from the segregation of the impurity atoms during the structure growth.

4. Selecting the parameters of the FPA on the basis of the GaAs/AlGaAs QWIP

While choosing a number of QW periods in the QWIP structure, the following must be taken into account: (A) an increase in a number of structure periods leads to an increase in the absorption coefficient, but, in turn, (B) the overall probability of capturing photoexcited electrons back into the QW, (C) the growth time of the heterostructure, (D) the mesa depth, and (E) the magnitude of the voltage applied to the structure increase as well. As a result, the sensitivity of the QWIP increases very weakly with an increasing number of the GaAs/AlGaAs layer periods [11]. Therefore, a heterostructure with 30 periods of the GaAs QW has been chosen.

A 2D diffraction grating with the parameters determined by the spectral range and properties of the dielectric applied is required to provide the absorption of a normal-incident light and increase the quantum efficiency of the QWIP [16]. The required etching depth of the lattice (d) is determined by the relation $d = \lambda/4n$, where λ is the radiation wavelength and n is the refractive index of the GaAs. These parameters are λ = 8–9 μm and d = 0.7–0.75 μm for the wavelength range applied. The period of the diffraction grating L = 2.8 μm was chosen so that the direction of the first diffraction order was an angle of 60° with the normal to the sample wafer.

The thickness of the upper contact layer of n+ doped GaAs was increased up to 1200 nm to adapt the fabrication technology for ohmic contacts. At a lower thickness of the contact layer, the metal may penetrate into the QW region as a result of diffusion during a long (5 minutes) annealing.

The multiplexer capacity and integration time of the signal were estimated by the dependency dark current upon the bias at the 77 K (**Figure 11**). The current at the bias voltage equal to 0.5 V is about 10^{-10} A, which, in the presence of the capacitance in the multiplexer equal to 6×10^6 electrons, allows integrating the signal during 10 ms.

5. The fabrication technology for the FPA on the basis of the GaAs/AlGaAs QWIP

The focal plane arrays on the basis of GaAs/AlGaAs QWIP structures were fabricated by the complex of coordinated technological operations based on photolithography processes using functional dielectric and metallic layers, etching processes, and chemical treatments of the heterostructures in the regimes determined by the physicochemical properties of the MQW layers. The technology development was carried out at the FPA with 384 × 288 elements with the 25 microns pitch. The structure control was carried out by the optical and scanning electron (LEO-1430, SU8220) microscopes.

5.1. Formation of the diffraction grating and mesa structure

A diffraction grating and mesa structures were formed by dry anisotropic etching of the GaAs layer in a remote gas-discharged plasma (ICP RF) of the BCl_3, argon, and nitrogen. The obtained diffraction grating holes are square shaped with rounded corners and vertical walls (**Figure 12A**). The view of the mesa structure walls formed by etching the GaAs at a given depth (2.4 μm) at the

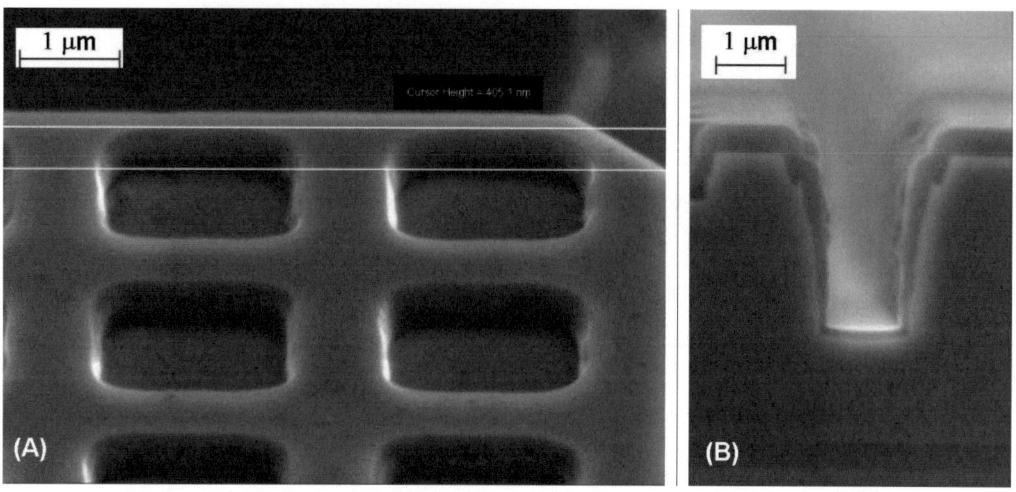

Figure 12. SEM images of (A) the diffraction grid and (B) the gap (2 μm) between the mesa structures of the FPA 384 × 288 elements obtained by dry etching with the 0.5-μm SiO_2 layer.

optimum ratios of reagent gases, the power of the RF and ICP generators, the reactor pressure, substrate temperature, heating, and etching time are shown in **Figure 12B.**

5.2. The fabrication of ohmic contacts

The ohmic contacts to the base and upper n+ doped GaAs layers (on the mesa surface) were fabricated by the Ge/Au/Ni/Au (20/20/20/100 nm) deposition [17] after the removal of the native oxide layer from the semiconductor surface by $HCl{:}H_2O$ (1:8) and annealing during 5 minutes at the 385°C in a hydrogen (**Figure 13**).

5.3. The mesa structure surface passivation

The SiO_2 dielectric layer was deposited by a low-pressure chemical vapor deposition (LP CVD) method at temperatures of 195–250°C to passivate and protect the mesa structure surface. The low temperature of this process excludes the disorder of the surface stoichiometric composition due to the evaporation at high temperatures of the fifth group element. Depending on the synthesis conditions, the layers of LP CVD SiO_2 (refractive index 1.46) have a dielectric constant of 5.9–6.5 and leakage currents of $6 \times \cdot 10^{-8}$–$9 \cdot \times 10^{-7}$A/cm^2 ($E = 2 \cdot 10^6$ V/cm) at the room temperature with a water content of 2.5–3.3 volume percent, respectively. The SiO_2 formation at the semiconductor surface results in coating the vertical walls of the mesa structures (**Figure 12B**).

5.4. Production of indium bumps on the FPA and silicon multiplexers

To assemble a FPA by cold welding, indium bumps were fabricated both on the GaAs/AlGaAs QWIP structure and silicon multiplexers. The Tl xLift photoresist was used to produce indium bumps (height of ~5 μm) by inverse photolithography. The view and cleavage of a separate mesa

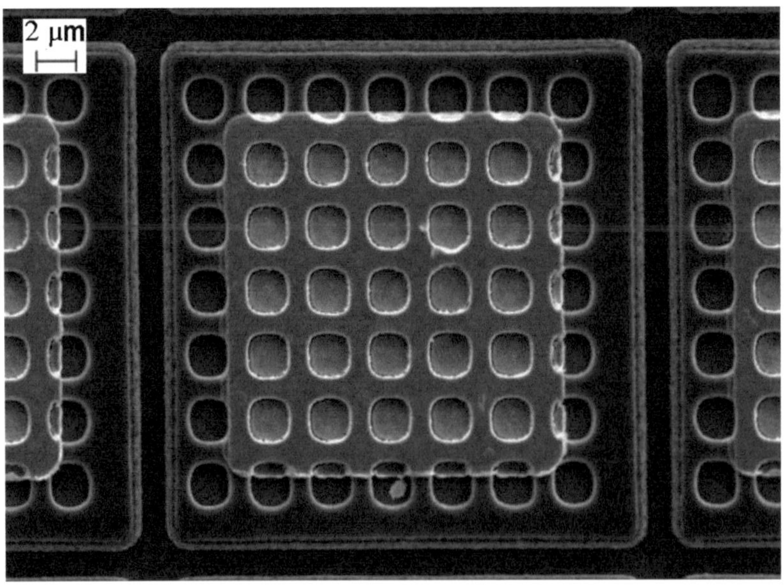

Figure 13. SEM image of the mesa structure of FPA 384 × 288 elements with a diffraction grating and the Ge/Au/Ni/Au (20/20/20/100 nm) metallization layer.

structure with an indium bump of the FPA of 384 ×·288 elements produced by the developed technology on the GaAs/AlGaAs heterostructures are shown in **Figure 14**.

5.5. Hybrid assembly of the FPA

The FPA modules were assembled by cold welding of the indium bumps under pressure [18]. The FPA and multiplexer crystals were docked on the M9 setup of Laurier company. The fusion of the indium contacts was performed by heating the module up to the indium melting temperature with the succeeding cooling. The surface autoplanarization is provided during the crystal compression process, which is achieved by installing polyamide stoppers along the perimeter of one of the module elements—an array or multiplexer. The maximum limit mechanical load is determined experimentally from the measurement of the curves of the plastic flow of the indium bumps, their height, and area. The pressure required for the plastic flow of the contacts ranges from 0.3 to 0.9 kg/mm^2. The total height of the indium bumps after the FPA assembly is 6–8 μm, which satisfies the requirements of the durability of the FPA hybrid assemblies [19].

5.6. The GaAs substrate removal from the FPA assembly

The technology of the substrate removal after the assembly of the FPA consisted of the successive processes of the mechanical grinding aimed at removing the main thickness of the GaAs substrate, chemical mechanical polishing and chemical dynamic polishing, in order to obtain a mirror-smooth surface of the array crystal. The processes of the chemical selective etching of the GaAs and heterostructure layers were applied to remove the GaAs substrate from the FPA surface completely (**Figure 15**) [20–22].

5.7. Technical characteristics of the multiplexer

The silicon multiplexers by Integral Joint Stock Company (IZ640FD format 640 × 512) made by the CMOS technology and meeting the QWIP requirements were used as a part of the FPA assembly [23].

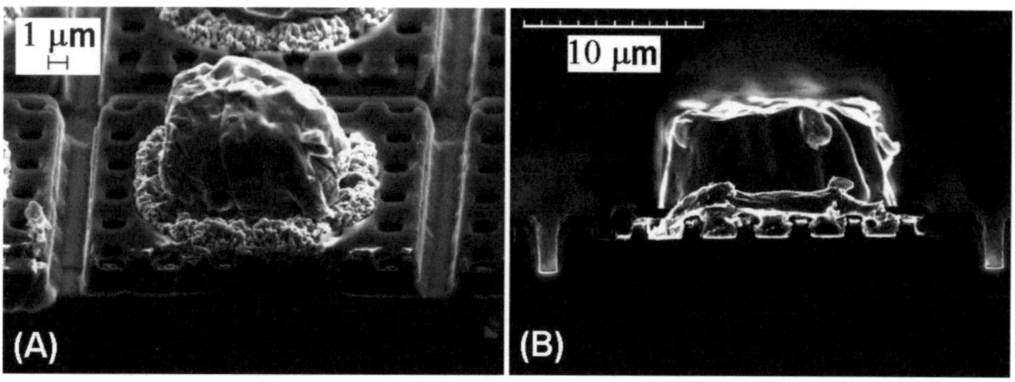

Figure 14. SEM images (A) of the mesa structure and (B) mesa structure cleavage of the FPA with 288 × 384 pixels on the basis of GaAs/AlGaAs QWIP structures with indium bumps.

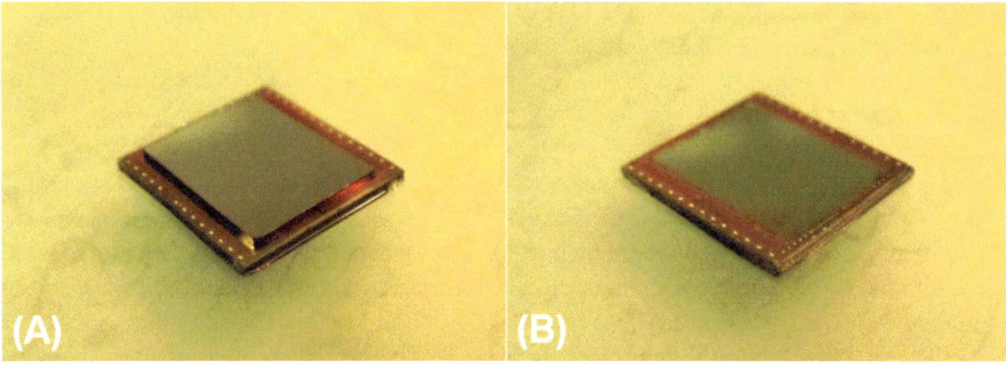

Figure 15. The photo of the FPA assembly with 384 × 288 elements (A) before and (B) after the removal of the GaAs substrate (650 μm).

The storage capacitance in every cell of the IZ640FD multiplexer was approximately 8 × 10^6 electrons with a reading noise of 1000 electrons. The adjustable integration time could vary from 100 microseconds to the entire duration of the frame scanning. The electric power consumed at a frame rate of 100 Hz did not exceed 120 mW. The electric power consumed at a frame rate of 100 Hz did not exceed 140 mW in the four output mode and 90 mW at a switch into one output mode, respectively.

The schematic diagram of the input block of the multiplexer is shown in **Figure 16**, where D is the photoresistor (detector), VD is the detector supply voltage, VB is the voltage specifying the detector bias voltage, VA is the voltage specifying the level of anti-burglary, VS is the skimming voltage, C1 is the integration capacitance in pixel, C2 is the storage capacity in pixel, C3 is the storage capacity in column, K1, K2, and K3 are the keys, A is the amplifier with a controlled gain, and B is the buffer. The signal integration occurs simultaneously on all the array elements, and then, the voltage from the capacitances C2 is line-by-line read out by connecting the key K2 to the column capacitance C3 and the column amplifier A.

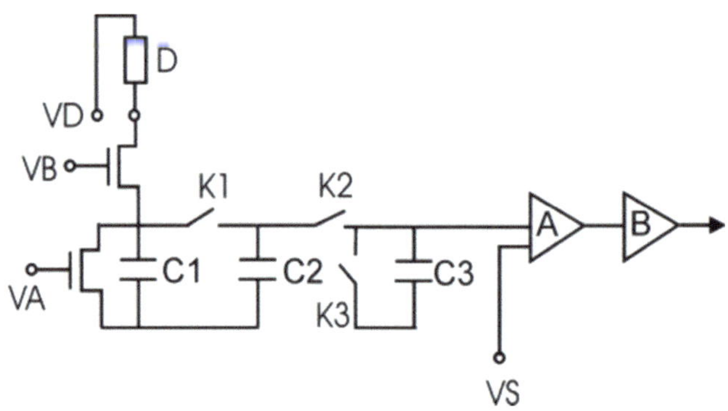

Figure 16. Schematic diagram of the input multiplexer for the FPA.

6. Opto-electronic characteristics of the FPA assembly

At the final stage, the opto-electronic characteristics of the fabricated FPA assembly were determined. For this purpose, the assembly was placed in a nitrogen cryostat with an entrance window made of the ZnSe. The operating temperature of 65 K was achieved due to the pressure pumped down by a vacuum pump. A cooled diaphragm provided a relative aperture of 1:2. To measure the sensitivity of the FPA, the module illumination was made by an extended-type absolutely black body. It should be noted that the high parameters of the FPA assembly both the absolute values of the signals and their homogeneity with respect to the array elements are supposed to be essential.

The distribution histogram of the total current (dark + photo signal from the 300 K background) for the FPA assembly thinned up to 170 μm is shown in **Figure 17A**. The integration time of the signal was chosen to be 9 ms. **Figure 17B** shows the distribution histogram of the temperature sensitivity of the S_T at the 300 K background upon the pixels of the FPA module BM20. Its average value is rather high and equal to 23.2 mV/K. **Figure 17C** shows

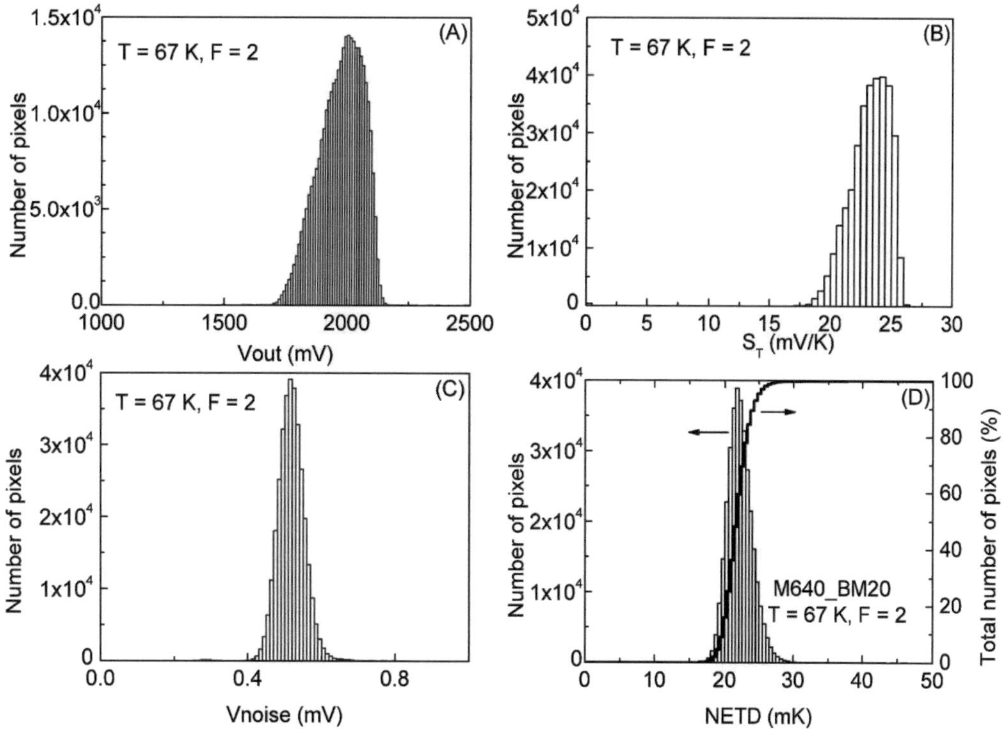

Figure 17. (A) The histogram of the total current distribution of the 640 × 512 FPA module BM20. (B) The histogram of the temperature sensitivity at a background of 300 K of the 640 × 512 FPA module BM20. (C) The histogram of the noise voltage of the 640 × 512 FPA module BM20. (D) Experimentally measured NEΔT histogram of the 640 × 512 FPA module BM20.

the distribution of the noise voltage V_n at the output of the photoreceptor module BM20 at a background of 300 K. All the histograms are rather narrow, which demonstrates the high uniformity of the array parameters. The noise equivalent temperature difference NETD = V_n/S_T of the FPA module BM20 is shown in **Figure 17D**. The average value of a NETD for nondefective pixels at the FPA temperature of 67 K is 22.2 mK. A number of defective elements with a NETD over 70 mK is 0.15%. A typical spectrum of photosensitivity of a 640 × 512 FPA is shown in **Figure 18**. An example of the IR image detected by a

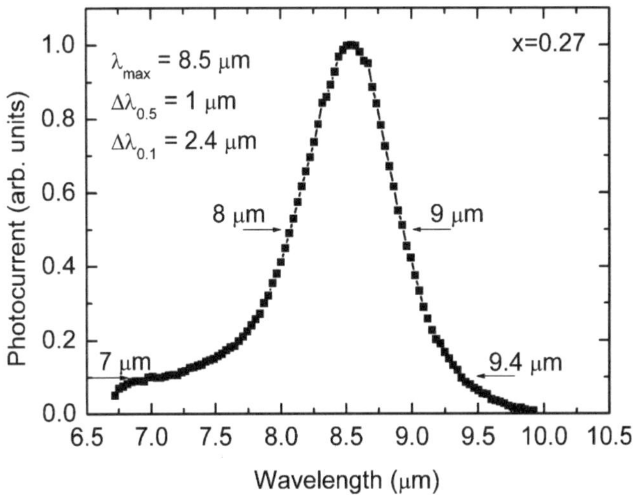

Figure 18. Photosensitivity spectrum of the FPA.

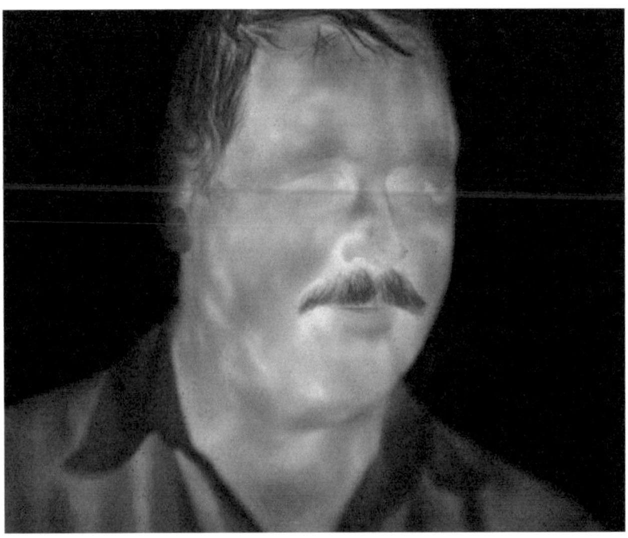

Figure 19. The example of an IR image. The temperature is 65 K. The integration time is 6 ms. The working elements are 99.6%.

640 × 512 FPA assembly module equipped by the germanium lens with D/F = 1:2 aperture is shown in **Figure 19**.

The developed technology for the FPA assembly is reproducible and has a rather high yield percentage of the suitable products well seen from **Figure 20** showing the scatter of the noise equivalent temperature differences and a number of defective elements for the 640 × 512 FPA assembly manufactured on the 5 MBE grown heterostructures.

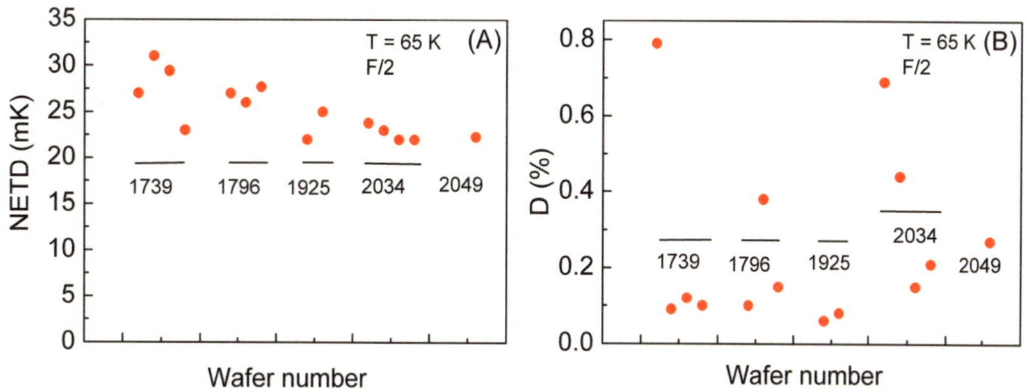

Figure 20. (A) Scatter of NETD and (B) a number of defective elements in the 640x512 FPAs produced on the five grown QWIP structures.

Figure 21. The photo of the QWIP IDCA.

	ISP SB RAS (Russia) BM20	Sofradir (France) SIRIUS-LW
Array format	640 × 512	640 × 512
Pixel pitch, μm	20	20
Peak sensitivity, μm	8.5–8.6	8.5
FWHM, μm	0.8–1	1
NETD, mK	<35	<35
Operability, %	>99.5	99.9%
Integration time, ms	6	7
Operating temperature, K	72	70–73

Table 1. Comparative characteristics of the GaAs/AlGaAs QWIP IDCA by the ISP of the SB RAS (Novosibirsk) and Sofradir (France).

6.1. Integrated detector cooler assembly

The fabricated and tested 640 × 512 FPA were installed in the body of a vacuum cryostat integrally coupled with a microcryogenic system. Resulted integrated detector cooler assembly (IDCA) is showed in **Figure 21**. A vacuum cryostat performs thermal isolation of the array from the environment to guarantee effective cooling of the photodetector up to the operating temperature (T = 68–72 K). The radiation from the detected objects is fed to the FPA through an input window made up of the germanium with the antireflection in the range of 8–10 μm and a cooled diaphragm with a relative aperture F/2 designed to reduce the background illumination. A low pressure in the vacuum cryostat is provided by a getter, whose reactivation is carried out by passing an electric current through it. Typical operating temperatures for the FPA based on the QWIP with the wavelength range 8–10 μm are 68–72 K. Thus, the powerful microcryogenic systems ensuring a cooling capacity at an operating temperature of 70 K not less than 0.4 W and the power consumption not more than 20 W are needed to provide the required temperature in a full range of climatic conditions.

As one can see from **Table 1**, the parameters of the developed 640 × 512 QWIP IDCA are comparable with those of the Sofradir products.

7. Conclusions

The technology of manufacturing the AlGaAs/GaAs QWIP FPA has been discussed. The parameters of a FPA of 640 × 512 format with a 20-μm pitch for a spectral range of 8–10 μm have been described. At an operating temperature of 72 K, the temperature resolution of the QWIP FPA is less than 35 mK. The frame rate is 100 Hz. A number of defect elements in the array do not exceed 0.5%. It is shown that the parameters of the QWIP FPA fabricated by Rzhanov Institute of Semiconductor Physics of SB RAS meet the world class standards.

Author details

Michael A. Dem'yanenko, Dmitry G. Esaev, Aleksandr I. Toropov, Natalia A. Valisheva, Sergey A. Dvoretsky, Dmitry V. Dmitriev, Dmitry V. Gulyaev, Vladimir A. Fateev, Igor V. Marchishin, Dmitry Yu Protasov, Anatoly P. Savchenko, Victor N. Ovsyuk and Konstantin Zhuravlev*

*Address all correspondence to: zhur@isp.nsc.ru

Rzhanov Institute of Semiconductor Physics of SB RAS, Novosibirsk, Russia

References

[1] West LC, Eglash SJ. First observation of an extremely large-dipole infrared transition within the conduction band of a GaAs quantum well. Applied Physics Letters. 1985;**46**:1156. DOI: 10.1063/1.95742

[2] Levine BF, Choi KK, Bethea CG, Walker J, Maliл RJ. New 10 μm infrared detector using intersubband absorption in resonant tunneling GaAlAs superlattices. Applied Physics Letters 1987;**50**:1092. DOI: http://dx.doi.org/10.1063/1.97928

[3] Rafol SB, Cho E, Lim W. Characterization of very large format 1K×1K LWIR QWIP focal plane array. Proceedings of SPIE. 2007;**6678**:6678X-1. DOI: http://dx.doi.org/10.1117/12.736137

[4] Robo JA, Costard E, Truffer JP, Nedelcu A, Marcadet X, Bois P. QWIP focal plane arrays performances from MWIR up to VLWIR. Proceedings of SPIE. 2009;**7298**:72980F-729801. DOI: http://dx.doi.org/10.1117/12.820470

[5] Reibel Y, Rubaldo L, Manissadjian A, Billon-Lanfrey D, Rothman J, de Borniol E, Destéfanis G, Costard E. High-performance MCT and QWIP IR detectors at Sofradir. Proceedings of SPIE. 2012;**8541**:85410A. DOI: http://dx.doi.org/10.1117/12.975671

[6] Gunapala SD, Bandara SV, Liu JK, Mumolo JM, Rafol SB, Ting DZ, Soibel A, Hill C. Quantum well infrared photodetector: Technology and applications. IEEE Journal of Selected Topics in Quantum Electronics. 2014;**20**(6):3802312. DOI: 10.1109/JSTQE.2014.2324538

[7] Martijn H, Gamfeldt A, Asplund C, Smuk S, Kataria H, Costard E. QWIPs at IRnova, a status update. Proceedings of SPIE. 2016;**9819**:981918. DOI: http://dx.doi.org/10.1117/12.2228348

[8] Davies JH. The Physics of Low-Dimensional Semiconductors: An Introduction. Cambridge: Cambridge University Press; 1998. 438 p

[9] Nextnano. Software for semiconductor nanodevices [Internet]. 2017. Available from: www.nextnano.de [Accessed: 2017-10-25]

[10] Schneider H, Liu HC. Quantum Well Infrared Photodetectors: Physics and Applications. Heidelberg: Springer; 2007. 248 p. DOI: 10.1007/978-3-540-36324-8

[11] Levine BF. Quantum well infrared photodetectors. Journal of Applied Physics. 1993;**74**(8): R1. DOI: http://dx.doi.org/10.1063/1.354252

[12] Seraphin O, Bottka N. Band-structure analysis from electro-reflectance studies. Physical Review. 1966;**145**:628. DOI: https://doi.org/10.1103/PhysRev.145.628

[13] Lee YR, Ramdas AK, Moretti AL, Chambers FA, Devane GP, Ram-Mohan LR. Piezomodulated reflectivity spectra of GaAs/AlGaAs single-parabolic-quantum-well heterostructures. Physical Review B. 1990;**41**:8380. DOI: https://doi.org/10.1103/PhysRevB.41.8380

[14] Vurgaftman I, Meyer JR, Ram-Mohan LR. Band parameters for III-V semiconductors and their alloys. Journal of Applied Physics. 2001;**89**:5815. DOI: http://dx.doi.org/10.1063/1.1368156

[15] Pollak FH, Shen H. Modulation spectroscopy characterization of MOCVD semiconductors and semiconductor structures. Journal of Crystal Growth. 1989;**98**:53. DOI: 10.1016/0022-0248(89)90185-1

[16] Anderssen JY, Lundqvist L and Paska ZF. Grating-coupled quantum–well infrared detectors: Theory and performance. Journal of Applied Physics 1992;**71**:3600. DOI: http://dx.doi.org/10.1063/1.350916

[17] Baca AG, Renb F, Zolpera JC, Briggsa RD, Peartonc SJ. A survey of ohmic contacts to III-V compound semiconductors. Thin Solid Films. 1997;**308-309**:599. DOI: 10.1016/S0040-6090(97)00439-2

[18] Klimenko AG, Voinov VG, Novoselov AR, Nedosekina TN, Vasilyev VV, Zakhariash TI, Ovsyuk VN. Soft indium bumps for indium bump bonding FPAs on CdHgTe. Optoelectronics Instrumentation and Data Processing C/C of Avtometriia. 1998;**4**:87

[19] Smith HI. Fabrication techniques for surface-acoustic-wave and thin-film optical devices. Proceedings of the IEEE. 1974;**62**:1361. DOI: 10.1109/PROC.1974.9627

[20] Zhao R, Lau WS, Chong TC, Li MF. A comparison of the selective etching characteristics of conventional and low-temperature-grown GaAs over AlAs by various etching solutions. Japanese Journal of Applied Physics. 1996;**35**:22. DOI: https://doi.org/10.1143/JJAP.35.22

[21] Spooner F, Quinn W, Hanes L, Woolsey S, Smith K, Maison J. A reproducible, high yield, robust wet etch etch-stop process using organic acid – peroxide solutions. [On-line Digest Table. The International Conference on Compound Semiconductor Manufacturing Technology (Sharing Ideas Throughout the Industry)]. [Internet]. Available from: http://csmantech.org/OldSite/Digests/2004/2004Papers/8.9.pdf [Accessed: 25 Oct 2017]

[22] Chang EY, Lai Y-L, Lee YS, Chen SH. A GaAs/AlAs wet selective etch process for the gate recess of GaAs power metal-semiconductor field-effect transistors. Journal of the Electrochemical Society. 2001;**148**:G4. DOI: 10.1149/1.1344555

[23] "INTEGRAL" Joint Stock Company [Internet]. Available from: HYPERLINK "http://www.integral.by/en" www.integral.by/en [Accessed: Oct 25, 2017]

Chapter 7

Type-II Superlattice Heterojunction Photodetector with Optoelectronic Characterization and Analytical/ Numerical Simulation

Sanghyun Lee and Kent J. Price

Additional information is available at the end of the chapter

http://dx.doi.org/10.5772/intechopen.71327

Abstract

This chapter focuses on characterization, modeling, and simulation about the type-II superlattices photodetector application. Despite dramatic improvements in type-II superlattices in the past 15 years, challenges still exist in InAs/GaSb and InAs/GaInSb superlattices: The diffusion current, Shockley-Read-Hall (SRH) recombination current, tunneling current, and surface leakage current at elevated temperature. To establish a set of modeling and simulation input parameters, in-depth materials and device characterization at different conditions are carried out for initial materials and device models. Based on input parameters, we will describe the development of analytical and numerical models of InAs/GaSb and InAs/GaInSb type-II superlattice-structured materials and device systems. At the end of this chapter, the fitting of modeled and simulated data will be performed to compare empirical data and modeling results at a set of temperature, which will provide guidance to achieve the higher performance.

Keywords: photodetector, superlattices, type-II, infrared detectors, modeling

1. Introduction

In the last decade, the type-II superlattice has been proposed as a promising candidate for the next generation of IR photodetectors in the long wavelength IR (LWIR) range. The InAs/GaSb and InAs/GaInSb type-II superlattices are considered to be equivalent to HgCdTe with superior cutoff wavelength beyond 15 μm which offers better stability and leverage [1, 2]. Moreover, InAs/GaSb and InAs/GaInSb type-II superlattices provide numerous advantageous optoelectronic properties such as high absorption coefficient, higher effective mass of electrons and holes, slower Auger recombination rate with a proper design of the valence band structure, a lower dark current, and higher operating temperatures. These make InAs/GaSb and InAs/GaInSb type-II superlattices comparable to established HgCdTe IR photodetectors with the high quantum efficiency [3–6].

The effective band gap of InAs/GaSb and InAs/GaInSb type-II superlattices can be engineered with great flexibility to match mid to far infrared photon energies by selecting appropriate thicknesses for the alternating InAs and GaSb (or GaInSb) layers during crystal growth [3]. Hence there are different performance levels which demonstrate the possibility of multiple band detectors with short wave infrared (SWIR), mid-wave infrared (MWIR), and long wave infrared (LWIR) detectors. In particular, MWIR InAs/GaSb superlattice photodiode application to a focal plane array (FPAs) shows the successful operation at 77 K [7]. The electronic band structure of InAs/GaSb and InAs/GaInSb superlattices enables a spatial separation of holes and electrons localized into different quantum wells. Furthermore, the alternating InAs and GaSb (or GaInSb) layers enable strain in InAs/GaSb and InAs/GaInSb type-II superlattice systems which reduces the Auger recombination rate and therefore, improves the detectivity of the photodetectors [8]. Since the current detector technologies including InAs/GaSb and InAs/GaInSb superlattices require significant cooling to be provided to the detectors, for example, down to 4 K, this implementation makes the photodetector system complicated as a whole which makes it impractical due to the high cost and, most of all, a major hurdle to be overcome before market commercialization. At room temperature, detectors can be used only in active mode with external source due to low sensitivity. Hence, the operation at higher temperatures with high quantum efficiencies remains a major challenge for long wavelength IR range detectors. The major sources that contribute to the dark current of InAs/GaSb and InAs/GaInSb are diffusion current, Shockley-Read-Hall (SRH) recombination current, tunneling current, and surface leakage current.

The goal of this chapter is to develop sustainable analytical and numerical models of device and materials which serve as a set of parameters for modeling and simulation. The established parameters will identify the entitled parameters to achieve maximized device performance. To optimize the performance of InAs/GaSb and InAs/GaInSb superlattices devices, the optoelectronic characterization methods will be carried out to provide basic materials input parameters and device key modeling parameters for the modeling and simulation. Based on established theoretical and empirical models, the baseline device and materials will be set up and provide the guidance for the new design concept to develop higher performance devices at elevated operating temperature.

2. Modeling

One of the major advantages of type-II InAs/GaSb and InAs/GaInSb superlattice systems is the ability to tailor the band gap by changing the thicknesses of InAs and GaSb (or GaInSb) layers. Additionally, the superlattices system provides the ability to hetero-engineer the band structure of the devices. As of now, a number of band gap engineered superlattices architectures, such as W-structure, M-structure for heterostructures (pMp or nBn) and complementary barrier infrared detector have been proposed with the improved performance compared to a homojunction design [9–12].

http://dx.doi.org/10.5772/intechopen.71327

Figure 1 shows the band structure of type-II InAs/GaInSb (also similar to InAs/GaSb) superlattices system grown by MBE on GaSb substrates [13]. The bottom of the conduction band of InAs is below the top of the valence band of GaInSb in a periodic manner. In this periodically alternating InAs and GaInSb layers, this typical band alignment separates the electrons and holes into these two different layers. The effects of coherent strain facilitate compressive stress in GaInSb and tensile stress in InAs layers, which results in splitting the light (LH) and heavy (HH) hole bands. Hence, the phenomena enhance the staggered band alignment. Consequently, the energy gap of the superlattice is formed between the localized hole states (E_{HH1}) in GaInSb and the electron states (E_{C1}) in InAs layers. The energy gap decreases monotonically while increasing the concentration of Indium and this is primarily due to increased strain (larger lattice constant of GaInSb).

The need to accurately predict type-II InAs/GaInSb photodetector performance through modeling and simulations becomes important due to the expensive fabrication cost and complexity of quantum sized materials and devices parameters, and lack of appropriate expertise and skill sets. The traditional choice of theoretical study for electronic transport modeling

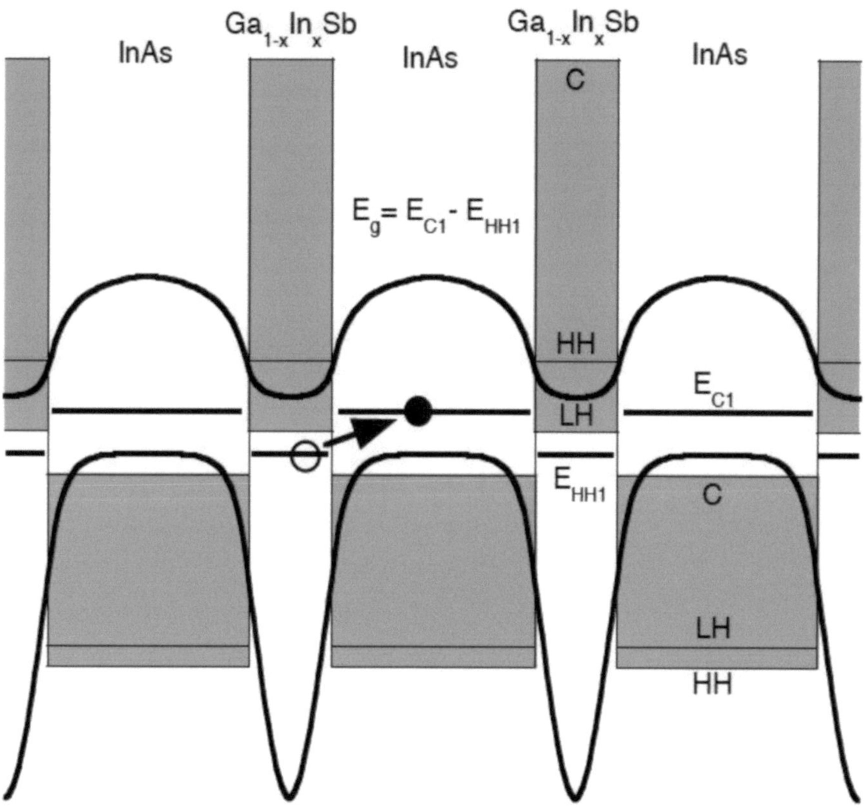

Figure 1. The superlattices energy gap Eg is E_{C1}–E_{HH1}. Energy band structure, electron (E_{C1}) and hole (E_{HH1}) states and corresponding wave functions (Ψ2) in a strained, InAs/GaInSb superlattices grown on GaSb substrate. Shaded parts indicate the forbidden gaps in InAs and GaInSb. The peaks of electron and hole wave functions are situated in different layers [13].

and simulation is governed by two important factors which are the critical spatial dimension of the simulated object (nano-scale alternating film layers) and the physical mechanisms of device operation such as SRH recombination, tunneling, thermionic emission, and temperature effect. The commonly used computational modeling and simulation for transport is described as below. Among these approaches, the drift-diffusion method obtained by using the first moment of the Boltzmann transport equation is the wide-spread methodology in commercially available technology computer aided design simulators (TCAD) such as Synopsys Sentaurus and Silvaco ATLAS. **Figure 2** shows the hierarchical overview of commonly used electronic transport solver for modeling and simulation for the TCAD. These prove an accurate description of carrier transport for the optoelectronic devices. However, the materials and device input parameters need to be precisely calibrated in order to predict the device behavior accordingly.

Over the last decade, there have been similar efforts in a photodetector community such as the k p method [15]. The k p method is used to investigate optical properties of quantum wells, superlattices, heterostructures, and direct-bandgap semiconductors for calculating the electronic structures [14]. The main advantages are to provide,

a. A reasonable compromise in time and model-complexity between simpler approaches such as effective-mass Hamiltonian and tight-binding, ab-initio;

b. Model parameters such as band gap, effective mass, optical matrix elements can be inferred from experiments;

c. An effective way to include strain and spin-orbit interactions;

d. Analytical description of band dispersions near high-symmetry points in Brillouin zone.

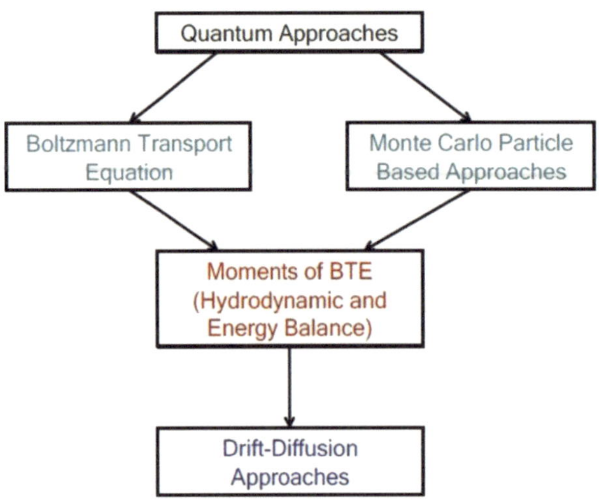

Figure 2. Hierarchical overview of commonly used electronic transport solver for modeling and simulation such as commercial technology computer aided design (TCAD) [14].

http://dx.doi.org/10.5772/intechopen.71327

For example, Grein et al. [15] described the analytical expression by using this k p method. As a similar effort, Roslan et al. [16] investigated InAs photodiode by using the advanced version of modeling and simulation with 2D Silvaco simulators. To accurately model and simulate, a set of accurate input materials and device parameters should be acquired by optoelectronic characterization such as lifetime, mobility, etc. through a multiple revision processes.

3. Fabrication and characterization methods

The initial step to establish the accurate models is to characterize a set of materials and device input parameters of type-II InAs/GaInSb superlattices, which will be the basis of design parameters for fabricating the device. To achieve their design parameters, there have been a number of approaches to growing InAs/InGaSb superlattices, focusing on the interface layer quality, growth temperature optimization, III–V flux ratios, and substrate effect, etc. [17]. Concurrently, there is a similar amount of efforts to improve GaSb wafers and polishing processes. Although various research groups try to grow the type-II superlattices on Silicon and GaAs substrates, GaSb substrates are able to provide less lattice mismatch related defects than other substrates.

The InAs/GaInSb superlattices samples are typically grown in a molecular beam epitaxy (MBE) system (or MOCVD) equipped with elemental In and Ga solid metal sources and valved As and Sb cracker sources providing $As_2$and Sb_2 [13, 15, 18]. The solid metal sources for In and Ga are SUMO cells designed to reduce spitting and the sources for As and Sb are valved cracker source. The cracker temperature for As and Sb is typically between 800 and 900°C to achieve As_2 and Sb_2 main fluxes over As_4 and Sb_4. The reason for valved cracker design is to ensure the control of As_2 and Sb_2 fluxes at the stage of superlattices growth without further tuning of the cell temperature, which enables the sharp interfaces and the reduction of cross contamination across the layers during switching source flux changes.

The other important processing parameter is the optimization of growth temperature. Based on extensive researches so far, the typical growth temperature is typically between 390 and 430°C to focus on the effects of substrate temperature on superlattices. In this temperature range, superlattices can be optimized to achieve better interface roughness, smoother interfaces at the higher temperature, and residual background carrier concentrations, and so on. To improve further, there have been attempts to utilize short in-situ anneal at higher temperature such as 450°C under a high As flux, which can possibly reduce point defects.

Another important parameter that affects film growth is the film layer growth rate which is very thin compared to other applications. Due to unique thin film growth of superlattices such as a multiple of monolayers (ML), the growth rate is required to be slow to ensure a good quality of the very thin film (<8 nm). Typical growth rates are less than 1 monolayer per second for each layer. With slow growth rates, precision control of the SL period and individual widths has been demonstrated. The samples are grown on n-type (100) GaSb substrates which consisted of a GaSb buffer layer, multiple periods (for example, 40) of superlattices layers, and a GaSb cap layer. The growth is performed with a reflective high-energy diffraction (RHEED) system on a static GaInSb surface [13, 15, 18].

High-resolution x-ray diffraction and atomic force microscopy can be used to confirm the structural quality, composition of the InAs/GaInSb superlattices layers, and the surface morphology. More importantly, the optical properties can be characterized with photoluminescence, absorption spectroscopy, electroluminescence, cathodoluminescence, temperature-dependent I-V, C-V, and deep level transient spectroscopy. Having all the required characterization of input parameters, the initial analytical/numerical modeling and simulation are carried out with MATLAB (k·p in Section 4.1), TCAD (Section 4.2), and Ab-initio methods.

4. Design and modeling of superlattices

The optoelectronic properties of type-II InAs/GaInSb superlattices depend on the energy gap determined by the bottom of the conduction band (CB) of InAs and the bottom of the valence band (VB) of $Ga_{1-x}In_xSb$, where CB maximum of InAs is lower than VB minimum of $Ga_{1-x}In_xSb$. Due to this typical band alignment, the band gap energy of superlattices is smaller than either constituent material, which separates electron (InAs) and holes ($Ga_{1-x}In_xSb$) into the different layers. Moreover, the unique periodic film structure of superlattices allows the separation of heavy and light hole bands by compressive stress ($Ga_{1-x}In_xSb$) and tensile strain (InAs). This staggered band structure of superlattices provides the smaller band gap and spatial separation of electron and hole, which requires spatial indirect transition across the band gap. To achieve good optical transition with sound oscillator strength (hence, good absorption coefficient), the wave functions of the first heavy hole state (E_{HH1}) and the first electron states (E_{C1}) are required to overlap each other. Consequently, the optical absorption of the superlattices is determined by the overlap of the wave function between the localized hole states (E_{C1} in $Ga_{1-x}In_xSb$) and electron states (E_{HH1} in InAs). The overlap of the wave function is strongest in the superlattices design structure due to the enhanced electron and hole wave function overlaps by the repetition of each thin layer. As the thickness of each layer increases, the wave function overlaps rapidly decreases.

In **Figure 3**, the critical thickness by the number of monolayers (ML) is calculated with respect to the Indium composition (x) at 400°C, based on the mechanical equilibrium theory [21]. The composition dependent-strain due to the lattices mismatch is inserted on the top. The critical layer thickness is the maximum strained $Ga_{1-x}In_xSb$ thickness which is dislocation-free. Consequently, the energy gap of superlattices can be modulated between 0 and 300 meV by adjusting the layer width and the alloy composition. In **Figure 4**, the energy gap changes are described as a function of the width of InAs monolayers with respect to Indium concentration at fixed 8 monolayers of GaInSb on the left. Similarly, the variation of energy gap is plotted against the width of GaInSb monolayers at different Indium concentration on the right.

Narrow energy band gap superlattices with good optical properties (in particular, infrared absorption) require designing of thin film layers with the corresponding higher concentration of Indium. This can facilitate the reduction of the layer widths and the energy band gap as well, which enables the application beyond a wavelength of 15 um or very long-wave infrared (VLWIR) detection [18]. The typical indium percentage in the GaInSb layer has not been less

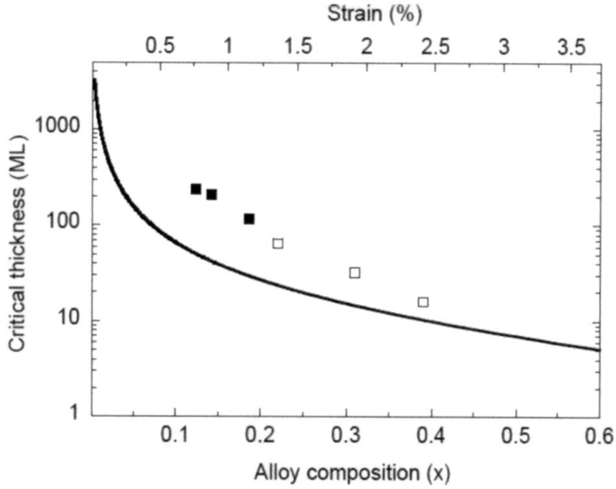

Figure 3. The calculated critical layer thickness in monolayers, ML (solid line) of $Ga_{1-x}In_xSb$ grown on GaSb at 400°C vs. indium concentration, x in $Ga_{1-x}In_xSb$ biaxial compression while InAs biaxial tension which lowers the conduction band and raises the heavy hole band. The experimental data are inserted in the figure (square) [19, 20].

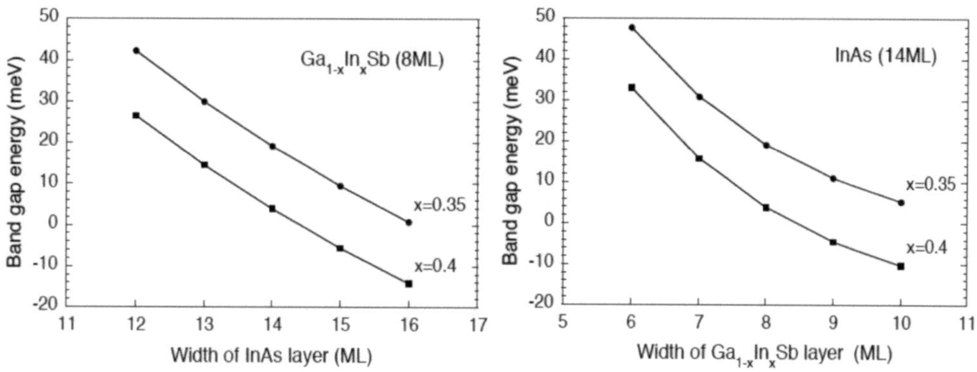

Figure 4. Band gap energy modification in InAs/$Ga_{1-x}In_xSb$ superlattices against the fixed width of $Ga_{1-x}In_xSb$ (left) and InAs (right) monolayers. The film thickness of $Ga_{1-x}In_xSb$ is fixed at 8 monolayers while varying the number of InAs monolayers at different indium concentration, x = 0.35 and 0.40 (left). For 14 monolayers of InAs, the film thickness of $Ga_{1-x}In_xSb$ changes with respect to indium x = 0.35–0.40 (right) [13].

than 30% to avoid the strain imbalance between these superlattices with high Indium concentration and GaSb lattice constant. The absorption coefficient spectra for various Indium concentration in superlattices are demonstrated in **Figure 5**.

4.1. Theoretical modeling: k·p methods

Since there are a number of input parameters that can be modulated to achieve designed superlattices at any given cutoff wavelength, it is imperative to provide a theoretical description

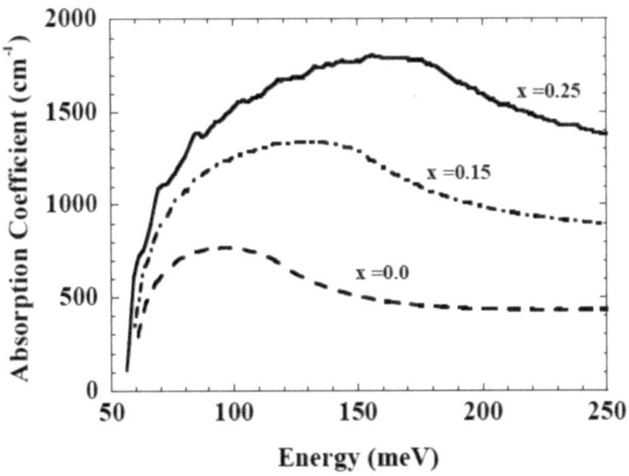

Figure 5. The calculated absorption coefficient spectra for various indium concentration in InAs/$Ga_{1-x}In_xSb$ superlattices design. The three different superlattices are designed to achieve the same narrow band gap with InAs (8.13 nm)/GaSb (2.66 nm), InAs (6.16 nm)/$In_{0.15}Ga_{0.85}Sb$ (2.28 nm), and InAs (5.27 nm)/$In_{0.25}Ga_{0.75}Sb$ (2.04 nm) [17].

by modeling chosen type-II superlattices system with the proper band gap energy, band structure, absorption spectra, and carrier lifetimes. The widely accepted method of the energy band description is the effective-mass approximation, which is also known as superlattice k·p method. It is a popular method for calculating the electronic structure of superlattices because of using a minimal set of parameters to describe and simulate heterostructure electronic properties independent of the number of atoms in the system [17]. The number of energy bands in the k·p Hamiltonian expands from the simple 4-band model to the 14-band model, which provides a continuation in the wave vector k of the energy bands near high-symmetry extremum points in the Brillouin zone such as Γ, X, L points. The simple 4-band model provides the accurate prediction of zone-center transition energies between the conduction and valence band whereas the 14-band model successfully describes the electronic structure throughout the Brillouin zone with the envelope function approximation, which discriminates crystal periodicity from the heterostructure energy envelope [17]. The k·p method is an efficient model to perform the simulation with material level properties up to the device-level properties without atomistic details.

The simulated energy band structure of 14 monolayers of InAs/8 monolayers of $Ga_{0.6}In_{0.4}Sb$ in [001] direction is demonstrated in **Figure 6**. To be consistent with the reported data, a set of material input parameters are taken from the Ref. 22 by using strain-dependent k·p model, including energy gaps, valence band offsets, deformation potentials, lattices constants, and elastic constants.

The superlattice is designed to target the energy gaps between 0 and 50 meV by considering the target layer thickness and alloy compositions. The 14 monolayers of InAs and 8 monolayers of $Ga_{0.6}In_{0.4}Sb$ strained layer superlattice is simulated and compared to the fabricated samples. The MBE grown samples are fed back to the superlattice modeling and design to achieve the target goal of energy gap around 9 meV. The superlattice energy gap is 4 meV,

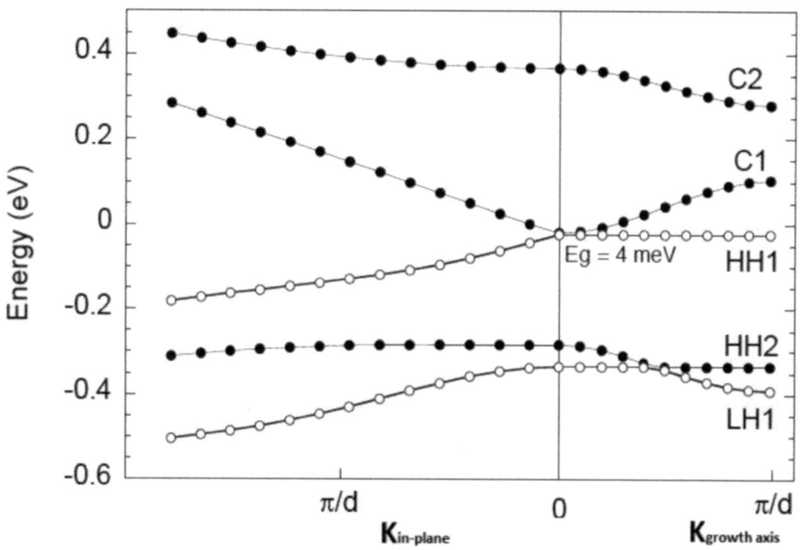

Figure 6. 8-band k·p model, calculating [001] in-plane (left) and growth direction (right) dispersion relations for the strained layer superlattice with 14 monolayers of InAs/8 monolayers of $Ga_{0.6}In_{0.4}Sb$. The period of the superlattices, d is 6.73 nm and the valence band edge of GaInSb happens to be located at zero energy level [13].

which is equivalent to the energy difference between the minimum of the electron energy state (E_{C1}) and the maximum of the heavy hole energy (E_{HH1}) at the zone center. As previously reported [8, 18], a large splitting between heavy hole and light hole energy states suppresses the hole–hole Auger recombination processes when compared to the energy gap along the growth direction (E_{HH1}–E_{LH1} < Eg) and thus, greatly improves the minority carrier lifetime, leading the improved device detectivity (D*) and operating temperature. Based on minimizing the Auger recombination, Grein et al. [15] calculated a strain-balanced VLWIR superlattice structure by using 14-band k·p method (a.k.a. superlattice k·p) to calculate states throughout the zone [8, 15], which is a reformulation of superlattice 14 band k·p method trying to explain Auger recombination and carrier lifetimes of type-II superlattice [15, 23]. In their calculation, a heterostructure restricted basis formalism is used by employing 14 bulk bands non-perturbative calculation [24], which successfully explained the optical properties, carrier recombination rates while incorporating interface bonding effects. The computed superlattice structure of 4.70 nm InAs/2.15 nm $Ga_{0.75}In_{0.25}Sb$ demonstrated a radiative lifetime of 140 ns whereas Auger-1 and Aguer-7 lifetimes were calculated to be more than 1 s, which could provide a theoretical limit of device detectivity of 6.0×10^{14} Johns at 40 K [18]. This long hole–hole Auger lifetime is caused by energy conservation forcing the most probable carriers into regions of lower occupation probability further band edges. **Figure 7** shows the simulated band structure of strained 4.70 nm InAs/2.15 nm $Ga_{0.75}In_{0.25}Sb$ superlattices, indicating the greater strain splitting between the uppermost valence bands, which increases the device detectivity in an order of magnitude. The positions of electrons (open circle) and holes (closed circle) indicates lower occupation probabilities at zone center which lowers the Auger recombination.

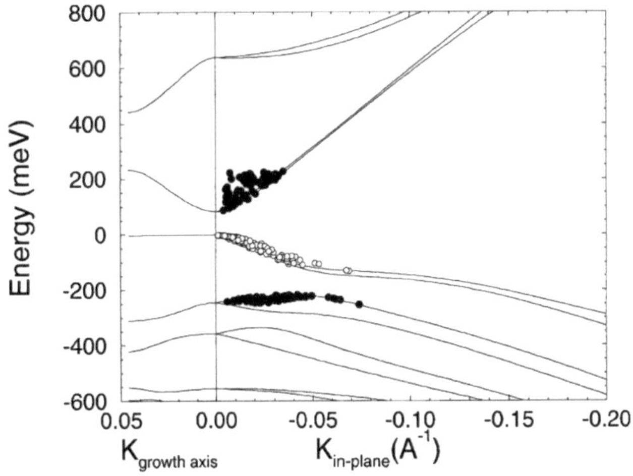

Figure 7. 14-band k·p model, calculating band structure of [001] strained 4.70 nm InAs/2.15 nm $Ga_{0.75}In_{0.25}Sb$ superlattices [001] in-plane (right) and growth direction (left). The positions of electrons (open circle) and holes (closed circle) is superimposed to highlight the lower occupation probabilities and slower Auger recombination at zone center [15].

4.2. Modeling of devices: TCAD

As 8-band k·p envelope function approximation model is used to calculate the band structure, the advantage is the flexibility of the model that can introduce an arbitrary number of layers per superlattice. After the analytical calculation, the resulting band structure can provide input parameters for the electrical device model such as band edge energy and effective masses. Furthermore, absorption coefficient spectra and tunneling coefficient can be modeled and simulated [25, 26]. With type-II InAs/GaSb superlattices, the goal of this section is to provide the TCAD simulation and computation of the electrical performance, in particular, dark current and complete device band structure with various sets of parameters such as composition, doping, and thickness of each layer. As reported in a recent report [27], choosing different period compositions for a given wavelength demonstrates the strong influence of the InAs/GaSb superlattice period thickness and composition on can boost the photodetector performance and material properties [27]. With an asymmetric superlattice period with thicker InAs layer than GaSb layer, the dark current can be reduced by a factor 4 compared to a symmetric superlattice period (equivalent thickness of both films), showing the same cut-off wavelength at 5 um at 77 K.

4.2.1. Modeling of dark current simulation

The main contributor to the dark current at moderate reverse bias is the generation-recombination at low temperature and diffusion current at high temperature. Furthermore, the other main contributors are the trap-assisted tunneling (TAT) via traps and residual dopant, the band-to-band tunneling currents (BTB), and Shockley-Read-Hall (SRH) Generation and Recombination (GR).

The diffusion of thermally generated minority carriers is expressed with a well-known drift-diffusion equation.

$$J_{n,p} = q(n,p)\,\mu_{n,p}\,E \pm q\,D_{n,p}\,\nabla_{n,p}, \tag{1}$$

where J is the current density, q is the charge, μ is the mobility, E is the electric field, D is the diffusion coefficient, and n, p subscripts are to clarify electron and hole components.

SRH recombination rate with trap states (E_t) and intrinsic energy level (E_i) is as below.

$$R_{SRH} = \frac{np - n_i^2}{\tau_p\left[n + n_i\,exp\left(\frac{E_t - E_i}{k\,T_L}\right)\right] + \tau_n\left[n + n_i\,exp\left(\frac{E_i - E_t}{k\,T_L}\right)\right]}, \tag{2}$$

where ni is the intrinsic carrier concentration, τ_n & τ_p are the electron and hole life times, and T_L is the lattice temperature.

The generation–recombination density is the integration along the simulating direction of thickness (W), y.

$$J_{GR} = q\int R_{SRH}\,dy,\ \textit{and integrate upon film thickness, W} \tag{3}$$

Estimating tunneling current is not trivial but if we follow Hurkx [28], the trap-assisted tunneling (TAT) current can be modeled with an additional term, a field effect factor Γ in SRH recombination equation in Eq. (2) [28]. If we modify the lifetime in SRH recombination, the Hurkx model about the field effect enhancement of the trap-assisted current is finally deduced with the net recombination rate (R_{SRH}) as below.

$$R_{SRH} = \frac{pn - n_i^2}{\frac{\tau_p}{1+\Gamma_p}\left[n + n_i\,exp\left(\frac{E_t - E_i}{k\,T_L}\right)\right] + \frac{\tau_n}{1+\Gamma_n}\left[n + n_i\,exp\left(\frac{-E_{Trap}}{k\,T_L}\right)\right]} \tag{4}$$

$$\Gamma_{n,p} = \frac{\Delta\,E_{n,p}}{k\,T_L}\int_0^1 exp\left(\frac{\Delta\,E_{n,p}}{k\,T_L}u - K_{n,p}\,u^{3/2}\right)du \tag{5}$$

where $\Delta E_{n,p}$ is the energy range where the tunneling of carriers happens and u is the integration term [29, 30]. $K_{n,p}$ is defined as below.

$$K_{n,p} = \frac{4}{3}\frac{\sqrt{2\,m_t\,\Delta E_{n,p}^3}}{q\,\frac{h}{2\pi}|E|}, \tag{6}$$

where m_t is tunneling mass, E is the electric field, and h is Plank's constant.

As you can see from the above equation, the field effect term considers the phonon-assisted enhancement of the carrier emission from a trap, which is enhanced under accelerated field circumstance. This model combined with GR and TAT current is a set of the simulation and computing in the following sections.

4.2.2. Input parameters for InAs/GaSb superlattices

The basic simulation structure is described in **Figure 8**. The simulated p-i-n device structure of InAs/GaSb superlattices is comprised of p-type GaSb buffer layer (200 nm), p-type 7 monolayers (ML) InAs/4 ML GaSb (1×10^{18} cm^{-3}), intrinsic 7 ML InAs/4 ML GaSb superlattices, n-type 7 ML InAs/4 ML GaSb superlattices (5×10^{17} cm^{-3}). The simulation is performed as a bulk material where the electronic transport inside mini-bands is ignored. To keep the consistency with the recent literature, modeling materials parameters for InAs and GaSb are listed as reported earlier [29–33].

4.3. Modeling and simulation results

Firstly, the band energy diagram is simulated at 77 K with the band gap energy and doping concentration (**Table 1**). As in the recent report [28, 30–33], residual doping is added as mentioned in the previous section. With reference to E_f at 0 eV location, the band diagram is demonstrated in **Figure 9**. The intrinsic doping level is experimentally determined as 2.8×10^{15} cm^{-3} [29].

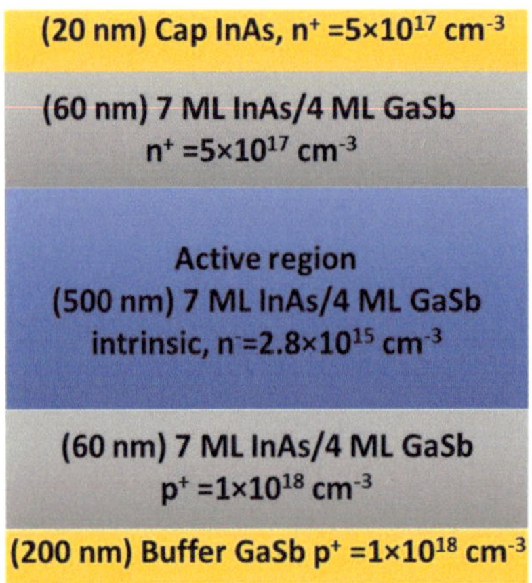

Figure 8. Simulated 7 monolayers (ML) InAs/4 ML GaSb superlattices p-i-n diode structure. Buffer is p-type GaSb, 1×10^{18} cm^{-3} and cap is n-type InAs, 5×10^{17} cm^{-3} for the simulation [29].

Parameter	Symbol	Value	Unit (or multiple of)
Electron effective mass	m_e	0.0254	m_0
Hole effective mass	m_h	0.245	m_0
Electron mobility	μ_e	2.6e4	cm^2/V-s
Hole mobility	μ_p	680	cm^2/V-s
Electron affinity	X	4.73	eV
Permittivity	ϵ	15.34	ϵ_0
Lifetime at 77 K	T	100	ns

Table 1. Input parameters for InAs/GaSb TCAD simulation.

The simulation of dark current curves is carried out at the various temperature from 77 to 237 K in **Figure 10** [29]. On the left hand side (**Figure 10a**), the simulated dark current curves are color-coded whereas empirical data are overlapped on the simulated curves with good matches, in particular, at the lower temperature. On the right hand side (**Figure 10b**), the

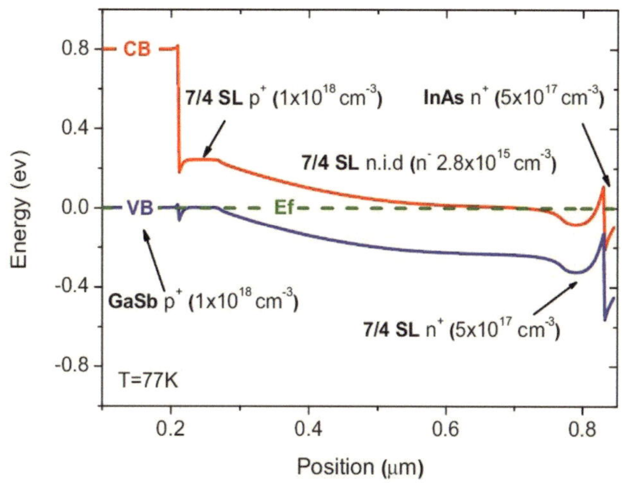

Figure 9. Energy band diagram calculated at 77 K with E_f located at 0 eV. The intrinsic InAs/GaSb superlattices are slightly n-type with 2.8×10^{15} cm^{-3} [29].

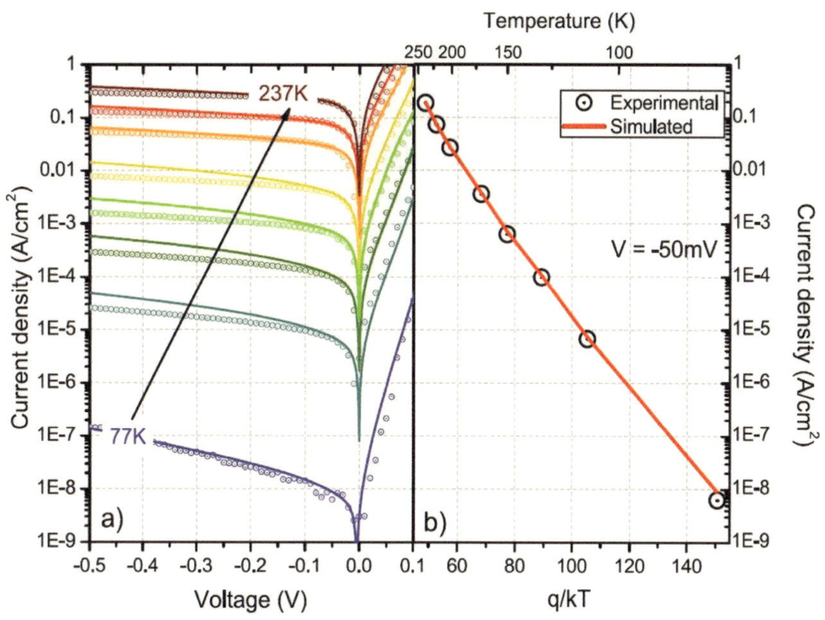

Figure 10. Empirical (open circle) and simulated dark currents (solid line) at different temperatures (a) and Arrhenius plot for the simulated data (solid line) at 50 mV reverse voltage, overlapped with empirical data (open circle) [29].

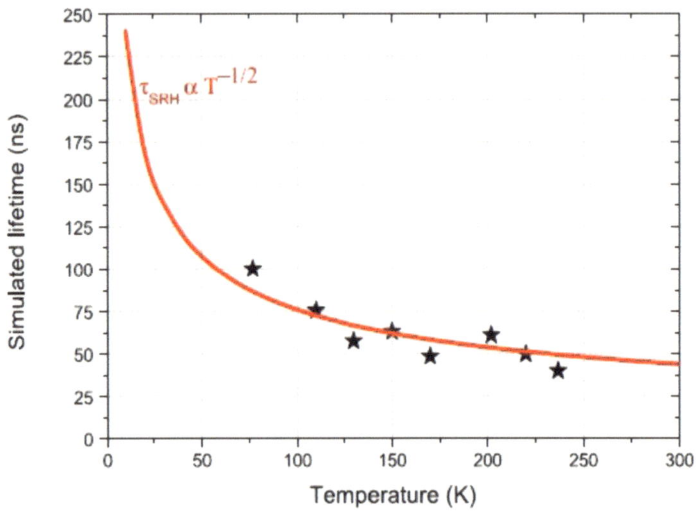

Figure 11. Fitting lifetime curves with power-law of $T^{-0.5}$ as a function temperature (star) [29].

Arrhenius plot of dark current at 50 mV reverse bias, considered to be a turn-on voltage is illustrated which well agrees well empirical data. The fitting of Arrhenius plot consists of the summation of diffusion component and generation-recombination plus trap-assisted tunneling components in the modeling and simulation. **Figure 11** shows the lifetime that is calculated from the simulation as a function of temperature. As you can see in the plot, the lifetime ranges from 50 nm to 100 ns from 77 to 230 K. The fitting of the lifetime is performed with temperature power-law, $T^{-0.5}$ which is the signature of SRH generation-recombination as mentioned by Connelly et al. [34].

5. Summary

In this chapter, we have discussed characterization, modeling, and simulation for the type-II superlattices photodetector application. By modeling with the k·p method, the superlattice band structure is modeled to serve as a set of input parameters for the following TCAD modeling and simulation. The theoretical modeling about Auger lifetime provides the insight about the dramatic reduction of Auger recombination. Furthermore, TCAD simulation and modeling are performed about the band the dark current. The dark current is theoretically characterized under the field effect enhancement of the trap-assisted current as well as SRH recombination and generation. The net recombination rate is finally deduced following a Hurkx model to compute the dark current. Having a tremendous amount of progress in the area of design, growth, and development over the past 15 years, has allowed the modeling and simulation of the type-II superlattices, this will provide new opportunities and guidance to the development of the next generation superlattice photodetector application.

Author details

Sanghyun Lee[1]* and Kent J. Price[2]

*Address all correspondence to: sanghyun.lee@indstate.edu

1 Department of Electronics and Computer Engineering Technology, Indiana State University, Terre Haute, IN, USA

2 Department of Physics, Morehead State University, Morehead, KY, USA

References

[1] Smith DL, Mailhiot C. Proposal for strained type II superlattice infrared detectors. Journal of Applied Physics. 1987;**62**:2545-2548

[2] Rogalski A, Antoszewski J, Faraone L. Third-generation infrared photodetector arrays. Journal of Applied Physics. 2009;**105**:091101

[3] Nguyen BM, Razeghi M, Nathan V, Brown GJ, Nguyen BM, Razeghi M, Nathan V, Brown GJ. Type-II M structure photodiodes: An alternative material design for mid-wave to long wavelength infrared regimes. Proceedings of SPIE. 2007;**6479**:64790S

[4] Kim HS, Plis E, Rodriguez JB, Bishop GD, Sharma YD, Dawson LR, Krishna S, Bundas J, Cook R, Burrows D, Dennis R, Patnaude K, Reisinger A, Sundaram M. Mid-IR focal plane array based on type-II InAs/GaSb strain layer superlattice detector with nBn design. Applied Physics Letters. 2008;**92**:183502P

[5] Delaunay, B. M. Nguyen, D. Hoffman, M. Razeghi. High-performance focal plane array based on InAs-GaSb superlattices with a 10-μm cutoff wavelength. IEEE Journal of Quantum Electronics. 2008;**44**(5):462-467

[6] Grein CH, Flatte ME, Olesberg JT, Anson SA, Zhang L, Boggess TF. Auger recombination in narrow-gap semiconductor superlattices incorporating antimony. Journal of Applied Physics. 2002;**92**:7311

[7] R. Rehm, M. Walther, J. Schmitz, J. Fleissner, J. Ziegler, W. Cabanski, R. Breiter. Dual-colour thermal imaging with InAs/GaSb superlattices in mid-wavelength infrared spectral range. Electronics Lett. 2006;**42**:577-578

[8] Grein CH, Young PM, Flatte ME, Ehrenreich H. Long wavelength InAs/InGaSb infrared detectors: Optimization fo carrier lifetimes. Journal of Applied Physics. 1995;**78**(12): 7143-7152

[9] Kim CS, Canedy CL, Aifer EH, Kim M, Bewley WW, Tischler JG, Larrabee DC, Nolde JA, Warner JH, Vurgaftman I, Jackson EM, Meyer JR. Molecular beam epitaxy growth of antimonide type-II "W" high-power interband cascade lasers and long-wavelength infrared photodiodes. Journal of Vacuum Science and Technology B. 2007;**25**:991

[10] Nguyen B-M, Hoffman D, Delaunay P-Y, Huang E, Razeghi M, Pellegrino J. Band edge tunability of M-structure for heterojunction design in Sb based type II superlattice photodiodes. Applied Physics Letters. 2008;**93**:163502

[11] Nguyen B-M, Bogdanov S, Pour S, Razeghi M. Minority electron unipolar photodetectors based on type II InAs/GaSb/AlSb superlattices for very long wavelength infrared detection. Applied Physics Letters. 2009;**95**:183502

[12] Maimon S, Wicks GW. nBn detector, an infrared detector with reduced dark current and higher operating temperature. Applied Physics Letters. 2006;**89**:2360235

[13] Patrashin M, Hosako I, Akahane K. Type-II InAs/GaInSb superlattices for terahertz range photodetectors. Proceedings of SPIE. 2001;**8188**:81880G-818801

[14] Ganapathi K. Tunneling in Low-Power Device-Design: A Bottom-Up View of Issues, Challenges, and Opportunities. Technical Report No. 2013;UCB/EECS-2013-164

[15] Grein CH, Lau WH, Harbert TL, Flatte ME. Modeling of very long infrared wavelength InAs/GaInSb strained layer Superlattice detectors. Proceedings of SPIE. 2002;**4795**:39-43

[16] Roslan P, Roslan PS, Ker PJ, Ahmad I, Pasupuleti J, Fam PZ. Modeling and simulation of InAs photodiode on electric field profile and dark current characteristics. Applied Physics Letters. 2016 IEEE-ICSE Proc. 2016;**1**:188-191

[17] Brown G. Type-II InAs/GaInSb Superlattices for infrared detection: An overview. Proceedings of SPIE. 2005;**5783**:5783-5783(13), Infrared Technology and Applications XXXI. DOI: 10.1117/12.606621

[18] Haugan H, Olson B, Brown G, Kadlec E, Kim J, Shaner E. Significantly enhanced carrier lifetimes of very long-wave infrared absorbers based on strained-layer InAs/GaInSb superlattices. Optical Engineering. 2017;**56**(9):091604

[19] Kuramochi E, Yoshifumi T. Strain effects in InGaSb/AlGaSb quantum wells grown by molecular beam epitaxy. Journal of Applied Physics. 1995;**77**(11):5706-5711

[20] Nilsen T, Brevik M, Selvig E, Fimland BO. Critical thickness of MBE-grown $Ga_{1-x}In_xSb$ ($x<0.2$) on GaSb. Journal of Crystal Growth. 2009;**311**:1688-1691

[21] Matthew JW, Blakeslee AE. Defects in epitaxial multilayers: I Misfit dislocations. Journal of Crystal Growth. 1974;**27**:118-125

[22] Vurgaftman J, Meyer R, Ram-Mohan LR. Band parameters for III–V compound semiconductors and their alloys. Journal of Applied Physics. 2001;**89**(11):5815-5875

[23] Youngdale ER, Meyer JR, Hoffman CA, Bartoli FJ, Grein CH, Young PM, Ehrenreich H. Auger lifetime enhancement in InAs-$Ga_{1-x}In_xSb$ superlattices. Applied Physics Letters. 1994;**64**:3160

[24] Olesberg JT. Design and Characterization of the Electronic and Optical Properties of Materials for Mid-Infrared Laser Diode Active Regions. [PhD thesis]. Iowa City, Iowa: University of Iowa; 1999

[25] Asplund C, Würtemberg RM, Lantz D, Malm H, Martijn H, Plis E, Gautam N, Krishna S. Performance of mid-wave T2SL detectors with heterojunction barriers. Infrared Physics & Technology. 2013;**59**:22-27

[26] Ashplund C, Wurtemberg RM, Hoglund L. Modeling tools for design of type-II superlattice photodetectors. Infrared Physics & Technology. 2017;**84**:21-27

[27] Taalat R, Rodriguez JB, Delmas M, Christol P. Influence of the period thickness and composition on the electro-optical properties of type-II InAs/GaSb midwave infrared superlattice photodetectors. Journal of Physics D: Applied Physics. 2014;**47**:015101

[28] Hurkx GAM, Degraff HC, Kloosterman WJ, Knuvers MPG. A new analytical diode model including tunneling and avalanche breakdown. IEEE Transactions on Electron Devices. 1992;**39**:2090

[29] M. Delmas, J-B Rodriguez, P. Christol. Electrical modeling of InAs/GaSb superlattice mid-wavelength infrared pin photodiode to analyze experimental dark current characteristics. Journal of Applied Physics. 2014;**116**:113101

[30] Callewaert F, Hoang AM, Razeg M. Generation-recombination and trap-assisted tunnelling in long wavelength infrared minority electron unipolar photodetectors based on InAs/GaSb superlattice. Applied Physics Letters. 2014;**104**:053508

[31] Svensson SP, Donetsky D, Wang D, Hier H, Crowne FJ, Belenky G. Growth of type II strained layer superlattice, bulk InAs an GaSb materials for minority lifetime characterization. Journal of Crystal Growth. 2011;**334**:103-107

[32] Murray LM, Lokovic KS, Olson BV, Yildirim A, Boggess TF, Prineas JP. Effects of growth rate variations on carrier lifetime and interface structure in InAs/GaSb superlattices. Journal of Crystal Growth. 2014;**386**:194

[33] Nguyen BM, Hoffman D, Delaunay PY, Huang E-K, Razeghi M. Very high performance LWIR and VLWIR type-II InAs/GaSb superlattice photodiodes with M-structure barrier. Proceedings of SPIE. 2008;**7082**:708205

[34] Connelly BC, Metcalfe GD, Shen H, Wraback M. Direct minority carrier lifetime measurements and recombination mechanisms in long-wave infrared type II superlattices using time-resolved photoluminescence. Applied Physics Letters. 2010;**97**:251117

Chapter 8

Graphene-Based Acousto-Optic Sensors with Vibrating Resonance Energy Transfer and Applications

Burhan Gulbahar and Gorkem Memisoglu

Additional information is available at the end of the chapter

http://dx.doi.org/10.5772/intechopen.73618

Abstract

Graphene as a two-dimensional planar material has numerous advantages for realizing high-performance nano-electromechanical systems (NEMS) such as nanoscale sensors including strain sensors, optical modulators or energy harvesters. Large Young's modulus (1 TPa for single layer graphene), ultra-low weight, low residual stress and large breaking strength properties are important properties as two-dimensional (2D) ultrathin resonators. Graphene resonators are recently utilized for low complexity design of nanoscale acousto-optic sensors based on a novel theoretical model describing vibrating Förster resonance energy transfer (VFRET) mechanism. Proposed system combines the advantages of graphene with quantum dots (QDs) as donor and acceptor pairs with broad absorption spectrum, large cross-sections, tunable emission spectra, size-dependent emission wavelength, high photochemical stability and improved quantum yield. Device structure supporting wide-band resonance frequencies including acoustic and ultrasound ranges promises high-performance applications for challenging environments. Remote sensors and acousto-optic communication channels are formed for in-body applications, wireless body area sensor networks (WBASNs), space and interplanetary systems, microfluidics and visible light communication (VLC)-based architectures.

Keywords: nano acousto-optic sensor, graphene resonator, nanoscale acousto-optic transduction, vibrating Förster resonance energy transfer, energy harvesting, interplanetary communications, in-body sensor network, microfluidics, visible light communications

1. Introduction

Graphene has important advantages in significantly many application areas of physical, chemical and biological sciences with advanced engineering system designs including sensors, modulators and two-dimensional (2D) ultrathin resonators. Mechanical designs utilizing graphene achieve high-performance nano-electromechanical systems (NEMS) with large Young's modulus (1 TPa for single layer graphene), ultra-low weight, low residual stress and large breaking strength, i.e., graphene strain allowing 25% without breaking [1]. Unique optical, electrical, physical and mechanical properties of graphene allow the designed resonators to be utilized in challenging applications, e.g., strain sensor, mass sensor, nanogenerators, transducers, photodetectors and novel NEMS devices [1–8].

Nanoscale photonic solutions with a novel design for sensing, energy harvesting or communication purposes are recently presented in [3–5, 8] with significant performances and rich set of applications. They utilize unique and unexploited features of graphene and quantum dots (QDs) combined with a special mechanical design. QDs are future promising as donor-acceptor couples in next generation nanoscale devices [3–5, 8, 9]. More specifically, passive acousto-optic nanoscale optical modulator design is presented by exploiting high performance mechanical properties of graphene resonators and radiating photon emission from QDs with a novel method denoted by vibrating Förster Resonance Energy Transfer (VFRET) [4]. It promises high performance applications in biomedical, space and microfluidic monitoring and tracking areas while utilizing energy harvesting, low complexity and all-in-one acousto-optic transducer mechanical design.

VFRET mechanism uniquely exploits important properties of graphene resonators having wideband spectrum covering acoustic and ultrasound frequencies combined with special features of QDs, and FRET which is a nanoscale energy transfer process between two molecules denoted as donors and acceptors [4]. FRET utilized in a rich set of biological, physical and chemical applications such as monitoring cellular activities has also been utilized for nanoscale communication channels [10–12]. Semiconductor nanocrystal QDs form a unique cooperation with graphene resonators by utilizing high performance properties such as sharp and broad absorption spectrum, large absorption cross-sections, tunable emission spectra and wavelength, photochemical stability and photoluminescence quantum yield [13, 14]. In [4], a modulator is designed to be utilized in challenging applications for communications, sensing and energy harvesting with a scalable range including both nanoscale and macroscale dimensions. Ambient light sources with low power levels are utilized instead of laser sources in a nanoscale hybrid acousto-optic platform while periodically modulating nanoscale distance between donors attached on vibrating graphene and acceptors fixed on a support. Therefore, a design combining flexibility without requiring special laser set-ups, energy efficiency with low power ambient sources and wideband resonance frequency including ultrasound levels is presented as a future promising tool. In addition, it is emphasized that graphene vibration can also be generated by thermo-acoustic or opto-acoustic methods by proposing the tool beneficial in significantly challenging environments.

VFRET system design promises important application areas as discussed in detail in Section 5 by combining unique properties of QDs and graphene including its biocompatibility. Various application areas are listed as follows presenting biomedical, space, microfluidics and communication purposes:

1. **Biomedical applications:**

   a. In-body: hybrid acousto-optic noninvasive communication channels inside body by utilizing external acoustic excitations.

   b. Wireless body area sensor networks (WBASNs): wearable sensors such as heart rate monitoring or any sensor transmitting data with optical sequences.

2. **Space applications:**

   a. Interplanetary remote sensing: sensing atmospheric events remotely.

   b. Optical wireless communications: realizing optical channels between devices in space environment.

3. **Microfluidics:** cell and particle monitoring and tracking.

4. **Visible light communications (VLC):** generating visible light sequences with advanced geometrical design to produce modulated data sequences.

In this Chapter, VFRET mechanism and nanoscale acousto-optic sensor design are briefly introduced in Sections 2 and 3, respectively, as fundamental novel tools described in [4]. Then, Section 4 discusses applications in biomedical sensing, interplanetary and space, microfluidics and communication networks. Finally, Section 5 concludes the Chapter with a brief summary of VFRET-based sensor design and its applications.

2. Vibrating FRET mechanism

An illustration describing the basics of VFRET mechanism is given in **Figure 1**. Vibrating multi-layer graphene membrane has height h, radius a, tunable donor-acceptor (D-A) distance of d_{AD} and resonance amplitude d_0. Three phases of VFRET are denoted by phase-a, phase-b and phase-c. Phase-a corresponds to the position at rest. FRET or VFRET occurs in phase-b at the shortest distance between D-A pairs and phase-c has the largest distance

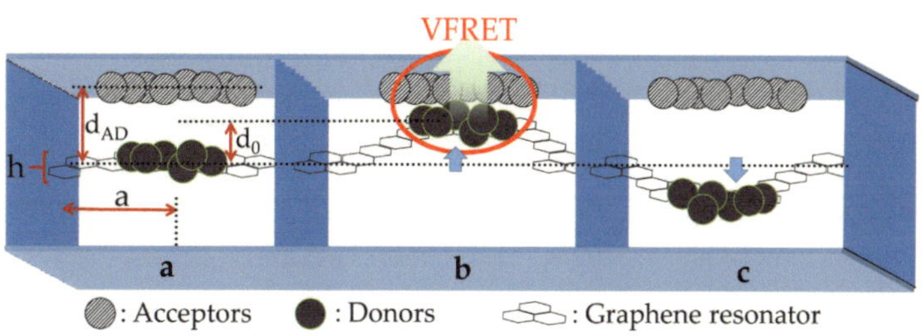

Figure 1. Nanoscale acousto-optic transduction mechanism with VFRET mechanism where mechanical vibration is converted to optical emission (adapted from [4]) with the phases denoted by a, b and c showing the relative positions of acceptors and donors in a single resonance period.

between the pairs without any FRET emission. QD donors are attached on graphene membrane firmly, while acceptor molecules are either the QDs or fluorescent dyes, and attached on top of the modulator frame.

Radiated light is controlled by modulating acoustic pressure on graphene resonator while donors are continuously excited by visible light sources. The number of photons emitted is modeled as follows [4]:

$$N_{FRET} = \Delta_t I_D D_B \sigma_D \Phi_A E N_D / \left(h_p f_D^a\right) \tag{1}$$

where Δt is the photon counting time interval which is much larger than the total time $t_p = t_D + t_{FRET} + t_A$ for acceptor emissions summing the donor excitation time denoted by t_D, FRET time t_{FRET} and acceptor emission time t_A, I_D and D_B are the light intensity in (W/m^2/nm) and bandwidth in (nm), respectively, for donor excitation, $\sigma_D = D_{ext} \times 3.825 \times 10^{-23}$ denotes absorption cross section of donor excitation in (m^2), D_{ext} is donor extinction coefficient in (M^{-1} cm^{-1}), Φ_A is the quantum yield of acceptors, E is FRET efficiency, N_D is the number of donor molecules, $f_D^a = c/\lambda_D^a$ is donor excitation frequency, $c = 3 \times 10^8$ (m/s), λ_D^a is the donor excitation wavelength, and $h_p = 6.62 \times 10^{-34}$ (J × s) is Planck's constant. FRET efficiency E is defined as follows [4]:

$$E = \frac{k_A R_0^6}{d^6 + k_A R_0^6} \tag{2}$$

where k_A denotes the number of acceptors corresponding to each donor, d is the donor-acceptor distance, and R_0 is Förster distance.

In the next section, proposed model is utilized to design an acousto-optic sensor composed of mainly vibrating multi-layer graphene membrane and D-A molecules.

3. Nanoscale acousto-optic sensor design

The proposed design is improved by including lenses, supports and special enclosure design to be utilized in challenging environments as shown in **Figure 2** as described in detail in [4]. Donors absorbing incoming light energy transfer the energy to acceptors while received and emitted light are filtered by a combination of optical lens and filter to prevent unintended acceptor emission and to intensify incoming and emitted light.

Graphene layer acoustically vibrates with minor change on the resonance frequency due to the weight of donor molecules. The distance between D-A pairs is tunable with respect to the desired vibration amplitude. An example is shown in **Figure 3** where multi-color emission capability is realized by including an array of resonators with different color emissions tuned to different sound pressure levels (SPLs). The set of D-A pairs in [4] is extended in **Table 1** to include multiple colors, and a diversity of D-A pair selections with different diameters and types, i.e., dyes, QDs and proteins [13, 14]. Excitation wavelengths of donors and acceptors,

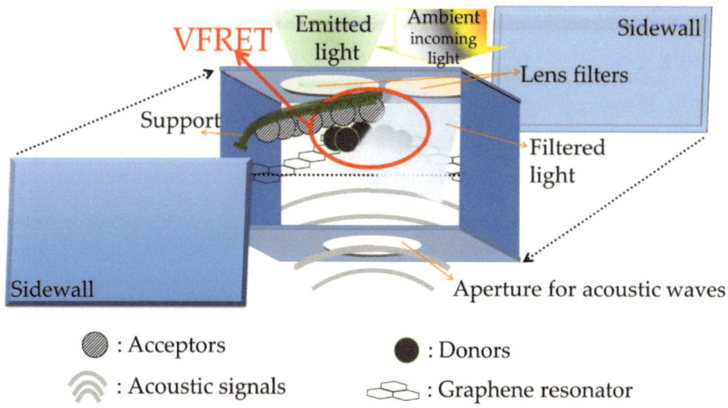

Figure 2. Physical structure of the acousto-optic transducer where QD donors on resonating graphene membrane transfer energy to QD acceptors on the support with VRET mechanism [4]. Acceptors have filtering interfaces for incoming and emitted light sources while resonators are vibrated by acoustic waves entering through aperture. Sidewalls and special geometrical design protect the device and provides durability (adapted from [4]).

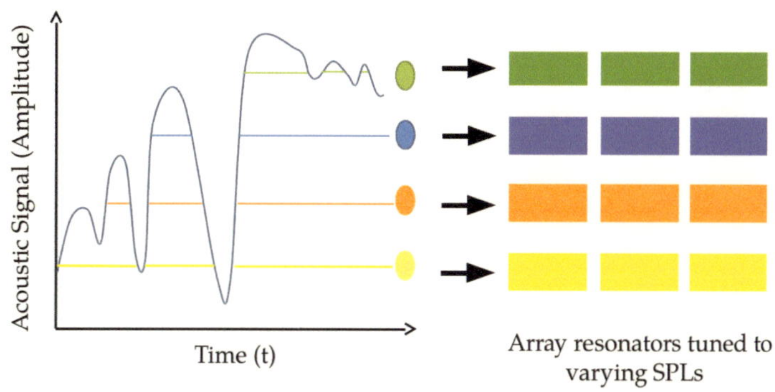

Figure 3. Resonators tuned to varying sound pressure level (SPLs) with different colors at the corresponding SPLs and arrays of resonators enhancing emission intensity.

Ref #	Donor and diameter	λ_D^a	λ_D^{PL}	Acceptor and diameter	λ_A^a	λ_A^{PL}	R_0	Φ_A
[13]	CdSe/ZnS (2.5 nm)	475	500	Dye (fluorescein27)	505	548 (green)	3.4	0.6
[14]	CdSe/ZnS (2.9 nm)	540	555	CdSe/ZnS (3.7 nm)	560	580 (yellow)	6.6	0.6
[14]	Cyan fluorescent protein	430	495	Yellow fluorescent protein	500	525 (green)	4.5	0.4
[14]	CdSe/ZnS (3.7 nm)	555	570	Fluorescent protein mCherry	575	615 (red)	5.8	0.6
[14]	CdSe/ZnS (3.3 nm)	510	550	CdSe/ZnS (3.3 nm)	555	590 (yellow)	6.1	0.6

CdSe/ZnS: cadmium selenide/zinc sulfide.

Table 1. Extended set of D-A pairs to utilize in acousto-optic sensor and modulator.

i.e., λ_D^a and λ_A^a, respectively, and emission wavelengths denoted by λ_D^{PL} and λ_A^{PL}, respectively, are tuned with changing diameters and types of molecules.

In addition to QDs, dyes and proteins are candidates to be utilized in various application environments depending with respect to the requirements. In the next section, applications of the proposed sensor and modulator devices are discussed.

4. Applications

The application areas of the proposed VFRET device include a diverse set of domains with various tasks of the device such as an acousto-optic sensor, modulator and hybrid communication transmitter. In-body hybrid sensors, WBASN applications with wearable sensors and transceivers, microfluidics, space, on-chip communication transceivers and visible light communications (VLC)-based applications are the most promising ones as discussed in the following sections.

Bio-compatibility of graphene and QDs gives opportunity for further improving the proposed design suitable for in-body applications [2, 15, 16]. More promising future applications include biomedical ultra-low power communication networks, acoustic sensors and transducers, and nanophotonics. In addition, array structures utilizing multiple resonators in horizontal and vertical dimensions improve light emission power.

VFRET-based nanoscale sensor/modulator architecture allows high performance future applications by exploiting important features of the special design with graphene and QDs. The proposed device brings unique advantages by exploiting the following features:

- Planar architecture to be attached on varying objects
- Lightweight design
- Broadband spectrum capability supporting ultrasound and acoustic waves
- Energy efficient design operating with low level of light sources such as ambient light levels for day time and LED sources for night time operation
- Passive and energy harvesting design by combining vibration and external light sources
- Hybrid acousto-optic architecture allowing utilizing ultrasound and acoustic signals for challenging environments such as creating optical communication links inside body by using external ultrasound vibrations
- Tuning to different frequencies, SPLs, light levels and applications by varying device geometry, graphene resonator and the number of QD molecules
- Flexible system design without requiring complicated laser set-ups having strict orientations
- Adaptability to different application areas including in-body communications, wireless body area networks (WBANs), nanoscale and microscale communication networks, microfluidic monitoring and tracking applications, space and aerial applications, and VLC.

Next, various application areas including biomedical sensing, interplanetary remote sensing and space applications, microfluidics and VLC are discussed in detail.

4.1. Biomedical sensing and communications applications

Biomedical sensing is an emerging and future promising topic with various applications including wearable sensors, wireless body area networks (WBAN), and in-body sensing and communications applications [16]. Proposed lightweight and energy efficient device structure is a low complexity and all-in-one alternative to the existing radio frequency (RF)-based WBAN sensor solutions [17]. Potential applications include heart tracking, mobility or any movement related monitoring of body parts. In **Figure 4**, an architecture is provided where acoustic vibrations of heart within specific frequency bands reaching hundreds of Hz are converted to optical emissions with tuned device geometry as indicator of heart rate signaling [18]. It provides a passive heart rate monitoring device without requiring battery and, separate modulator and coding blocks with complicated architectures.

In disaster scenarios, heart rate monitoring device provides a method to track and to monitor the health condition of the people remotely. In addition, multiple people including highly crowded groups can be tracked by monitoring emitted periodic sequences. A simple and remote personal monitoring device is proposed.

Besides that, it allows to realize communication channels inside body, e.g., for hybrid communication links for intra-bone communication architectures as discussed in [19] and as shown in **Figure 5**. Remote signal transmission and communication with in-body region by using transceivers outside the body are highly difficult with RF and optical signals. On the other hand, ultrasound signals easily penetrate through body. A hybrid communication channel is formed by modulating the device inside the body, e.g., bones, with an external ultrasound excitation while emitting optical signals inside the bone region for optical communication. Therefore, optical communication channels anywhere inside the body are easily formed by exciting externally with ultrasound waves.

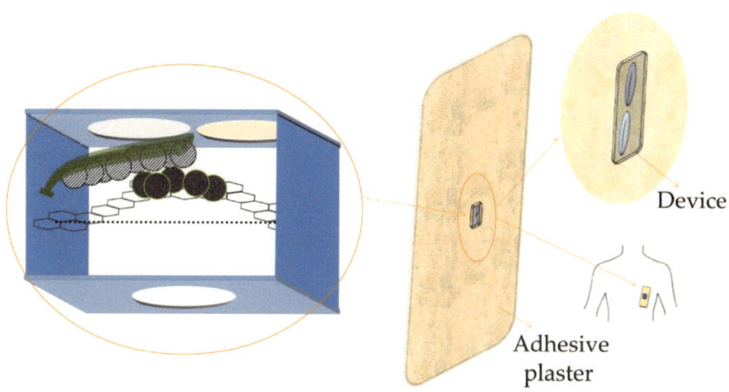

Figure 4. Biomedical sensing for heart rate monitoring where acousto-optic transducer device is attached on the surface of the heart with an adhesive plaster.

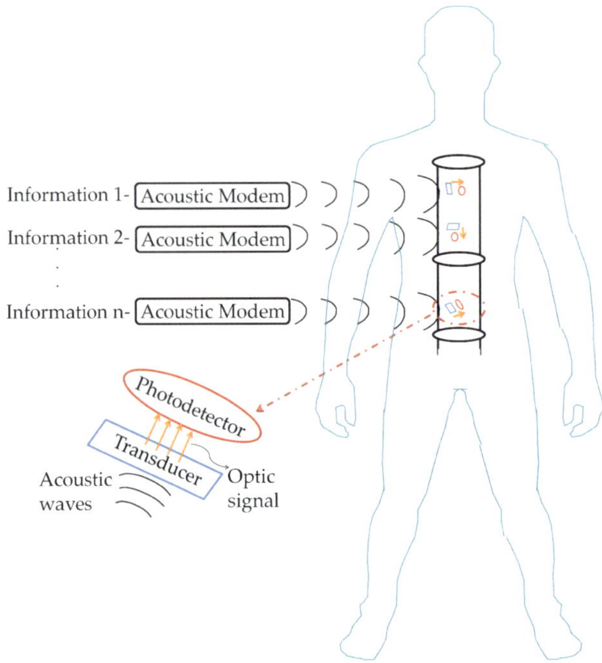

Figure 5. Hybrid in-body sensing and communication scenario for realizing optical communication links inside the body with external acoustic excitation. Photodetectors detect acousto-optic signals inside the bones as a challenging medium to realize communication networks.

In-body applications are realized by exploiting acousto-optic transduction properties, energy harvesting, low power operation and unique features of graphene and QDs. Hybrid communication links are formed such that graphene resonator is vibrated remotely without requiring contact by using ultrasound waves for creating optical in-body networks. Noninvasive design promises to improve opto-genetics with capability to operate inside different parts of the body [20]. In-body system design requires low power LEDs to be available with the proposed device. This is feasible compared with high power laser operations requiring strict orientations and complicated set-ups. Scalability of the geometrical design, planar structure, flexible and simple system design, and all-mechanical structure make VFRET design with graphene resonators more promising for challenging optical in-body nanonetworks.

Information is externally generated and transmitted to photodetectors inside the body by using external acoustic modulation and hybrid acousto-optic transduction in the VFRET device as a solution for significant challenges of the optical or RF signals to penetrate inside the body [16].

4.2. Interplanetary remote sensing and space applications

Long range acoustical sensors are utilized for passive sensing of vibrations, flutter, atmospheric turbulence, and terrestrial and planetary sounds remotely [21]. Utilization of reflective surfaces with the light shining on the diaphragm limits the practicality of these devices. On the other hand, proposed acousto-optical system architecture allows nanoscale architectures, flexible design with a controlled light source without any reflection-based positioning requirements and multi-color light modulation. Atmospheric events and space-based environmental sensing applications of vibrations are easily tracked with optical signal detection where the light energy is harvested from high power and line-of-sight radiation of the sun as shown in **Figure 6**. Hard-to-reach areas where vibratory events occur are monitored with a sensor network composed of energy harvesting transducers with significantly high lifetimes, durability and low complexity design allowing long duration space travel.

On the other hand, energy harvesting nature and optical emission capability are utilized for creating optical wireless communications channels in space. Simple device mechanism and array forming capability provide a unique opportunity to utilize in challenging environments in space.

4.3. Microfluidic applications

Multi particle tracking (MPT) or single particle tracking (SPT) with labeling by observing fluorescent molecule emission are common tools for monitoring cellular processes for in-vivo nano-biological and in-vitro microfluidic systems [3, 5]. Special fluorescing tags attached on molecules are tracked and digital image processing tools allow nanometer (nm) resolution in positioning. Image processing methods and algorithms are developed in significantly many studies for particle and cell tracking in microscopic platforms [22, 23]. However, there are important challenges for collection capability and analyzing high complexity imaging data. Tags do not have signaling capability to support a signaling-based tracking compared with traditional imaging-based tracking systems like fluorescence lifetime imaging for cell tracking.

Recently, a novel nanoscale acousto-optic radar and particle tracking system design using chirp spread spectrum (CSS) sequences with special geometrical design of acousto-optic transducers [5]. The design denoted with the name *CSSTag* promises signaling-based tagging in microfluidic environments by generating unique tags as shown in **Figure 7**. CSSTag system utilizes spread

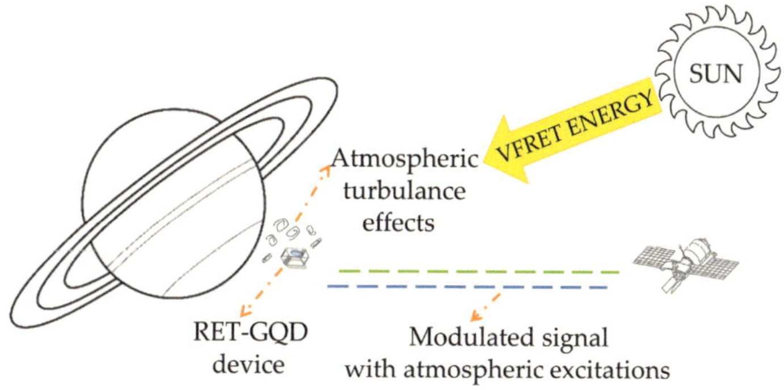

Figure 6. Space application for interplanetary remote sensing where energy is harvested from sun to track acoustic vibrations of atmospheric effects in space or planetary environments.

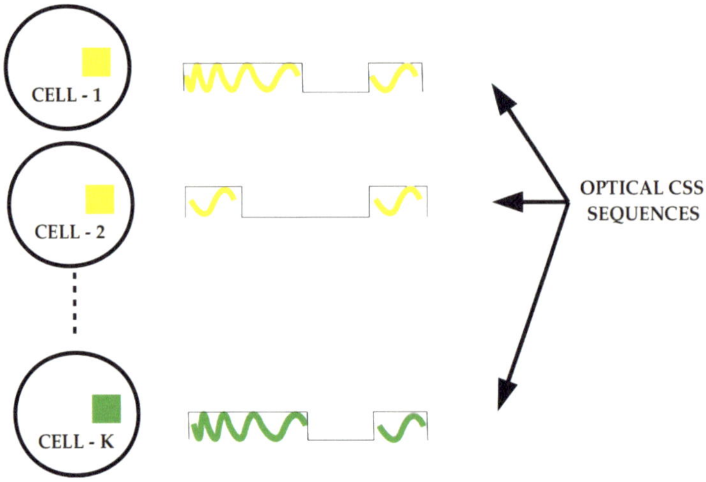

Figure 7. Multiple optical sequence generation for tagging different cells where the first two cells are tagged with the same color but different sequences while Kth cell is tagged with same sequence with the first cell but with different color.

spectrum modulation, time difference of arrival (TDOA)-based positioning and cell specific signaling sequences by exploiting the theoretical design providing nanoscale acousto-optic transduction with VRET and graphene resonators. Ultrasound carrier vibrated graphene resonators with specially placed multi-layer geometrical design to produce different optical emission sequences which are detected by a set of photodetector arrays with color filters.

4.4. Visible light communications

The capability to generate modulated optical signals in visible frequency band allows to realize nanoscale VLC transmitter at the resonance frequency of the membrane utilizing various modulation structures [24]. Multi-color emission capability and spatial diversity support generation of more advanced modulation methods such as color shift keying (CSK) as one of the important standard methods in VLC. The proposed device is embedded either into on-chip platforms for intra-chip hybrid communications or can be utilized externally in free space communication channel.

Future works promise novel modulation/demodulation methods for generating data by optimizing geometry and material selection with specific colors and donor-acceptor pair arrays.

In addition, nanophotonics is another important area where the proposed device as a VLC modulator promises novel optical devices such as in applications utilizing single photon devices, quantum communications, plasmonics and electro-optic modulator structures. A simple, nanoscale and hybrid acousto-optic VLC modulator is designed without complicated device architecture and potential with high speed modulation capability [25]. The proposed system design is a future promising VLC transmitter with significantly many application areas.

5. Conclusion

In this chapter, VFRET mechanism-based nanoscale acousto-optic sensor, acousto-optic wireless communication modulator and transducer converting the vibrations of multi-layer graphene resonator to multi-color photon emission are described. Graphene resonators achieve vibration frequencies ranging from acoustic to ultrasound ranges while significant advantages of graphene such as lightweight and 2D planar structure, large Young's modulus, low residual stress and large breaking strength. The device structure combined with QDs includes unique advantages of broad absorption spectrum, large cross-sections, tunable emission spectra, size dependent emission wavelength, high photochemical stability and improved quantum yield. Hybrid nanoscale acousto-optic system design promises important applications in sensing and communications for in-body networks and WBANs, space and interplanetary applications, microfluidics and VLC-based systems.

Acknowledgements

This work is supported by Vestel Electronics Inc., Manisa, Turkey.

Author details

Burhan Gulbahar[1]* and Gorkem Memisoglu[2]

*Address all correspondence to: burhan.gulbahar@ozyegin.edu.tr

1 Department of Electrical and Electronics Engineering, Applied Research Center of Technology Products, Ozyegin University, Istanbul, Turkey

2 Vestel Electronics Inc., Organize Sanayi Bolgesi, Manisa, Turkey

References

[1] Castallenos-Gomez A, Singh Van der Zant HS, Steele GA. Mechanics of freely suspended ultrathin layered materials. Annalen der Physik. 2015;**527**:27-44. DOI: 10.1002/andp.201400153

[2] Ferrari AC et al. Science and technology roadmap for graphene, related two-dimensional crystals, and hybrid systems. Nanoscale. 2015;**7**:4598-4810. DOI: 10.1039/C4NR01600A

[3] Memisoglu G, Gulbahar B. Acousto-optic tagging and identification, PCT/EP409508, Patent Application. June 2017

[4] Gulbahar B, Memisoglu G. Nanoscale optical communications modulator and acousto-optic transduction with vibrating graphene and resonance energy transfer. In: Proc. of the IEEE ICC 2017 Selected Areas in Communications Symposium Molecular, Biological and Multi-Scale Communications Track; 21-25 May 2017; France. IEEE; 2017. DOI: 10.1109/ICC.2017.7997036. http://ieeexplore.ieee.org/document/7997036/

[5] Gulbahar B, Memisoglu G. CSSTag: Optical nanoscale radar and particle tracking for inbody and microfluidic systems with vibrating graphene and resonance energy transfer. IEEE Transactions on NanoBioscience. 2017;**16**:905-916. DOI: 10.1109/TNB.2017.2785226

[6] Gulbahar B, Akan OB. A communication theoretical modeling of single-layer graphene photodetectors and efficient multireceiver diversity combining. IEEE Transactions on Nanotechnology. 2012;**11**:601-610. DOI: 10.1109/TNANO.2012.2187068

[7] Gulbahar B. Theoretical modeling of KHz to THz simultaneous energy harvesting and magneto-inductive communications with molecular magnets on vibrating graphene [Internet]. 2017. Available from: https://faculty.ozyegin.edu.tr/burhangulbahar/publications/ [Accessed: 2017-12-06]

[8] Gulbahar B, Memisoglu G. Acousto-optic transducer, array and method. PCT/EP2017/054408, Patent Application. July 2017

[9] Sevim S, Memisoglu G, Varlikli C, Dogan LE, Tascioglu D, Ozcelik S. An ultraviolet photodetector with an active layer composed of solution processed polyfluorene:Zn0.71Cd0.29S hybrid nanomaterials. Applied Surface Science. 2014;**305**:227-234. DOI: 10.1109/LPT.2014.2384532

[10] Ha T et al. Probing the interaction between two single molecules: Fluorescence resonance energy transfer between a single donor and a single acceptor. Proceedings of the National Academy of Sciences. 1996;**93**:6264-6268. DOI: PMC39010

[11] Kuscu M, Akan OB. Multi-step FRET-based long-range nanoscale communication channel. IEEE Journal on Selected Areas in Communications. 2013;**31**:715-725. DOI: 10.1109/JSAC.2013.SUP2.1213004

[12] Wojcik K, Solarczyk K, Kulakowski P. Measurements on MIMO-FRET nano-networks based on Alexa Fluor dyes. IEEE Transactions on Nanotechnology. 2015;**14**:531-539. DOI: 10.1109/TNANO.2015.2415201

[13] Shivkumar MA, Inamdar (Doddamani) LS, Rabinal MHK, Mulimani BG, Advi Rao GM, Inamdar SR. FRET from CdSe/ZnS core-shell quantum dots to fluorescein 27 dye. Open Journal of Physical Chemistry. 2013;**3**:40-48. DOI: 10.4236/ojpc.2013.31006

[14] Chou KF, Dennis AM. Förster resonance energy transfer between quantum dot donors and quantum dot acceptors. Sensors. 2015;**15**:13288-13325. DOI: 10.3390/s150613288

[15] Rosenthal SJ, Chang JC, Kovtun J, McBride R, Tomlinson ID. Biocompatible quantum dots for biological applications. Chemistry and Biology. 2011;**18**:10-24. DOI: 10.1016/j.chembiol.2010.11.013

[16] Gulbahar B. Theoretical analysis of magneto-inductive THz wireless communications and power transfer with multi-layer graphene nano-coils. IEEE Transactions on Molecular, Biological, and Multi-Scale Communications. 2017;**3**:60-70. DOI: 10.1109/TMBMC.2017.2655022

[17] Huang L et al. Ultra-low power sensor design for wireless body area networks: Challenges, potential solutions, and applications. JDCTA. 2009;**3**:136-148. DOI: 10.1.1.653.293

[18] Kanai H et al. Transcutaneous measurement and spectrum analysis of heart wall vibrations. IEEE Transactions on Ultrasonics, Ferroelectrics, and Frequency Control. 1996; **43**:791-810. DOI: 10.1109/58.535480

[19] Gulbahar B, Memisoglu G. Implementable medical device and intra-bone wireless communication system and methods. EP17172207.7, EPO Patent Application. May 2017

[20] Deisseroth K. Optogenetics. Nature Methods. 2011;**8**:26-29. DOI: 10.1038/nmeth.f.324

[21] Slater D, Ridenoure RW. Passive remote acoustic sensing in aerospace environments. In: AIAA SPACE 2015 Conference and Exposition; 2015. p. 4661. DOI: 10.2514/6.2015-4661

[22] Smal I, Loog M, Niessen W, Meijering E. Quantitative comparison of spot detection methods in fluorescence microscopy. IEEE Transactions on Medical Imaging. 2010;**29**:282-301. DOI: 10.1109/TMI.2009.2025127

[23] Shen H, Tauzin LJ, Baiyasi R, Wang W, Moringo N, Shuang B, Landes CF. Single particle tracking: From theory to biophysical applications. Chemical Reviews. 2017;**117**:7331-7376. DOI: 10.1021/acs.chemrev.6b00815

[24] Rajagopal S, Roberts RD, Lim SK. IEEE 802.15.7 visible light communication: Modulation schemes and dimming support. IEEE Communications Magazine. 2012;**50**(3):72-82. DOI: 10.1109/MCOM.2012.6163585

[25] Kim YT et al. Synthesis of a CdSe-graphene hybrid composed of CdSe quantum dot arrays directly grown on CVD-graphene and its ultrafast carrier dynamic. Nanoscale. 2013; **5**:1483-1488. DOI: 10.1039/c2nr33294a

Chapter 9

Increasing Light Absorption and Collection Using Engineered Structures

Yasir J. Noori, Christopher S. Woodhead and
Robert J. Young

Additional information is available at the end of the chapter

http://dx.doi.org/10.5772/intechopen.73079

Abstract

In recent years we have witnessed an explosion of interest in two dimensional (2D) materials, due to their unique physical properties. Excitement surrounds the promise of replacing conventional bulk photodetectors with devices based on 2D materials, allowing better integration, flexibility and potentially improving performance. However, the low inherent light absorption of 2D materials is an outstanding issue to be solved. In this chapter we review two independent approaches to tackling this problem, which have the potential to be combined to find a robust solution. The first approach involves patterning the substrate with a rod-type photonic crystal (PhC) cavity structure, which is shown to increase the light absorption into a 2D material flake coupled spatially to the cavity mode. Secondly, we review 2D–compatible solid immersion lenses (SILs) and their ability to increase both the optical magnification of the structures they encapsulate, and the longevity of the material. SILs have been shown to reduce the requirements for complex optics in the implementation of 2D materials in optoelectronic devices, and also in preserving the photodetector's optical performance over long periods of time. Finally, we show how by combining rod-type PhC cavities with SILs, we can improve the performance of 2D material-based photodetectors.

Keywords: rod-type photonic crystals, quality factor, solid immersion lenses, epoxy

1. Introduction

Two-dimensional transition metal dichalcogenides (TMDs) are a class of semiconducting materials, which can be exploited for a range of diverse applications [1]. Their ultra-thin dimensions and novel properties can provide unique advantages in fields such as energy [2, 3], computing [4–6] optoelectronics [7, 8], sensing [9, 10] and security [11, 12]. Current bottlenecks in commercialising devices formed from TMDs are their relatively poor quantum yield, low optical absorption and high rate of degradation in ambient conditions. The quantum yield of TMDs, namely the number of excitons generated divided by the number of photons impinged on the device, is typically between 0.01 and 6% [13], which is exceptionally low in comparison to other emitters such as III-V based structures that can have quantum yields reaching close to 100% [14, 15].

TMD monolayers such as MoS_2 have excellent absorption characteristics [16], and high photoresponsivity [9] relative to their dimensions. However, their extreme thinness of just 0.65 nm means that the absolute absorption is usually relatively low (typically less than 6% at 450 nm) [17] severely limiting the efficiency of TMDs in real implementations such as detectors and solar cells. This value can be increased by using multiple monolayers, [17] or by careful preparation of the incident light [18]; however such methods may not be practical for specific optoelectronic applications.

Multiple existing approaches have been developed to circumvent these limitations. One method involves the use of superacid treatment to obtain near a 100% photoluminescence quantum yield from these materials [13, 19]. Currently, reports of superacid treatment have all involved using Bis(trifluoromethane) sulfonimide as the superacid material for treating MoS_2. The photoluminescence intensity of MoS_2 monolayers have been shown to increase up to 190-fold after treatment [13]. While the explanation of the enhancement in light emission as a result of superacid treatment is still not very clear, recent studies demonstrated that enhancement could be caused by the transformation of trions into neutral excitons combined with a reduction in the density of mid-gap trap states for CVD grown monolayers [20].

Improved monolayer absorption can be achieved by coupling TMD monolayers to an engineered structure that confines plasmonic modes such as gold nanostructures [21, 22], nanoparticles [23] and nanorods [24, 25]. This phenomenon stems from the fact that plasmonic resonance couples to both excitation and emission fields, hence boosting the light-matter interaction at the nanoscale. For example, Johnson et al. have observed a giant enhancement in the luminescence intensity of tungsten diselenide (WSe_2) by coupling it to plasmonic structures, resulting in an increased light absorption [26]. They report achieving an enhancement factor of up to 200 using silver nanotriangle arrays coupled to the monolayers. Sobhani et al. in [23] have observed that by tuning plasmonic core-shell nanoparticles to the direct bandgap of monolayer MoS_2 and depositing them sparsely onto the monolayer's surface, the photocurrent achieved through the monolayer increases 3-fold, hence promising a model for more sensitive TMD based photodetectors. Other structures such as nanocubes [27] and bowtie antennae [28, 29] have also been utilized for enhancing light emission, resulting in up to 2000 fold increase in material's absorption.

Absorption of TMD monolayers can also be increased when they are coupled to photonic crystal (PhC) structures such as cavities, which can greatly aid in improving the efficiency of photodetectors [30, 31]. Like plasmonic structures, the strong localization of electromagnetic radiation in photonic cavities increases the light-matter interaction, leading to higher absorption efficiency. For example, by monolithically integrating graphene with a Fabry-Perot microcavity, the optical absorption can be enhanced by 26-fold, reaching absorption values greater than 60% [32]. Combining graphene with hole-type PhC cavities, has been shown to increase light reflectivity from graphene by 4.0x [33]. A photodetector can be realized by fabricating metal contacts above sheets, through which a circuit current can be amplified and measured when the device is illuminated.

An alternative method for enhancing monolayer light absorption using PhCs was proposed by Noori et al. in [31]. In this method, a rod-type PhC is used to achieve an increased light to monolayer coupling. This PhC can be combined with a high refractive index solid immersion lens (SIL) [34, 35], which is placed directly above the cavity. This has two effects: the first is to act as a reflective interface to redirect any leaking light back into the cavity, increasing the chance of light absorption by the monolayer. The second is to increase the coupling of light into/out of the cavity, increasing the probability that an applied photon gets into the cavity, which SILs are considered especially effective at [36, 37]. SILs are commonly made out of glass [36], which can pose a problem when integrating into delicate structures such as PhCs and 2D materials. Recent work forming SILs out of photopolymers, has shown progress in solving this issue [38–40].

For the rest of this chapter, we will focus on the concept of a rod-SIL photodetector. We will review the rod-based PhC structure and compare it to the more common method of coupling monolayers to uncapped hole-type PhCs. We will then look at photopolymer based solid immersion lenses, and their effectiveness at both coupling light into/out of a material, as well as protecting it from the ambient environment. Finally, we discuss how the combination of the two structures can provide an efficient optical package for use in photodetection.

2. Enhancing light absorption using photonic crystal cavities

2.1. Coupling 2D materials to photonic crystals

In 1987, Yablonovitch proposed the theoretical concept of using periodic optical structures to create photonic bandgap systems, now popularly known as photonic crystals [41]. A PhC structure comprises of a series of periodic changes in refractive index that creates a range of disallowed states for a photon, effectively forming a photonic bandgap. This effect is very similar to electronic bandgaps experienced by electrons in an atomic lattice, thus PhCs are sometimes describes as 'optical lattices'.

Experimental realization of manmade PhCs occurred a decade later [42, 43]. Ever since that time, PhCs became a potential platform for making integrated photonic components such as cavities, waveguides, mirrors and wavelength/polarization multiplexers [44, 45]. This allows the realization of technologies such as single photon sources, lasers, filters, interferometers, modulators and slow light waveguides.

Coupling 2D materials to photonic cavities can lead to substantial improvements in the material's absorption efficiency, opening the doors for their use in the development of ultrathin but highly efficient detectors. TMD monolayers have also been coupled to hole-type PhC cavities, resulting in lasing and enhancement in the spontaneous emission rate for light emitted from these 2D materials [46–50]. Hole-type PhC cavities can have high Q-factors due to their reduced mode losses in the vertical direction due to the refractive index contrast at the bridge-air interface. However, the hole-type PhCs have distinct disadvantages when being used with non-embedded emitters such as 2D materials, because the cavity mode is confined within the bridge structure; a region where the TMD monolayer cannot realistically be placed. This causes reduced coupling between the photo-absorber monolayer and the cavity mode's maximum, leading to a reduction in light-matter interactions. Secondly, the large dielectric-to-air volume ratio that exists in the structures of hole-type PhCs results in undesirable light absorption by the dielectric material. This becomes critical when a high absorption material, such as GaAs, is used with an operating wavelength that lies in the visible regime. However, a lot of these issues can be solved by changing the PhC structure from a hole-type to a rod-type [31].

2.2. Photonic crystal cavity design

PhC structures are commonly formed from a hexagonal array structure, rather than a square array because, they are easier to fabricate in practice [51]. For rod-type cavities, square lattice arrays exhibit a transverse magnetic (TM) bandgap with Q-factors that may exceed 1000. However, confining light in the visible wavelength range using the square lattice PhC requires the structure to have feature sizes that may be difficult to fabricate using conventional lithography techniques. For example, assume a square lattice PhC is designed for the optical bandgap of monolayer molybdenum disulfide (MoS_2), i.e. approximately 660 nm [52]; Villeneuve et al. in [51] showed that for a rod-type PhC lattice with material index $n = 3.4$, $a = 1$ and $r = 0.2a$, a H1 cavity would have its cavity mode at $\lambda = 2.56a$. In other words, a PhC cavity with a mode at $\lambda = 660\ nm$, requires its lattice constant to be, $a = 258\ nm$ and the radius to be, $r = 52\ nm$, with rod heights of at least 660 nm. Designing rods with such dimensions is challenging and requires extreme controllability of the fabrication process. Furthermore, small fabrication uncertainties that change the rods' radius, the lattice constant, sidewall roughness and/or their vertical height can shift the cavity mode and change its Q-factor. An H1 hexagonal lattice PhC cavity of similar dimensions has its cavity mode at a much smaller wavelength. Hence, a cavity designed to have a mode wavelength at 660 nm will have larger dimensions, making it less dependent on fabrication limitations. The cavity that will be discussed in this chapter can achieve this goal using $a = 595\ nm$ and $r = 95\ nm$, which are easier to achieve in practice.

The PhC studied here consists of a hexagonal lattice of rods surrounded by air for, optimum index contrast. Since MoS_2 has an absorption wavelength at 660 nm, due to its direct bandgap, rods made from common substrates such as silicon or GaAs can have a refractive index that exceeds 3.5 at this wavelength. To make the photonic cavity, a rod is omitted from the PhC lattice allowing a photonic state to be created within the lattice's photonic bandgap. The confined wavelength of the cavity is also known as the cavity mode. In previous work by Kay et al. [53], 2D materials have been shown to dip when transferred onto hollowed regions such as etched trenches on a substrate. Through exploiting the flexibility of TMD monolayers, spatial coupling between the cavity and the material is possible by suspending the monolayer over the rods; **Figure 1** illustrates this concept. Having a sufficient lattice separation, the cavity region, can allow the suspended flake to sag within the cavity such that the flake's topological minimum spatially matches the cavity mode's antinode. Optimized alignment in both position and energy optimizes absorption by the 2D material. The PhC hexagonal lattice is surrounded by air for optimum index contrast. MoS_2 is the material considered here for this work because of its large neutral and charged exciton binding energies, making it a great candidate for room temperature photodetection. Nevertheless, the PhC design is universal, as the normalized parameters may be scaled and adjusted to match any light absorptive TMD. It consists of rods with a lattice constant, $a = 1$, and a radius, $r = 0.165a$, where the rods are made by etching the silicon substrate.

2.3. Cavity performance

2D PWE simulations of the PhC structure (**Figure 2a**), have shown that a bandgap exists in the frequency range $1.05a < \lambda < 1.17a$. Once the photonic band diagram was simulated, a H1 cavity was formed in the centre of the array. Subsequently, FDTD is commonly used to test the cavity's performance. 2D simulations can confirm the presence of the cavity mode and used to analyze its central wavelength. Performing 2D simulations at this point can usually save a lot of computation time and power that may be spent with exhaustive 3D simulations. Q-factor measurements from 2D simulations typically show astronomical values since leakage in the vertical direction is not included.

3D FDTD simulations of the cavity mode are shown in **Figure 2b** and **c**. Initially, an E_x polarised source was placed at the center of the cavity, with a central wavelength λ of $1.1a$ and a full-width at half maximum (FWHM) of $d\lambda = 0.1a$. Using Meep, a visual image of the light propagation can be produced. After allowing enough time to pass following initialization of the dipole, to allow edge states and propagating modes to leak, the cavity standing mode's field was observed on its own, as shown in **Figure 3a**. The cavity is later simulated with an E_y

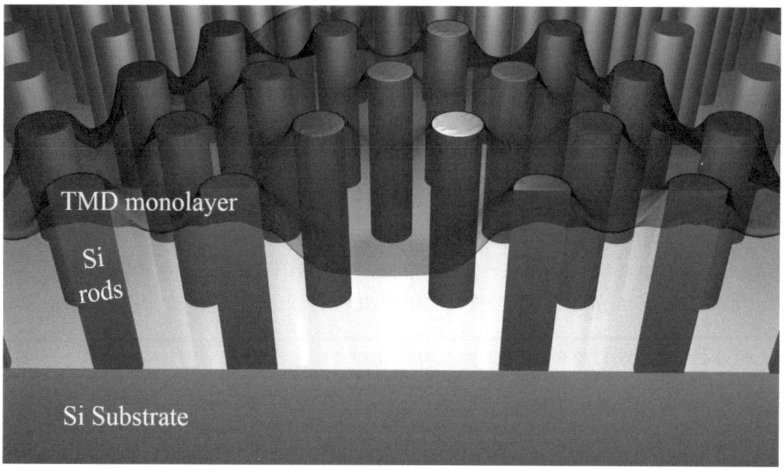

Figure 1. Cross-sectional illustration of a silicon rod PhC cavity with a monolayer transferred on top. Declining of the monolayer within the cavity region can increase coupling between excitons and the cavity modes.

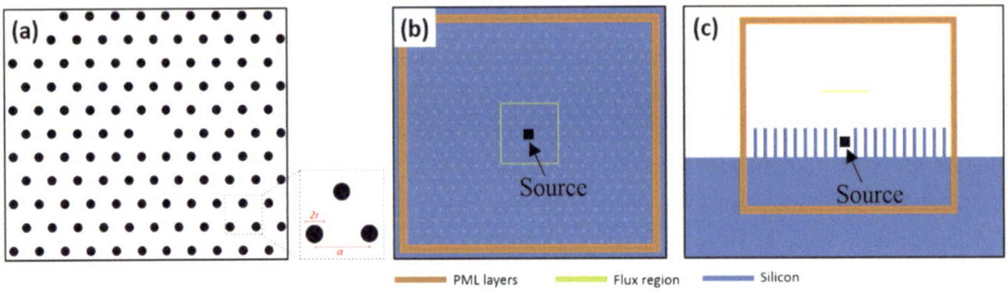

Figure 2. (a) An x-y cross-section of the simulated PhC cavity structure. x-y (b) and y-z (c) screenshots of the simulated structure from Lumerical FDTD package showing the PM layers and flux region.

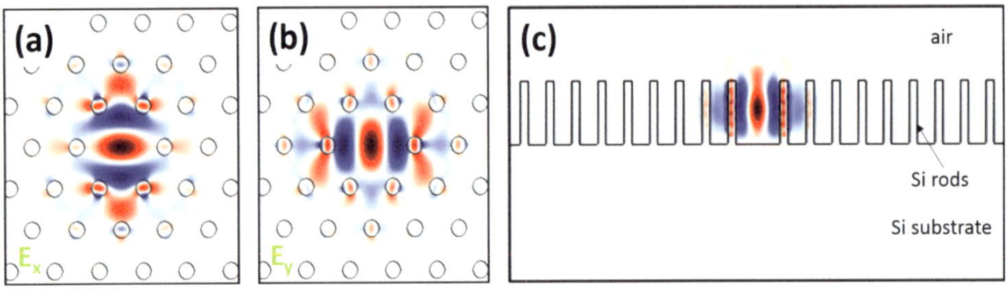

Figure 3. 3D FDTD simulation of the photonic cavity mode showing time slice images of the confined E_x (a) and E_y (b) components. (c) Cross section of the E_y field within the microcavity. Red (blue) represents positive (negative) components of the electric field.

polarised dipole, which showed an electric field distribution for the cavity mode, as shown in **Figure 3b**. Using the harmonic inversion (harminv) tool embedded in Meep, the cavity mode was found to have a wavelength $\lambda = 1.12a$ with a Q-factor of 300. On the other hand, **Figure 3c** shows the field distribution for a cross-sectional slice along the x-direction through the cavity. Unsurprisingly, the PhC rod's height also influences the mode's confinement. If the rods are too short, the cavity mode's shape can extend into the air above the PhC. This results in a reduction in the spatial interface between the mode and the rods, lowering the light collection ratio. Conversely, if the rods are too long, higher order and propagating modes can form within the cavity structure. Furthermore, etching high aspect ratio rods will always require complex dry etching techniques to achieve the desired degree of anisotropy. It has been revealed that maximum gap size for a square-lattice rod-type PhC is achieved for rod heights of approximately $2.3a$, corresponding to approximately two cavity modes wavelengths [51].

The Q-factor of a cavity can be defined as the ratio of the energy stored in the resonating cavity to the energy dissipated per cycle. For a cavity with a Q-factor of 300, such as the rod-type PhC cavity discussed here, a resonating photonic mode inside a cavity is expected to oscillate 300 times before it leaks half of its energy outside the cavity. For a MoS_2 monolayer with an absorption coefficient of 0.05, coupled to a rod-type PhC cavity the absorption of the monolayer can be enhanced to almost unity [52]. This can be calculated using the following equation:

$$\alpha_c = 1 - (\alpha_m)^{Q_c} \quad (1)$$

where α_c is the absorption of a monolayer that is coupled to a cavity, α_m is the absorption of a monolayer that is not coupled to a cavity and Q_c is the cavity's Q-factor. Note that in this approximation, absorption due to the dielectric host material (in this case considered silicon) was not taken into account. It is expected that absorption due to the material can reduce the cavity's Q, however this becomes less critical for cavities designed to have low-mid Q-factors such as the ones shown here where light is allowed to leak into the cavity to enhance absorption for optimized photodetection.

A series of simulations were carried out with a source inside a PhC cavity containing rods with different radii. The aim is primarily to investigate the robustness of this design to fabrication imperfections that are likely to occur due to the relatively small structure. Causes of shifts in the radius of the fabricated rods could be due to the inaccurate selection of exposure dosage in the electron-beam (e-beam) writing process. Other reasons could be due to non-anisotropic sidewall profile development of the exposed regions of the resist, due to temperature variations of the developing solution and/or the samples themselves. Hence, the radius was varied between $0.155a$ and $0.170a$. **Figure 2b** and **c** show screenshots taken directly from Lumerical's user interface for the simulation cell. The mesh resolution used in this simulation is 20 elements per one lattice constant, *a*, in every dimension. Corresponding to approximately one element per 30 nm in real units. To measure the cavity mode, a flux region was setup above the cavity, collecting light from the cavity mode radiated vertically upward. A maximum Q-factor of 341 was achieved for a PhC cavity having rods with a radius, $r = 0.161a$, where the cavity mode's central wavelength is $\lambda = 1.104a$, as shown in **Figure 4**. As the radius deviates from $r = 0.161a$, the collected power flux starts to decrease, which results from a decrease in the Q-factor, recording a minimum of 155 at $\lambda = 1.138a$ for $r = 0.170a$. It is clear from **Figure 4** to note that reducing the radius of the rods tends to blue-shift the cavity mode's energy, hence reducing its wavelength. This is expected, as the relationship between the PhC's cell radius (hole-type or rod-type) and the cavity mode's wavelength was anticipated through previous studies [54]. **Figure 4** plots a comparison between the obtained fluxes for two cases. The first case is when a flake is exfoliated directly onto a flat substrate. The second case is when the flake is transferred on top of the PhC structure, dipping to half the height of the rods where the cavity's field maximum exists for rods with $r = 0.161a$. Later, the effect of having the source away from the cavity's centre will be discussed.

Throughout the scope of this work, the simulations were made with the assumption that the flake is dipped to about half the height of the rods. This is expected to be the ideal case since the flake will be coupled to the field's antinode, creating maximal spatial coupling to the cavity. Performing simulations to anticipate the amount of dipping of a monolayer over a typical structure depends on many factors. These include the monolayer material, the rods' surface roughness, temperature, etc. These typically require setting up finite element simulations while taking into account the van der Waals, Coulombic and gravitational forces, which is very exhaustive. Hence, a series of simulations were run to investigate the effect of dipping of the monolayer emitter on the monolayer-cavity coupling. These simulations are done

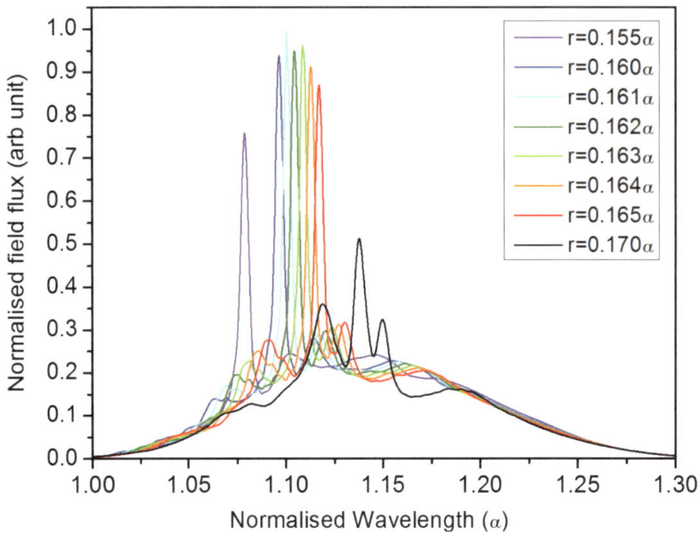

Figure 4. The effect of changing the rods' radii r from 0.155a to 0.170a on the cavity mode, showing a maximum Q for r = 0.161a and an increase in the cavity's central wavelength when r is increased, and vice versa.

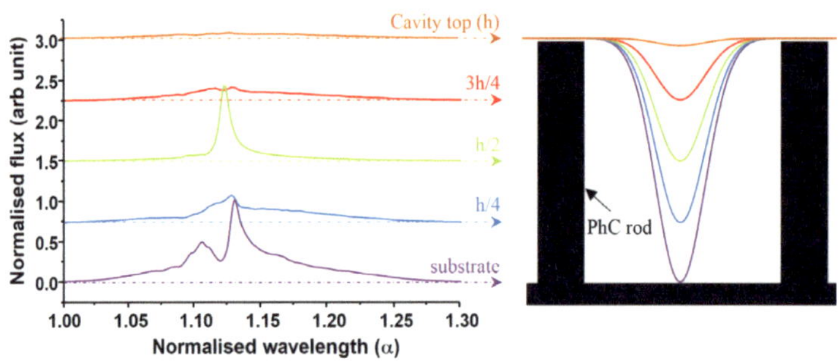

Figure 5. Comparison of the flux spectra for emission from the cavity as the dipping of the material is varied along the z-axis. The drawing on the right illustrates the degrees of bowing of the 2D material from the surface of the structure into the cavity for each of the cases shown on the left.

by placing the dipole at different heights along the vertical direction inside the cavity. Even though the source position in the vertical direction was changed along the height of the cavity, it is important to mention that the flux region height above the photonic structure was not changed. The cavity spectra for the different heights are shown in **Figure 5**. It is expected that maximum absorption of the monolayer should be achieved when the source is spatially aligned with the cavity mode's electric field maximum. This exists at the center of the cavity, one wavelength above the substrate, as was previously shown in **Figure 3c**. The intensity of the cavity mode was found to decrease as the source is moved toward the top of the rods due to reduced spatial coupling between the dipole source and the cavity mode's antinode, which exists in the center of the cavity.

http://dx.doi.org/10.5772/intechopen.73079

Compared to H1 cavities, L3 cavities are usually favored for better coupling to PhC waveguides and tend to have lower confinement quality factors of roughly 150. In L3 cavities, the Q-factor is expected to reduce due to increased coupling of the cavity mode to radiative modes.

Having an engineered structure that can improve absorption of monolayers is essential, in the next section we will describe how a different structure such as a micro-lens can help increase the light collection efficiency to these monolayers/cavity systems.

3. Enhancing light collection using photopolymer lenses

A solid immersion lens (SIL) is an optical element with a high refractive index that can be placed on the surface of a semiconductor to increase the optical extraction efficiency of a surface/subsurface emitter. SILs are typically formed from high index glass [55] and are placed directly on the surface of an optical structure to increase the light being coupled into and/or out from it. SILs can also be formed from photopolymers such as UV-curable epoxy [40], which gives the SIL unique properties such as easy tunability and mounting. The rest of this section will focus on these types of SILs.

Two different geometries of SIL have been predominantly studied; those with a hemispherical shape (h-SILs) and those with a Weierstraß shape (s-SILs) [55], as shown **Figure 6**. S-SILs have a higher magnification than h-SILs, scaling as the refractive index of the SIL squared as opposed to a direct linear relationship for a h-SIL. In subsurface emitters such as quantum dots (QD's), s-SILs have a higher input/output coupling efficiency over h-SILs, due to them being able to refract a higher number of rays at the SIL-air boundary, thus collecting/delivering a greater solid angle of light (θ_S) to and from a device. In the case of a surface emitter such as a 2D material this SIL-air boundary refraction in the s-SIL is essential to increase the input/output coupling, as there is no contribution to light coupling from the substrate-SIL boundary.

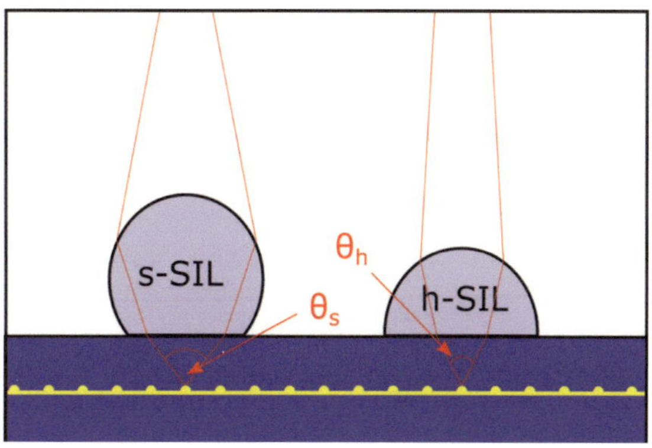

Figure 6. Diagram showing the difference in shape between a s-SIL and h-SIL on a sample with embedded emitters (quantum dots) in the above case (yellow features). The dashed lines highlight how a larger angle of emitted light is collected in the s-SIL relative to the h-SIL.

3.1. SIL fabrication

The process of creating an epoxy SIL onto a 2D material is described in this chapter. The sample containing the 2D material is immersed into a glycerol bath, which provides the liquid phase medium needed to enable the formation of droplets with high contact angles, such as the SILs those shown in **Figure 6**. This arises due to a modification of the surface tension experienced by the droplet, and can be explained by considering the Young equation given in (2) [56] and illustrated in **Figure 7**.

$$\cos\theta_y = \frac{\gamma_{sf} - \gamma_{sl}}{\gamma_{lf}} \quad (2)$$

where γ_{sf}, γ_{sl} and γ_{lf} are the solid-filler, solid–liquid and liquid-filler surface tensions, respectively. When the filler solution is air, γ_{sf} is greater than both γ_{sl}, and γ_{lf}. This makes $\cos(\theta y) > 0$, resulting in a small equilibrium contact angle (θy). However, when a filler solution such as glycerol is used, γ_{sf} reduces dramatically, allowing $\cos(\theta y) < 0$; this allows droplets with a contact angle over 90 degrees to form. Glycerol is an ideal filler solution due to it being relatively inert with respect to the epoxy. Other filler solutions such as water are less ideal as they are known to be absorbed by the epoxy, leading to a significant reduction in the SIL's transparency [40].

The dispensed UV-curable polymer can be tuned, using one of two methods. The first and most simple is to exploit the polymer's relatively strong attraction to the substrate to create a larger than required SIL with the correct sized base. The SIL can then be tuned by withdrawing epoxy from the center of the dispensed droplet, allowing the volume to reduce whilst the base's dimensions stay relatively constant. Another method of tuning the shape involves applying a bias between the needle and sample, and using the principle of electro-wetting to change the droplets shape [38–40]. Once the desired shape is obtained the SIL can be cured by exposure to UV light, permanently fixing its location, and shape. The sample can then be removed from the glycerol bath and washed with deionized water to remove any remaining glycerol. Further details on the fabrication process of an epoxy SIL, can be found from [39].

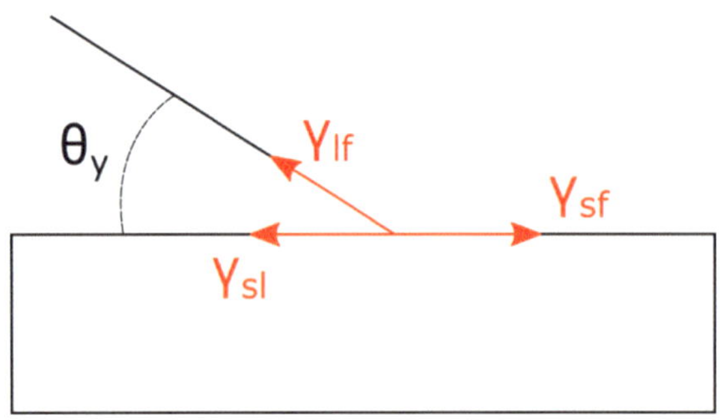

Figure 7. Diagram illustrating Young's equilibrium contact angle (θ y), due to the balancing of the solid–liquid (γ sl), liquid-filler (γ lf) and solid-filler (γ sf) surface tensions.

3.2. Enhanced light coupling

Figure 8a shows how a SIL enhances the coupling of light out of a TMD based device. The SIL on the RHS of the blue dashed line refracts the light rays produced by the TMD at the SIL-air boundary closer to the normal leading to more rays entering the lens (shown as a double ended red arrow).

The increase in the coupling of light can be used to increase the light output of a TMD by increasing the solid angle of light that can be observed rather than lost to the environment. **Figure 9a** shows the measured photoluminescence (PL) spectra for a monolayer at 10 μW of excitation power. Comparing the integrated intensity of the flake both before and after the application of a SIL, an increase in PL intensity of 4.0× is observed. This enhancement arises from the SIL refracting light at the SIL-air boundary. A SIL with the dimensions shown in **Figure 8b** should increase the solid angle of light emitted vertically by 1.33×, the SiO_2 layer underneath it will also reflect light back, creating a virtual source which will, in turn be enhanced by the SIL. Calculating these values, we find that the theoretical solid angle of the reflected light would be increased by 3.15×, which when scaled to take account of the percentage of light that would be reflected and added to the vertical emission we get a total enhancement of 2.0 × .

The theoretical enhancement value calculated is only half the power of the experimental results, however, this only considers the coupling of light out of the SIL, and does not consider what happens to the excitation source entering the SIL. A beam of light travelling in

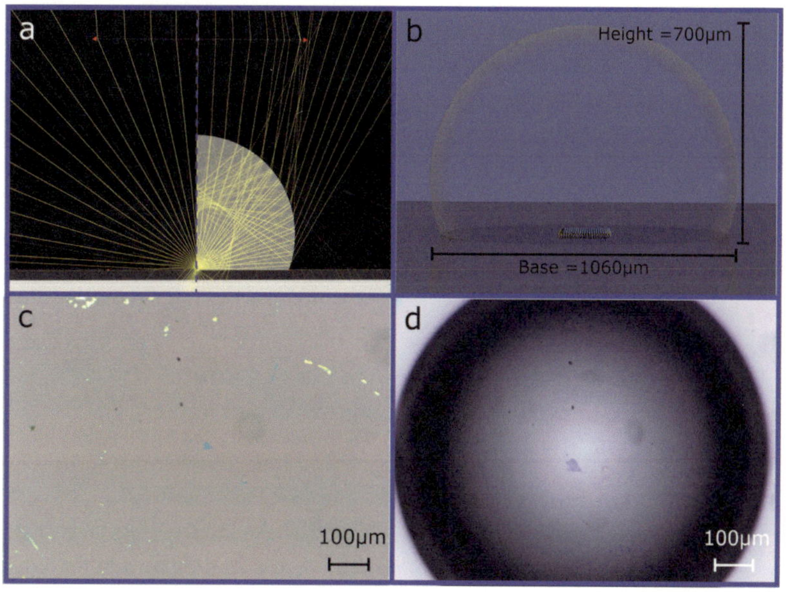

Figure 8. (a) A ray trace simulation demonstrating how the SIL increases the coupling of light, (b): A render of the SIL in c and d from the side with dimensions, (c and d): Microscope images showing an isolated TMD flake both before (c), and after (d) being magnified by a SIL.

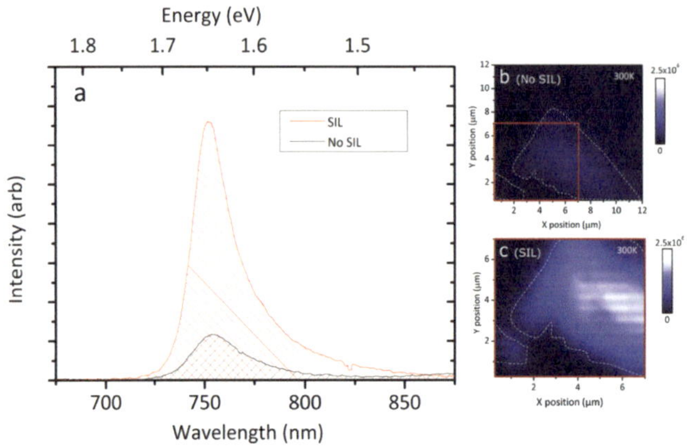

Figure 9. (a) Graph showing the room temperature photoluminescence of a WSe_2 monolayer both with and without a SIL, demonstrating the influence of the SIL in improving light collection efficiency. (b and c) Photoluminescence map of the same flake as (a) both without a SIL (b) and with a SIL (c), the intensity scale in these maps are identical, and the orange square in b shows the area in c that has been magnified by the SIL.

air is normally diffraction limited in terms of its size, and can be described by the Rayleigh criterion. When considering a beam being focused through a lens onto a surface we can use the following equation to describe the half-width at half-maximum (FWHM) of the resultant airy pattern:

$$FWHM = \frac{0.52\lambda}{n\,NA_{obj}} \quad (3)$$

where n is refractive index of the medium above the TMDC, NA_{obj} is the numerical aperture of the μPL system and λ is the excitation wavelength [57]. For a 532 nm laser travelling through the air we get a FWHM of 0.42 nm; in contrast a SIL with refractive index of 1.56 (such as the one demonstrated in **Figure 8**) has a FWHM of 2.7 nm. Assuming light is not lost due to scattering/reflection from the SIL then the incident optical power is unchanged, meaning that the power density of the laser spot increases due to the same optical power being focused into a smaller area. This change in power density for the aforementioned SIL translates to a 2.4× increase in light per unit area. Unfortunately, this will not lead to a 2.4× increase in light output of the monolayer WSe_2, as the quantum efficiency of WSe_2 is typically low [13, 58, 59]. However, the increased excitation will lead to an increase in PL intensity (providing the excitonic ground state does not become saturated), and may explain the difference observed between theory and experimental observations.

Figure 8c and **d** show a 2D flake before and after being magnified through the application of a SIL. This magnification increase arises from the SIL creating an optical lever effect [60] (i.e. moving the focal position laterally across the SIL produces a smaller lateral movement under the SIL). The magnification allows maps to be made with a higher number of measurements per unit area. This can be easily observed from **Figure 9b** and **c** which show PL maps of the emission of a flake without and with a SIL, respectively. Note that both maps have the same number of pixels despite **Figure 9c** showing a map of a smaller area. This increased resolution is especially important, when pushing the limits of micro-photoluminescence, and can have many optoelectronic applications. In the example of a photodetector, the increased magnification can enable the size of the detector to be reduced, whilst maintaining the same collection area, giving potential reductions in response times and jitter.

The magnification increase from **Figure 8c–d** was found to be 1.8×, indicating that the SIL has a shape between that of a hemisphere (linear dependence with n giving 1.56×) and an ideal super-sphere (quadratic dependence with n giving 2.43×). This result shows that SILs in between the h and s-SIL geometries (shown in **Figure 6**), can give optical properties that are a combination of the two, with the studied SIL showing a greater magnification than an h-SIL without introducing strong chromatic aberrations.

4. Improved device longevity

We have up to this point considered the properties of an epoxy SIL in terms of its optical performance, which although considerable is limited by the epoxy's refractive index. SILs formed from materials like glass can have higher refractive indices, and thus can cause better optical enhancement, and indeed are well suited to coupling light into/out of embedded structures with a flat surface over them. However, due to being solid and having a large hardness, glass SILs are unsuitable for coupling light into/out of sensitive photo-absorbers/emitters like 2D materials, as their presence risks damage to the flake and is likely to leave an air gap that will significantly degrade its optical performance. Epoxy formed SILs do not suffer from this problem as the SIL can be formed over the sensitive structure as a liquid, then hardened into a solid to fully encapsulate the emitter. This hardening process causes no damage to the emitter and fully seals the emitter away from any physical or chemical harm, giving the epoxy SILs a significant advantage over other types of SILs.

Encapsulation can be essential for materials that have finite lifetimes in an ambient environment. The photoluminescence of emitters such as colloidal quantum dots (CQD's) [61, 62] and monolayer TMD's are strongly suppressed when exposed to oxidants and organic contaminants present in the air [63–65]. In the case of TMD's this strong suppression of PL is accompanied by a large change in the flake's structure and morphology especially at grain boundaries [66]. Strategies such as encapsulation with hexagonal boron nitride (h-BN) have been shown to be effective at preventing this oxidization [67] however, this form of encapsulation adds complexity and cost to the potential manufacturing processes. Photopolymers such as UV cured epoxies are commonly used in the mass-manufacture of products [68] and should also seal the TMD from oxygen. To test this hypothesis, a monolayer of WSe_2 was encapsulated within a SIL and its appearance and PL intensity measured over several months in ambient conditions (a time period known to cause a reduction in PL [69]). The results of this experiment are shown in **Figure 10**, the total integrated PL of the peak was found to stay constant within error over 5 months. In addition, the inset microscope images show that there are no visible changes in the monolayers appearance, suggesting that the SIL encapsulation helps to prolong the emissive lifetime of a monolayer in ambient conditions.

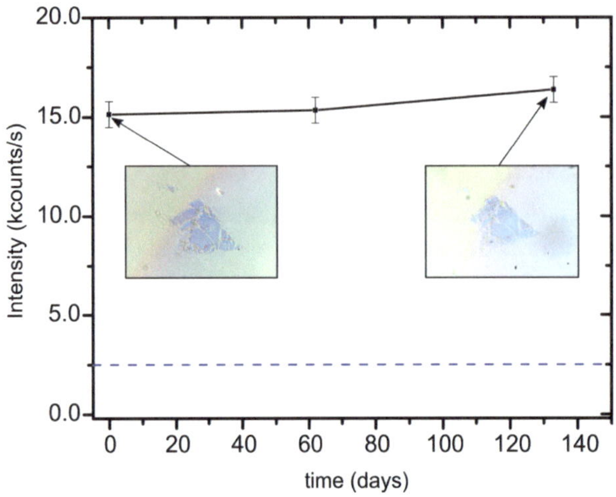

Figure 10. Graph showing the change in photoluminescence intensity with time, for a flake of WSe_2 immersed within an epoxy SIL (the dashed blue line shows the peak intensity before application of the SIL).

The enhanced light coupling provided by a SIL combined with the tunability and protective nature of the liquid epoxy that forms makes it an efficient, cheap and scalable solution for the capping of 2D TMD based optoelectronics. Their obvious use is to improve the emission of light emitting devices such as TMD based LED's [8]. Their high magnification and ability to focus light into a structure as well as out of a structure make them valuable for photodetector applications. PhC cavities as we have previously explored are ideal for coupling light into a 2D material based detector; however, they require high NA objective lenses to focus incoming light to a tight enough spot that they can be effectively coupled into the crystal by the output coupler. Mounting a SIL on-top of the PhC, and centered on the coupler can allow much simpler optics to be used as the SIL is providing some of the required magnification.

5. Conclusions and outlook

2D materials have shown great promise for replacing conventional bulk semiconductors in optoelectronic applications such as photodetectors and light emitters [8, 70]. Unfortunately, their low absorption due to their inherent atomic thickness presents limitations in adapting the material for photodetection applications. Here we have reviewed a rod-type PhC cavity structure which increases the light absorption of a 2D material flake coupled spatially to a cavity mode. Coupling the flake to a cavity with a Q-factor of 300 has been shown to be able to increase the material absorption to almost unity which has the potential to eliminate the absorption limitation of 2D material based photodetectors.

Directing collimated light to such a small structure for photodetection applications requires complicated optics such as expensive, high-NA objective lenses. SILs formed using UV-cured epoxy have been shown to magnify a 2D material based structure, this can enable simpler coupling optics to be utilized. The higher refractive index of the SIL can also help reduce the diffraction limit of a focused spot, helping to couple more light into one area. Hence for optimum efficiency, our proposed scheme for a 2D material based photodetector involves the transfer of a flake on top of a rod-type PhC cavity, and spatial coupling of the monolayer and the cavity mode maximum via the flake's declining into the cavity structure. A UV-curable epoxy SIL is then dispensed on top of the cavity. Dispensing the SIL above the cavity provides three advantages. The first is to help collimate light into a tighter focal spot and magnify the 2D flake to simplify coupling optics. The second is providing an encapsulation of the 2D material to isolate it from the external environment, thereby increasing its longevity and improving its optical performance. The third advantage of dispensing a UV-cured epoxy SIL over the PhC cavity is enhancing the Q-factor of the cavity by reflecting vertically leaking light back into the cavity due to the refractive index contrast between the air and epoxy.

Recent work by Bie et al. [30] has shown that TMDs can be used to form a waveguide-integrated light source and photodetector based on a p-n junction. This novel scheme enables the photoresponsivity of the detector to be tuned and the scheme reversed for realizing a 2D material coupled to hole-PhC LED. The advantage of this scheme is that one device can perform two roles. Using this concept with rod-type PhC coupled to SILs can provide an increase in material's absorption for a photodetection mode as well as greatly increasing the light output of the LED mode. Opening new doors to niche applications and technologies.

Acknowledgements

The work was supported by the Royal Society through a University Research Fellowship (UF110555) held by Robert J. Young. The material is supported by the Air Force Office of Scientific Research under award number FA9550-16-1-0276. This work was also supported by grants from the Engineering and Physical Sciences Research Council in the UK (grant numbers Ep/K50421X/1 and EP/L01548X/1). We also acknowledge the support of The Wolfson Foundation, the J P Moulton Charitable Foundation and Garfield Weston Foundation in establishing the IsoLab facility. The authors would like to thank Dr. Yameng Cao, for his valuable knowledge on 2D materials and his preliminary data on placing them on rod-type photonic crystals. The authors would also like to acknowledge the efforts of Benjamin Astbury and James Fong in proofreading this chapter.

Author details

Yasir J. Noori*, Christopher S. Woodhead and Robert J. Young

*Address all correspondence to: yasir.noori@gmail.com

Department of Physics, Lancaster University, Lancaster, UK

References

[1] Ferrari AC, Bonaccorso F, Fal'ko V, et al. Science and technology roadmap for graphene, related two-dimensional crystals, and hybrid systems. Nanoscale. 2015;**7**(11):4598-4810. DOI: 10.1039/c4nr01600a

[2] Quesnel E, Roux F, Emieux F, et al. Graphene-based technologies for energy applications, challenges and perspectives. 2D Materials. 2015;**2**(3):030204. DOI: 10.1088/2053-1583/2/3/030204

[3] Furchi MM, Pospischil A, Libisch F, et al. Photovoltaic effect in an electrically tunable van der Waals heterojunction. Nano Letters. 2014;**14**(8):4785-4791. DOI: 10.1021/nl501962c

[4] Radisavljevic B, Radenovic A, Brivio J, et al. Single-layer MoS2 transistors. Nature Nanotechnology. 2011;**6**(3):147-150. DOI: 10.1038/nnano.2010.279

[5] Lee J, Mak KF, Shan J. Electrical control of the valley hall effect in bilayer MoS2 transistors. Nature Nanotechnology. 2016;**11**(5):421-425. DOI: 10.1038/nnano.2015.337

[6] Wachter S, Polyushkin DK, Bethge O, et al. A microprocessor based on a two-dimensional semiconductor. 2017;**8**:14948. DOI: 10.1038/ncomms14948

[7] Fiori G, Bonaccorso F, Iannaccone G, et al. Electronics based on two-dimensional materials. Nature Nanotechnology. 2014;**9**:768. DOI: 10.1038/nnano.2014.207

[8] Withers F, Del Pozo-Zamudio O, Mishchenko A, et al. Light-emitting diodes by band-structure engineering in van der Waals heterostructures. Nature Materials. 2015;**14**(3):301-306. DOI: 10.1038/nmat4205

[9] Lopez-Sanchez O, Lembke D, Kayci M, et al. Ultrasensitive photodetectors based on monolayer MoS2. Nature Nanotechnology. 2013;**8**:497. DOI: 10.1038/nnano.2013.100

[10] Sarkar D, Liu W, Xie X, et al. MoS2 field-effect transistor for next-generation label-free biosensors. ACS Nano. 2014;**8**(4):3992-4003. DOI: 10.1021/nn5009148

[11] Yameng C, Alexander JR, Abdullah A, et al. Optical identification using imperfections in 2D materials. 2017;**4**(4):045021. DOI: 10.1088/2053-1583/aa8b4d

[12] Alharbi A, Armstrong D, Alharbi S, et al. Physically unclonable cryptographic primitives by chemical vapor deposition of layered MoS2. ACS Nano. 2017. DOI: 10.1021/acsnano.7b07568

[13] Amani M, Lien D-H, Kiriya D, et al. Near-unity photoluminescence quantum yield in MoS2. Science. 2015;**350**(6264):1064. DOI: 10.1126/science.aad2114

[14] Schnitzer I, Yablonovitch E, Caneau C, et al. Ultrahigh spontaneous emission quantum efficiency, 99.7% internally and 72% externally, from AlGaAs/GaAs/AlGaAs double heterostructures. Applied Physics Letters. 1993;**62**(2):131-133. DOI: 10.1063/1.109348

[15] Gauck H, Gfroerer TH, Renn MJ, et al. External radiative quantum efficiency of 96% from a GaAs/GaInP heterostructure. Applied Physics A. 1997;**64**(2):143-147. DOI: 10.1007/s003390050455

[16] Tonndorf P, Schmidt R, Böttger P, et al. Photoluminescence emission and Raman response of monolayer MoS2, MoSe2, and WSe2. Optics Express. 2013;**21**(4):4908-4916. DOI: 10.1364/OE.21.004908

[17] Kang K, Xie S, Huang L, et al. High-mobility three-atom-thick semiconducting films with wafer-scale homogeneity. Nature. 2015;**520**:656. DOI: 10.1038/nature14417

[18] Ansari N, Moradi M. Optical absorption in air/monolayer MoS2/S (SSiO2 or Si) trilayer stacks at oblique incidence. Superlattices and Microstructures. 2017:**104**(Supplement C):104-111. DOI: 10.1016/j.spmi.2017.02.010

[19] Kim H, Lien D-H, Amani M, et al. Highly stable near-unity photoluminescence yield in monolayer MoS2 by fluoropolymer encapsulation and superacid treatment. ACS Nano. 2017;**11**(5):5179-5185. DOI: 10.1021/acsnano.7b02521

[20] Alharbi A, Zahl P, Shahrjerdi D. Material and device properties of superacid-treated monolayer molybdenum disulfide. Applied Physics Letters. 2017;**110**(3):033503. DOI: 10.1063/1.4974046

[21] Zeng S, Hu S, Xia J, et al. Graphene–MoS2 hybrid nanostructures enhanced surface plasmon resonance biosensors. Sensors and Actuators B: Chemical. 2015;**207**:801-810. DOI: 10.1016/j.snb.2014.10.124

[22] Wang Z, Dong Z, Gu Y, et al. Giant photoluminescence enhancement in tungsten-diselenide-gold plasmonic hybrid structures. Nature Communications. 2016;**7**. DOI: 10.1038/ncomms11283

[23] Sobhani A, Lauchner A, Najmaei S, et al. Enhancing the photocurrent and photoluminescence of single crystal monolayer MoS2 with resonant plasmonic nanoshells. Applied Physics Letters. 2014;**104**(3):031112. DOI: 10.1063/1.4862745

[24] Lee KCJ, Chen Y-H, Lin H-Y, et al. Plasmonic gold nanorods coverage influence on enhancement of the photoluminescence of two-dimensional MoS2 monolayer. Scientific Reports. 2015;**5**:16374. DOI: 10.1038/srep16374

[25] Kim J, Son H, Cho DJ, et al. Electrical control of optical plasmon resonance with graphene. Nano Letters. 2012;**12**(11):5598-5602. DOI: 10.1021/nl302656d

[26] Johnson AD, Cheng F, Tsai Y, et al. Giant enhancement of defect-bound exciton luminescence and suppression of band-edge luminescence in monolayer WSe2–Ag plasmonic hybrid structures. Nano Letters. 2017;**17**(7):4317-4322. DOI: 10.1021/acs.nanolett.7b01364

[27] Akselrod GM, Weidman MC, Li Y, et al. Efficient nanosecond photoluminescence from infrared PbS quantum dots coupled to plasmonic nanoantennas. ACS Photonics. 2016;**3**(10):1741-1746. DOI: 10.1021/acsphotonics.6b00357

[28] Duan H, Fernández-Domínguez AI, Bosman M, et al. Nanoplasmonics: Classical down to the nanometer scale. Nano Letters. 2012;**12**(3):1683-1689. DOI: 10.1021/nl3001309

[29] Emani NK, Chung T-F, Ni X, et al. Electrically tunable damping of plasmonic resonances with graphene. Nano Letters. 2012;**12**(10):5202-5206. DOI: 10.1021/nl302322t

[30] Bie Y-Q, Grosso G, Heuck M, et al. A MoTe2-based light-emitting diode and photodetector for silicon photonic integrated circuits. Nature Nanotechnology. 2017. DOI: 10.1038/nnano.2017.209

[31] Noori YJ, Cao Y, Roberts J, et al. Photonic crystals for enhanced light extraction from 2D materials. ACS Photonics. 2016;**3**(12):2515-2520. DOI: 10.1021/acsphotonics.6b00779

[32] Furchi M, Urich A, Pospischil A, et al. Microcavity-integrated graphene photodetector. Nano Letters. 2012;**12**(6):2773-2777. DOI: 10.1021/nl204512x

[33] Majumdar A, Kim J, Vuckovic J, et al. Electrical control of silicon photonic crystal cavity by graphene. Nano Letters. 2013;**13**(2):515-518. DOI: 10.1021/nl3039212

[34] Mansfield SM, Kino GS. Solid immersion microscope. Applied Physics Letters. 1990;**57**(24):2615-2616. DOI: 10.1063/1.103828

[35] Yoshita M, Sasaki T, Baba M, et al. Application of solid immersion lens to high-spatial resolution photoluminescence imaging of GaAs quantum wells at low temperatures. Applied Physics Letters. 1998;**73**(5):635-637. DOI: 10.1063/1.121931

[36] Vamivakas AN, Younger RD, Goldberg BB, et al. A case study for optics: The solid immersion microscope. American Journal of Physics. 2008;**76**(8):758-768. DOI: 10.1119/1.2908186

[37] Zwiller V, Björk G. Improved light extraction from emitters in high refractive index materials using solid immersion lenses. Journal of Applied Physics. 2002;**92**(2):660-665. DOI: 10.1063/1.1487913

[38] Jin X, Guerrero D, Klukas R, et al. Microlenses with tuned focal characteristics for optical wireless imaging. Applied Physics Letters. 2014;**105**(3):031102. DOI: 10.1063/1.4890967

[39] Woodhead CS, Roberts J, Noori YJ, et al. Increasing the light extraction and longevity of TMDC monolayers using liquid formed micro-lenses. 2D Materials. 2017;**4**(1):015032. DOI: 10.1088/2053-1583/4/1/015032

[40] Born B, Landry EL, Holzman JF. Electrodispensing of microspheroids for lateral refractive and reflective photonic elements. IEEE Photonics Journal. 2010;**2**(6):873-883. DOI: 10.1109/JPHOT.2010.2076800

[41] Yablonovitch E. Inhibited spontaneous emission in solid-state physics and electronics. Physical Review Letters. 1987;**58**(20):2059-2062. DOI: 10.1103/PhysRevLett.58.2059

[42] Astratov VN, Bogomolov VN, Kaplyanskii AA, et al. Optical spectroscopy of opal matrices with CdS embedded in its pores: Quantum confinement and photonic band gap effects. Il Nuovo Cimento D. 1995;**17**(11):1349-1354. DOI: 10.1007/BF02457208

[43] Krauss TF, Rue RMDL, Brand S. Two-dimensional photonic-bandgap structures operating at near-infrared wavelengths. Nature. 1996;**383**:699. DOI: 10.1038/383699a0

[44] Busch K, von Freymann G, Linden S, et al. Periodic nanostructures for photonics. Physics Reports. 2007;**444**(3):101-202. DOI: 10.1016/j.physrep.2007.02.011

[45] Hennessy K, Badolato A, Winger M, et al. Quantum nature of a strongly coupled single quantum dot-cavity system. Nature. 2007;**445**:896. DOI: 10.1038/nature05586

[46] Wu S, Buckley S, Schaibley JR, et al. Monolayer semiconductor nanocavity lasers with ultralow thresholds. Nature. 2015;**520**:69. DOI: 10.1038/nature14290

[47] Liu T, Qiu H, Yin T, et al. Enhanced light-matter interaction in atomically thin MoS2 coupled with 1D photonic crystal nanocavity. Optics Express. 2017;**25**(13):14691-14696. DOI: 10.1364/OE.25.014691

[48] Sanfeng W, Sonia B, Aaron MJ, et al. Control of two-dimensional excitonic light emission via photonic crystal. 2D Materials. 2014;**1**(1):011001

[49] Gan X, Gao Y, Mak KF, et al. Controlling the spontaneous emission rate of monolayer MoS2 in a photonic crystal nanocavity. Applied Physics Letters. 2013;**103**(18):181119. DOI: 10.1063/1.4826679

[50] Ye Y, Wong ZJ, Lu X, et al. Monolayer excitonic laser. Nature Photonics. 2015;**9**(11):733-737. DOI: 10.1038/nphoton.2015.197

[51] Villeneuve PR, Fan S, Joannopoulos JD. Microcavities in photonic crystals: Mode symmetry, tunability, and coupling efficiency. Physical Review B. 1996;**54**(11):7837-7842. DOI: 10.1103/PhysRevB.54.7837

[52] Paton KR, Coleman JN. Relating the optical absorption coefficient of nanosheet dispersions to the intrinsic monolayer absorption. Carbon. 2016;**107**(Supplement C):733-738. DOI: 10.1016/j.carbon.2016.06.043

[53] Kay ND, Robinson BJ, Fal'ko VI, et al. Electromechanical sensing of substrate charge hidden under atomic 2D crystals. Nano Letters. 2014;**14**(6):3400-3404. DOI: 10.1021/nl500922h

[54] Johnson SG, Fan S, Villeneuve PR, et al. Guided modes in photonic crystal slabs. Physical Review B. 1999;**60**(8):5751-5758. DOI: 10.1103/PhysRevB.60.5751

[55] Qian W, Ghislain LP, Elings VB. Imaging with solid immersion lenses, spatial resolution, and applications. Proceedings of the IEEE. 2000;**88**(9):1491-1498. DOI: 10.1109/5.883320

[56] Frieder M, Jean-Christophe B. Electrowetting: From basics to applications. Journal of Physics. Condensed Matter. 2005;**17**(28):R705. DOI: 10.1088/0953-8984/17/28/R01

[57] Moehl S, Zhao H, Don BD, et al. Solid immersion lens-enhanced nano-photoluminescence: Principle and applications. Journal of Applied Physics. 2003;**93**(10):6265-6272. DOI: 10.1063/1.1567035

[58] Mak KF, Shan J. Photonics and optoelectronics of 2D semiconductor transition metal dichalcogenides. Nature Photonics. 2016;**10**(4):216-226. DOI: 10.1038/nphoton.2015.282

[59] Mak KF, Lee C, Hone J, et al. Atomically thin MoS_2: A new direct-gap semiconductor. Physical Review Letters. 2010;**105**(13):136805

[60] Serrels KA, Ramsay E, Dalgarno PA, et al. Solid immersion lens applications for nanophotonic devices. Journal of Nanophotonics. 2008;**2**(1):021854-021829. DOI: 10.1117/1.3068652

[61] Shirasaki Y, Supran GJ, Bawendi MG, et al. Emergence of colloidal quantum-dot light-emitting technologies. Nature Photonics. 2013;**7**(1):13-23. DOI: 10.1038/nphoton.2012.328

[62] Igor LM, Uyeda HT, Ellen RG, et al. Quantum dot bioconjugates for imaging, labelling and sensing. Nature Materials. 2005;**4**(6):435. DOI: 10.1038/nmat1390

[63] Sang-Yul P, Hyo-Sun K, Jeseung Y, et al. Long-term stability of CdSe/CdZnS quantum dot encapsulated in a multi-lamellar microcapsule. Nanotechnology. 2015;**26**(27):275602. DOI: 10.1088/0957-4484/26/27/275602

[64] Roberto CL, Rafik A, Santosh KC, et al. Intrinsic air stability mechanisms of two-dimensional transition metal dichalcogenide surfaces: Basal versus edge oxidation. 2D Materials. 2017;**4**(2):025050. DOI: 10.1088/2053-1583/aa636c

[65] Gao J, Li B, Tan J, et al. Aging of transition metal dichalcogenide monolayers. ACS Nano. 2016;**10**(2):2628-2635. DOI: 10.1021/acsnano.5b07677

[66] Rong Y, He K, Pacios M, et al. Controlled preferential oxidation of grain boundaries in monolayer tungsten disulfide for direct optical imaging. ACS Nano. 2015;**9**(4):3695-3703. DOI: 10.1021/acsnano.5b00852

[67] Ahn S, Kim G, Nayak PK, et al. Prevention of transition metal dichalcogenide photodegradation by encapsulation with h-BN layers. ACS Nano. 2016;**10**(9):8973-8979. DOI: 10.1021/acsnano.6b05042

[68] Crivello JV, Reichmanis E. Photopolymer materials and processes for advanced technologies. Chemistry of Materials. 2014;**26**(1):533-548. DOI: 10.1021/cm402262g

[69] He Z, Wang X, Xu W, et al. Revealing defect-state photoluminescence in monolayer WS2 by cryogenic laser processing. ACS Nano. 2016;**10**(6):5847-5855. DOI: 10.1021/acsnano.6b00714

[70] Koppens FHL, Mueller T, Avouris P, et al. Photodetectors based on graphene, other two-dimensional materials and hybrid systems. Nature Nanotechnology. 2014;**9**:780. DOI: 10.1038/nnano.2014.215

Chapter 10

Ultrafast Dynamics in Topological Insulators

Phuoc Huu Le and Chih-Wei Luo

Additional information is available at the end of the chapter

http://dx.doi.org/10.5772/intechopen.74918

Abstract

Ultrafast dynamics of carriers and phonons in topological insulator Bi_2Se_3, $Cu_xBi_2Se_3$ (x = 0, 0.1, 0.125) single crystals were studied by time-resolved pump-probe spectroscopy. The coherent optical phonon (A_{1g}^{1}) is found via the damped oscillation in the transient reflectivity changes ($\Delta R/R$) for $Cu_xBi_2Se_3$. The observed red shift of A_{1g}^{1} phonon frequency suggests the intercalation of Cu atoms between a pair of the quintuple layers of Bi_2Se_3 crystals. Moreover, the relaxation processes of Dirac fermion near the Dirac point of Bi_2Se_3 are studied by optical pump and mid-infrared probe spectroscopy through analyzing the negative peak of the $\Delta R/R$. The Dirac fermion-phonon coupling strength was found in the range of 0.08–0.19 and the strength is reduced as it gets closer to the Dirac point. The ultrafast dynamics and fundamental parameters revealed by time-resolved pump-probe spectroscopy are important for designing the optoelectronics in the mid-IR and THz ranges.

Keywords: topological insulators, ultrafast dynamics, pump-probe spectroscopy, Dirac fermion

1. Introduction

Recently, topological insulators (TIs) [1–8] and two-dimensional (2D) materials such as graphene [9], MoS_2, WS_2, and $MoSe_2$ [10] are of great interests because of their unique physical properties and applications. These materials have a band structure that is linearly dispersed with respect to momentum, in which the transportation of electrons in these materials is essentially governed by Dirac's (relativistic) equation with zero rest mass and an effective "speed of light"—$c^* \approx 10^6$ m/s [9]. In TIs, a novel electronic state called the topological surface state (TSS) has been predicted and observed [1–8]. Unlike the trivial insulator, TIs have a spin degenerate and fully gapped bulk state but exhibit a spin polarized and gapless electronic state on the surface [8]. This metallic surface state has a linear energy-momentum dispersion relation in the low-energy region, which is known as a Dirac cone. Unlike the Dirac cone of graphene, the Dirac cone of a TI is protected by the time-reversal symmetry. This robust TSS can survive under time-reversal invariant perturbations, such as surface pollution, crystalline defects, and distortions of the surface [6]. Additionally, because of the fully spin-polarized characteristics of the surface state, TIs have a high potential for the development of spintronic devices and quantum computation [6, 11].

The optoelectronic properties of TIs are important subjects for the development of optoelectronic devices. Therefore, the issues associated with electron–phonon interaction, carrier lifetime, carrier dynamics, energy loss rate, and low-energy electronic responses are very important for optimizing device performance. These ultrafast dynamic properties of the materials can be resolved by pump-probe spectroscopy. This chapter provides a brief introduction to the materials, time-resolved pump-probe spectroscopy, and some ultrafast dynamic properties of Bi-based topological insulators.

2. Bismuth-based topological insulators

Bismuth chalcogenide compounds (Bi_2Ch_3, Ch = Se, Te) have been extensively investigated in material science and condensed-matter physics because of their intriguing properties regarding thermoelectricity [12–14] and three-dimensional TIs [15–18]. Bi_2Ch_3 is a narrow bandgap semiconductor with a rhombohedral crystal structure belonging to the $D_{3d}^{5}(R\bar{3}m)$ space group. The Bi_2Ch_3 crystal structure is constructed from repeated quintuple layers (QLs) arranged along the *c*-axis. The unit lattice cell of a Bi_2Ch_3 crystal is composed of three QLs. Each QL is stacked in a sequence of atomic layers Ch(1)-Bi-Ch(2)-Bi-Ch(1) and is weakly bonded to the next QL via Van der Waals interaction. The crystal structures of Bi_2Se_3 and Bi_2Te_3 are shown in **Figure 1**. For convenience, these crystal structures are also described by a hexagonal lattice, where the *a*-axis and *c*-axis lattice constants of Bi_2Se_3 (Bi_2Te_3) are 4.138 Å (4.384 Å) and 28.64 Å (30.487 Å), respectively [19].

In 2009, Zhang et al. predicted that the Bi_2Ch_3 crystal is a strong TI [15]. A calculation of the electronic structure with spin-orbit coupling in the Bi_2Se_3 crystal has also been performed [15]. By tuning the spin-orbit coupling in the system, band inversion occurred around the Γ point. As these two levels, which are closest to the Fermi energy, have opposite parity, the inversion between them drives the system into a TI phase [15]. **Figure 2** shows the calculated energy and momentum dependence of the local density of states (LDOS) for Sb_2Se_3, Sb_2Te_3, Bi_2Se_3, and Bi_2Te_3. All of these materials have the same rhombohedral crystal structure with the space group $D_{3d}^{5}(R\bar{3}m)$. Zhang et al. predicted that Bi_2Se_3, Bi_2Te_3, and Sb_2Se_3 are candidates for a TI, whereas Sb_2Se_3 is not because the spin-orbit coupling effect of Sb_2Se_3 is not strong enough to induce band inversion [15]. Following this prediction, Xia et al. [20] and Hsieh et al. [4] investigated the existence of the TSS in Bi_2Se_3, Bi_2Te_3, and Sb_2Se_3 through angle-resolved photoemission spectroscopy (ARPES).

Figure 3(a) and **(b)** shows the ARPES results of the surface electronic structure on a Bi_2Se_3 (111) surface [20]. Around the $\bar{\Gamma}$ point, the clear V-shaped band is observed to approach the

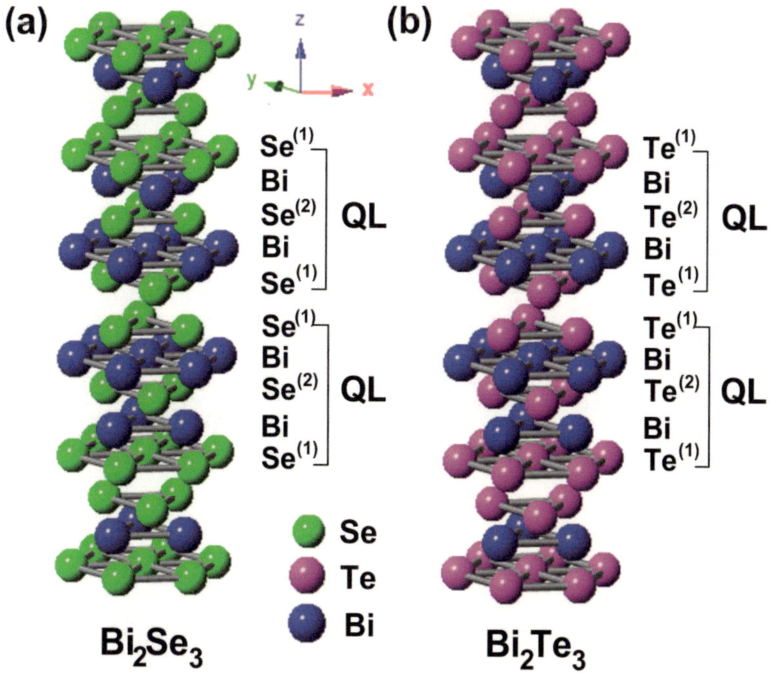

Figure 1. (a) The crystal structures of (a) Bi_2Se_3 and (b) Bi_2Te_3, in which 5-atomic-layer-thick lamellae of –($Se^{(1)}$–Bi–$Se^{(2)}$–Bi–$Se^{(1)}$)– or –($Te^{(1)}$–Bi–$Te^{(2)}$–Bi–$Te^{(1)}$)– is called a quintuple layer (QL).

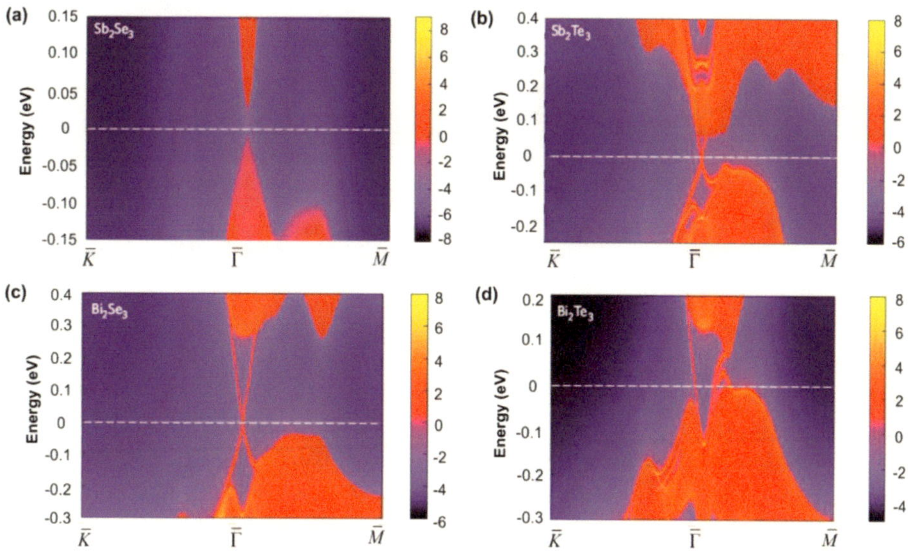

Figure 2. The calculated energy and momentum dependence of the LDOS for (a) Sb_2Se_3, (b) Sb_2Te_3, (c) Bi_2Se_3, and (d) Bi_2Te_3 on the (111) surface. The TSSs are clearly seen around the Γ point as a red line in the Sb_2Te_3, Bi_2Se_3, and Bi_2Se_3 graphs. No TSS exists in Sb_2Se_3 [15].

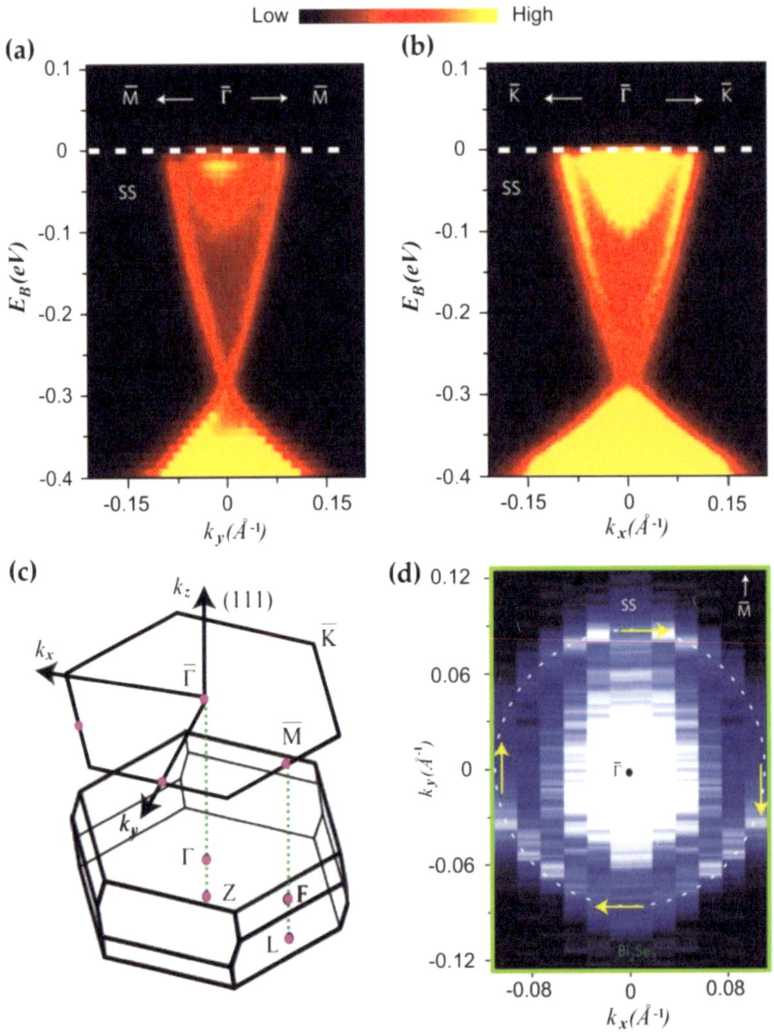

Figure 3. The ARPES measurements of the electronic structure on Bi_2Se_3 near the $\bar{\Gamma}$ point along the (a) $\bar{\Gamma}-\bar{M}$ and (b) $\bar{\Gamma}-\bar{K}$ directions. (c) The bulk 3D Brillouin zone and the surface 2D Brillouin zone of the projected (111) surface. (d) The Fermi surface of the surface state [20].

Fermi level. The slopes of this V-shaped band along the $\bar{\Gamma}-\bar{M}$ and $\bar{\Gamma}-\bar{K}$ directions are nearly equivalent [20]. The U-shaped bands near the Fermi level and below the V-shaped band are the bulk conduction band (BCB) and bulk valence band (BVB) of Bi_2Se_3. This result matches the prediction that the surface state exists between the BCB and the BVB. The detail of the surface state is shown in **Figure 3(d)**. The ring-like Fermi surface formed by the Dirac cone-like surface state is centered at the Γ point. The unique spin-momentum lock behavior can also be observed in this figure.

http://dx.doi.org/10.5772/intechopen.74918

3. Principle of femtosecond spectroscopy

3.1. Degenerate pump-probe spectroscopy

Highly temporal resolution is one of the unique characteristics in femtosecond optics. By the pump-probe technique, the photoexcited carrier dynamics and phonon dynamics in solid state materials can be clearly resolved. Additionally, the interband and intraband relaxation processes can be also obtained.

The basic understanding of time-resolved pump-probe spectroscopy is introduced as follows. The pump pulses are served as a perturbation which leads to the changes of the electronic population in materials. The probe pulses are used for the detection of the optical property changes of the materials. By controlling the time interval between the pump and probe pulses, the transient changes of the optical properties can be recorded. In pump-probe spectroscopies, the transient reflectivity changes ($\Delta R/R$) or transient absorption changes ($\Delta A/A$) can be measured.

Here, we explain more experimental details about the detection of $\Delta R/R$. As shown in **Figure 4**, the pump-induced reflectivity changes are plotted as $R(t)$. The $f_{\text{modulation}}$ is the modulation frequency of the chopper. The f_{laser} is the repetition rate of femtosecond laser pluses. The period of the $R(t)$ is correlating to the period of the $I_{\text{pump}}(t)$ pulse train. Since the reflectivity of material, $R(t)$, is modulated by the $I_{\text{pump}}(t)$ pulse train, the intensity of reflective probe beam $I_{R,\text{probe}}(t)$ is also modulated by the $I_{\text{pump}}(t)$ pulse train. Thus, the $I_{R,\text{probe}}(t)$ can be described by the superposition of the DC intensity signal $I_{R,0}$ and AC intensity signal ΔI_R with the specific frequency f (= $f_{\text{modulation}}$). Typically, the ΔI_R is much smaller than the $I_{R,0}$ in the order of 10^{-3}–10^{-6}. By using the phase-lock technique, the amplitude of AC intensity signal ΔI_R can be extracted out by providing the reference frequency $f_{\text{modulation}}$ for the lock-in amplifier. Because the ΔI_R is small compared to the $I_{R,0}$, the $\langle I_{R,probe}(t)\rangle = I_{R,0} + \Delta I_R/2 \approx I_{R,0}$. Thus, the $I_{R,0}$ can be obtained by using a multimeter for the measurements of $\langle I_{R,probe}(t)\rangle$. The measured $\Delta R/R$ can directly relate to the $\Delta I_R/I_{R,0}$ via the following relationship.

$$\frac{\Delta R}{R} = \frac{\Delta I_R / I_{i,\text{probe}}}{I_{R,0} / I_{i,\text{probe}}} = \frac{\Delta I_R}{I_{R,0}} \tag{1}$$

where the $I_{i,probe}$ is the intensity of the incident probe beam. Thus, the time evolution of $\Delta R(t)/R$ can be measured by swapping the time interval Δt between the pump pulse and the probe pulse.

3.2. Optical pump and mid-infrared probe spectroscopy

The plasma edge of the doped n-type semiconductor usually lies in the mid-infrared (MIR) regime. By measuring the reflectivity around the plasma edge, many characteristics of carriers such as scattering rate and carrier concentration can be obtained [21]. The development of a pulsed mid-infrared light source provides the opportunities for understanding the dynamics of carriers. The mid-infrared pump-probe spectroscopy has been already applied on various

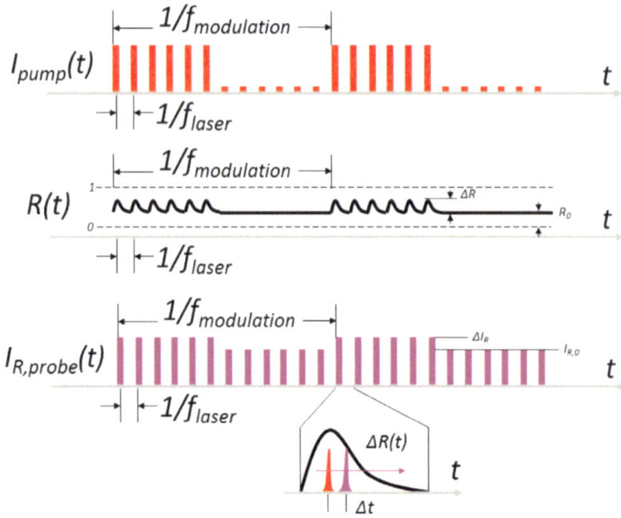

Figure 4. The scheme for the principle of the pump-probe technique. The details are described in the text.

materials (i.e., oxides, semiconductors, superconductors, graphene, and topological insulators) [22–28]. In the reflection-type mid-infrared pump-probe spectroscopy, the effect of multiple reflections should be considered in the analysis, and the dynamical characteristics of carriers can be further obtained through modeling the measured data with the Drude-Lorentz model.

Figure 5 shows a schematic diagram of our optical pump and mid-infrared probe (OPMP) spectroscopy. The light source of the pump-probe system is a regenerative amplifier with 800 nm

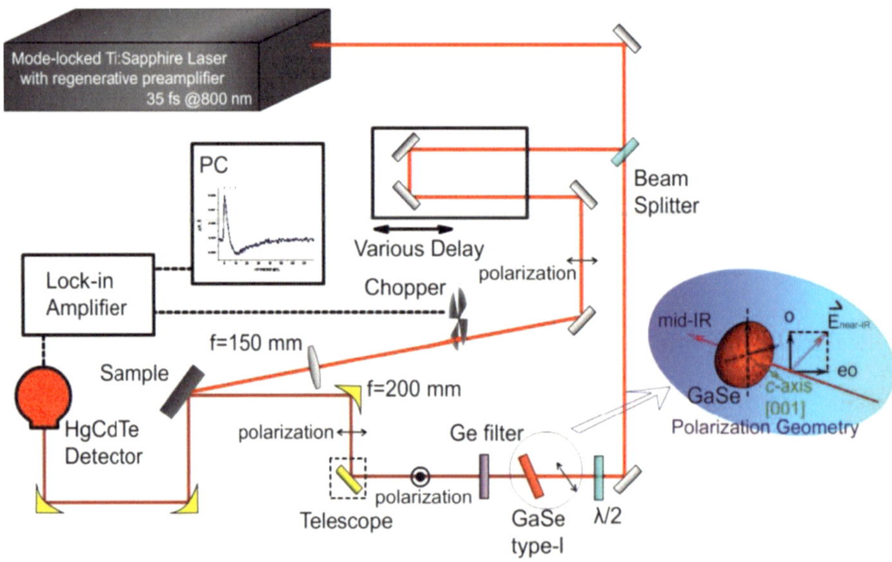

Figure 5. Schematic diagram of the optical pump and mid-infrared probe (OPMP) system.

central wavelength, 5 kHz repetition rate, and 30 fs pulse duration. The beam is split into a pump beam (40% of the incident light) and a probe beam (60% of the incident light). The probe beam passes through a 0.7-mm-thick GaSe crystal to generate mid-infrared (MIR) pulses, in which the MIR wavelength can be tuned from 9.0 μm (138 meV) to 14.1 μm (88 meV) through differential frequency generation (DFG). The optical pump beam with the fluence of 68 μJ/cm^2 and a spot size of 485 μm (in diameter) is focused on the sample using a 150 mm lens. An Au-coated off-axis parabolic mirror with f = 200 mm is employed to focus the probe beam on the sample surface with a spot diameter of 392 μm. It is ensured that the spot size of the pump beam is larger than that of the probe beam. The probe beam is further collimated and refocused onto a MIR detector (e.g., liquid nitrogen-cooled HgCdTe) using an Au-coated off-axis parabolic mirror (f = 50 mm).

4. Ultrafast dynamics in topological insulators

4.1. Time-resolved spectroscopy in a topological insulators

The dynamic properties of photoexcited TIs have attracted a great deal of attention. For example, the relaxation behavior of a carrier near the Fermi surface has been observed by the time-resolved angle-resolved photoemission spectroscopy (Tr-ARPES) [29–32]. **Figure 6(c)** shows that the 1.55 eV photons excite the electrons from the bulk valence band to a higher-lying state in the bulk materials. Then, the photoexcited carriers fall into the bulk conduction band (BCB) and the surface state within 1 ps [31]. In **Figure 6(a)**, we can see the rise time of curve 10 is ~1 ps. This means that after photoexcitation, the carriers in the higher lying band are rapidly relaxed into the BCB, then cooled to the bottom of the BCB via intraband scattering. These interband transitions and intraband scattering are shown in **Figure 6(d)** and **(e)** [31].

Furthermore, the relaxation time of curve 10 in **Figure 6(a)** is longer than 10 ps. This slow relaxation indicates the metastable behavior of the population of carriers in the BCB [30, 31]. Meanwhile, as curves 6–9 shown in **Figure 6(b)**, the population of surface states also exhibits an unusually long-lived existence [31]. Here, the relaxation bottleneck is attributed to the scattering processes between the BCB and the surface state [31]. As **Figure 6(f)** shows, the photoexcited carriers first relax via surface-bulk scattering and then cooling via surface-state intraband scattering. This scattering channel is mainly in response to the acoustic phonon-mediated surface-bulk coupling and the acoustic phonon scattering of the surface-state Dirac fermions [32]. The Tr-ARPES can directly deliver information about the population changes of the electronic state near the Fermi level. However, reports on the transition processes occurring in the early stages after photoexcitation are rare. To fully understand the photoexcited carrier dynamics, studies for the interband transition and the intraband cooling are needed, which can be revealed using optical pump/optical probe spectroscopy (OPOP) and optical-pump/mid-infrared probe (OPMP) spectroscopy.

4.2. Interband relaxations in topological insulators

The interband relaxation of photoexcited carriers in topological insulator (TI) single crystals is examined by the optical pump and optical probe spectroscopy [33]. In this section, we present the phonon and carrier dynamics in doped TI $Cu_xBi_2Se_3$ (x = 0, 0.1, 0.125) single crystals.

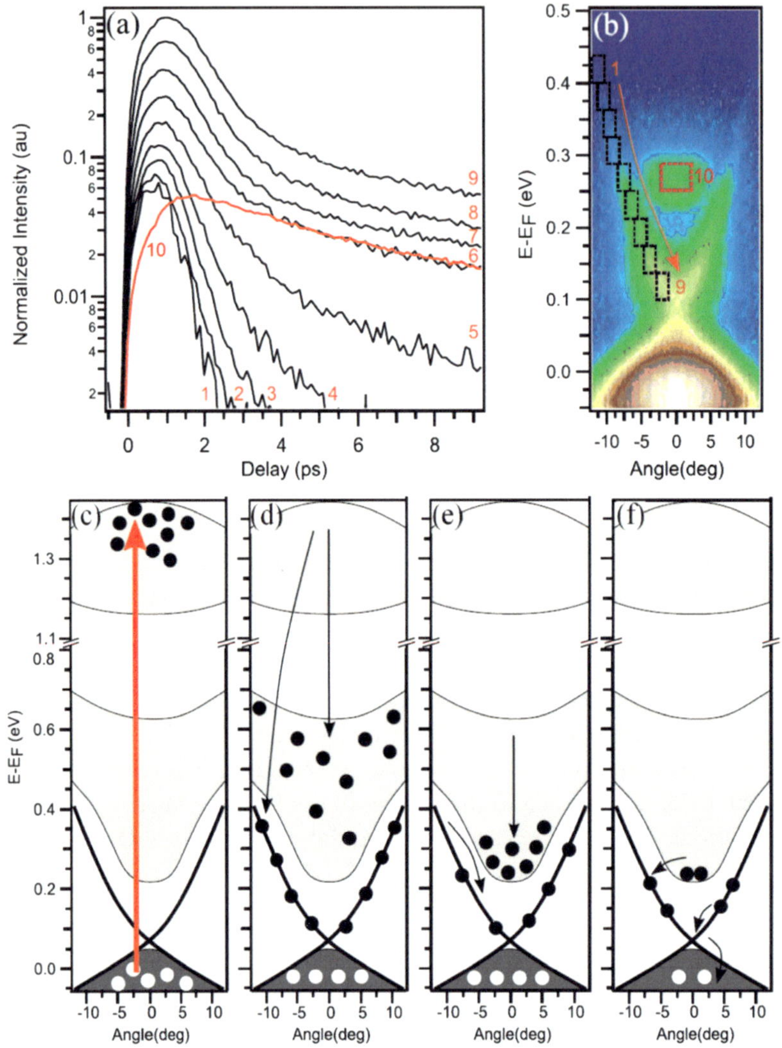

Figure 6. (a) The normalized population within the integration windows indicated in (b). (b) The integration windows over the BCB and the surface state. (c)–(f) Schematics of carrier dynamics over the transition energy range [31]. (c) $t = 0$ ps. (d) $t \sim 0.5$ ps. (e) $t \sim 2.5$ ps. (f) $t > 5$ ps.

Figure 7(a) shows the typical $\Delta R/R$ signals as a function of delay time for $Cu_xBi_2Se_3$ crystals at room temperature. Generally, different energy-transfer processes can be unambiguously extracted from the time evolution of $\Delta R/R$ curve. After pumping, the thermalization between electrons and optical phonons which occurred in a sub-picosecond timescale is characterized by the fast component in $\Delta R/R$. A subsequent slow component in a timescale of several picoseconds is assigned to the thermalization between electrons and acoustic phonons [34]. After these electron-lattice relaxation processes, the heat diffusion out of the illuminated area on the sample is further revealed by the quasi-constant component in $\Delta R/R$ [35]. Furthermore, all of the $\Delta R/R$ curves show two damped oscillation components with different periods.

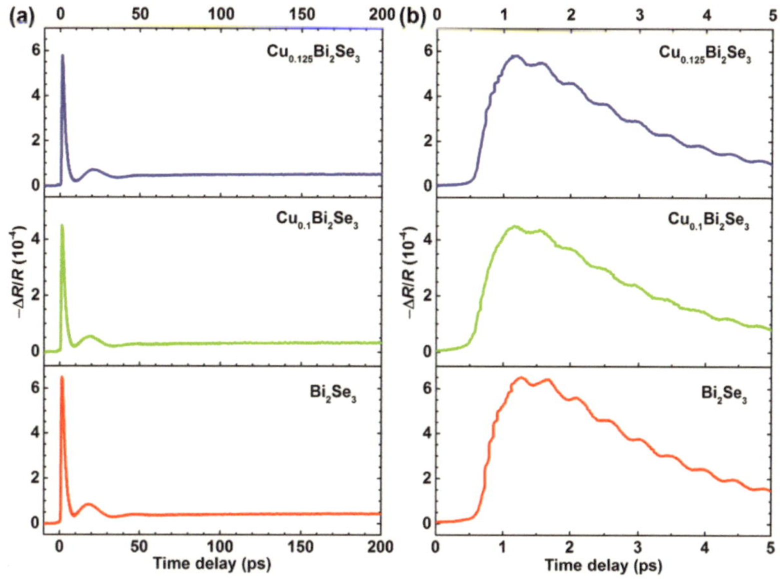

Figure 7. Temporal variations of $\Delta R/R$ signals for $Cu_xBi_2Se_3$ crystals (x = 0, 0.1, 0.125) at room temperature by using the 1.55 eV degenerate pump-probe spectroscopy, shown (a) in the full timescale and (b) in short timescale.

The slow oscillation components, as shown in **Figure 7(a)**, are attributed to the coherent acoustic phonons (CAPs) generated by ultrafast laser pulses. This damped slow oscillation in $\Delta R/R$ is generated by the interference between two probe beams, respectively, reflected from the sample surface and the strain pulse that propagate longitudinally with the sound velocity. The relationship between the period τ_{CAP} of the slow oscillation and the longitudinal sound velocity v_s is $\tau_{CAP} = \lambda/(v_s\sqrt{n^2 - \sin^2\theta})$, where λ is the probe wavelength, n is the refractive index at λ, and θ is the incident angle of the probe beam [33]. Consequently, the sound velocity can be estimated by measuring the CAP oscillations when the refractive index of the material is known. The frequency of the CAP for the Bi_2Se_3 crystals is ~0.033 THz (~ 30 ps in period). Additionally, it completely decays within ~60 ps. The disappearance of the CAP (slow oscillation) around 60 ps, according to the strain pulse model, is determined by the penetration depth of an 800 nm probe beam in Bi_2Se_3 crystals. Taking the refractive index of Bi_2Se_3 crystals reported in [36], the sound velocity is estimated to be 1996 m/s at room temperature [33]. **Figure 7(a)** also reveals that the periods of the slow oscillations in $Cu_xBi_2Se_3$ (x = 0, 0.1, 0.125) crystals vary slightly from 29.9 to 30.2 ps.

The fast oscillation components of $Cu_xBi_2Se_3$ crystals are presented in **Figure 7(b)**, which can be extracted by removing the relaxation background from the $\Delta R/R$ signals. The results are presented in **Figure 8(a)**. The frequency of the component is 2.148 THz, which can be further assigned as the $A_{1g}{}^1$ coherent optical phonon (COP) mode of Bi_2Se_3, based on comparison with the steady-state Raman spectroscopy [37]. Interestingly, the frequencies of the fast oscillations considerably vary with Cu content (x) of the $Cu_xBi_2Se_3$ samples and are associated with the changes in the chain length of the QL and in the lattice constant of *c*-axis.

Figure 8(a) shows the fast oscillation component for $Cu_xBi_2Se_3$ (x = 0, 0.1, 0.125) crystals. In order to quantitatively analyze these oscillations, a damped oscillation function, $A_{osc}\cos(2\pi f_{osc} t + \varphi)\, e^{-t/\tau_{dephasing}}$,

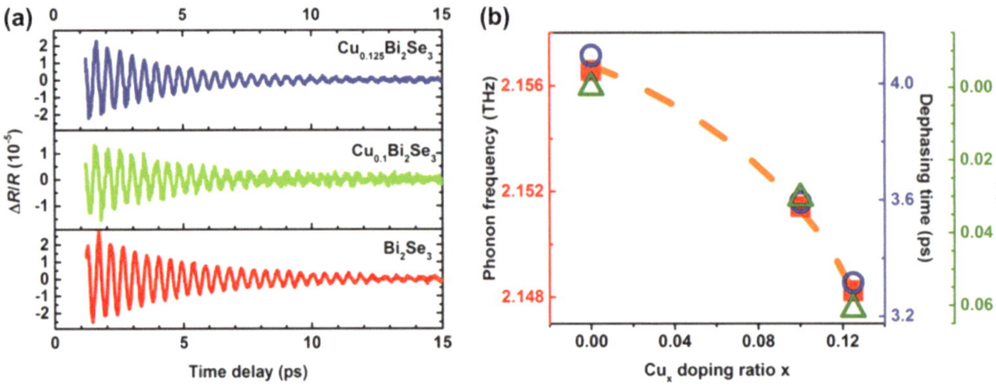

Figure 8. (a) High-frequency temporal variations of $\Delta R/R$ signals for $Cu_xBi_2Se_3$ (x = 0, 0.1, 0.125) crystals extracted from Figure 7(b). (b) Phonon frequency (red squares), dephasing time of A_{1g}^{1} phonon mode (blue circles) and changes in lattice constant of *c*-axis (green triangles) in $Cu_xBi_2Se_3$ crystals as a function of Cu doping concentrations.

was used to fit the original data in **Figure 8(a)** to get the dephasing time ($\tau_{dephasing}$) and the phonon frequency (f_{osc}) for the $Cu_xBi_2Se_3$ crystals. As shown in **Figure 8(b)**, both dephasing time and phonon frequency shrink as Cu concentrations increase (x). This indicates that an additional Cu atom deforms the Se-Bi-Se-Bi-Se chain in $Cu_xBi_2Se_3$ crystals. Furthermore, the lattice constant of *c*-axis increases slightly with increasing Cu concentrations [**Figure 8(b)**], implying that the QL chain in $Cu_xBi_2Se_3$ is stretched by introducing Cu atoms. Thus, the scenario of stretching the QL chain length is that the Cu atoms (form a mediated layer) are intercalated between QLs to strengthen the interaction between QLs. Moreover, the QLs are further deformed by these intercalated Cu atoms.

4.3. Intraband relaxations in topological insulators

The femtosecond snapshots of the relaxation processes and Dirac fermion-phonon coupling strength of 3D TI Bi_2Se_3 were revealed by OPMP spectroscopy [26]. In this study, several selected Bi_2Se_3 single crystals with a wide range of carrier concentrations (n) from 51.5×10^{18} to 0.25×10^{18} cm^{-3} were studied. **Table 1** summarizes the doping levels of samples (#1: n = 51.5 × 10^{18} cm^{-3}, #2:

Code	E_F-$E_{\text{Dirac point}}$ (meV)	Carrier concentration		$n_{surface}/(n_{surface}+n_{bulk}\cdot d)$
		n_{bulk} (10^{18} cm^{-3})	$n_{surface}$ (10^{13} cm^{-2})	
#1	422	−51.5 ± 0.84	−1.45	0.11
#2	325	−13.9 ± 0.26	−0.83	0.20
#3	284	−5.58 ± 0.25	−0.72	0.35
#4	260	−0.25 ± 0.01	−0.47	0.89

Table 1. Fermi energy and carrier concentrations of bulk and surface states of various Bi_2Se_3 single crystals. All samples are ***n***-type. "***d*** = 23.5 nm" is the penetration depth of 800 nm pumping light.

$n = 13.9 \times 10^{18}$ cm^{-3}, #3: $n = 5.58 \times 10^{18}$ cm^{-3}, and #4: $n = 0.25 \times 10^{18}$ cm^{-3}). The OPMP spectra and the corresponding ARPES images of the samples are shown in **Figure 9(a)** and **(b)** [26]. The OPMP spectra clearly show a positive $\Delta R/R$ peak for high $n \geq 13.9 \times 10^{18}$ cm^{-3} (Bi_2Se_3 #1 and #2). In contrast, this positive peak gradually diminishes as n decreases, while an additional negative peak appears for the cases of $n = 5.58 \times 10^{18}$ cm^{-3} and $n = 0.25 \times 10^{18}$ cm^{-3}.

Based on the ARPES image and the energy band structure of TI Bi_2Se_3, a model is proposed [in **Figure 10(a)**] for the optical pumping (1.55 eV) and mid-infrared probing processes to elucidate the origins of both positive and negative signals. The band gap of Bi_2Se_3 is ~300 meV, as shown in the ARPES images of **Figure 9(b)**, which is much larger than the probe photon energy (87~153 meV) of the mid-infrared (mid-IR). Thus, it does not allow the occurring of the interband transitions between the valence band (VB) and the conduction band (CB) of the bulk. Meanwhile, the free-carrier absorption in the CB [the probe (1) in **Figure 10(a)**] and Dirac cone surface state [the probe (2) in **Figure 10(a)**] will dominate the probe processes, which can be assigned to the positive and negative peaks in $\Delta R/R$, respectively. To reveal the physical meanings of the positive peak in $\Delta R/R$, the photon energy dependence of $\Delta R/R$ for #1

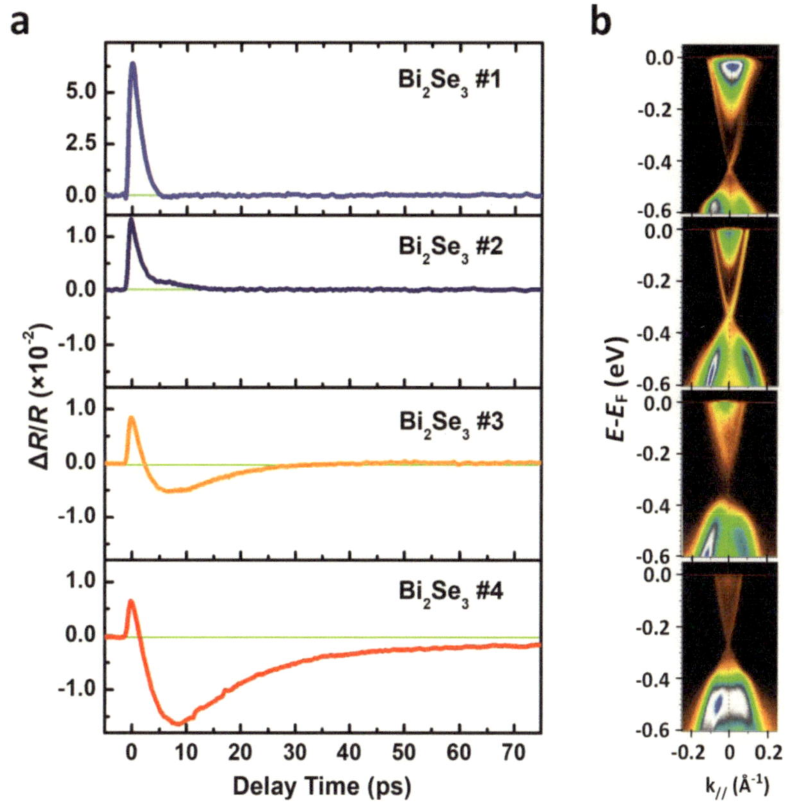

Figure 9. Carrier concentration (n) dependence of the transient changes in reflectivity $\Delta R/R$ in Bi_2Se_3 single crystals. (a) $\Delta R/R$ of samples #1 ($n = 51.5\times10^{18}$ cm^{-3}), #2 ($n = 13.9\times10^{18}$ cm^{-3}), #3 ($n = 5.58\times10^{18}$ cm^{-3}), and #4 ($n = 0.25\times10^{18}$ cm^{-3}) with a fluence of 34 μJ/cm^2 for pumping and a photon energy of 141 meV for probing. (b) ARPES images on samples of (a) [26].

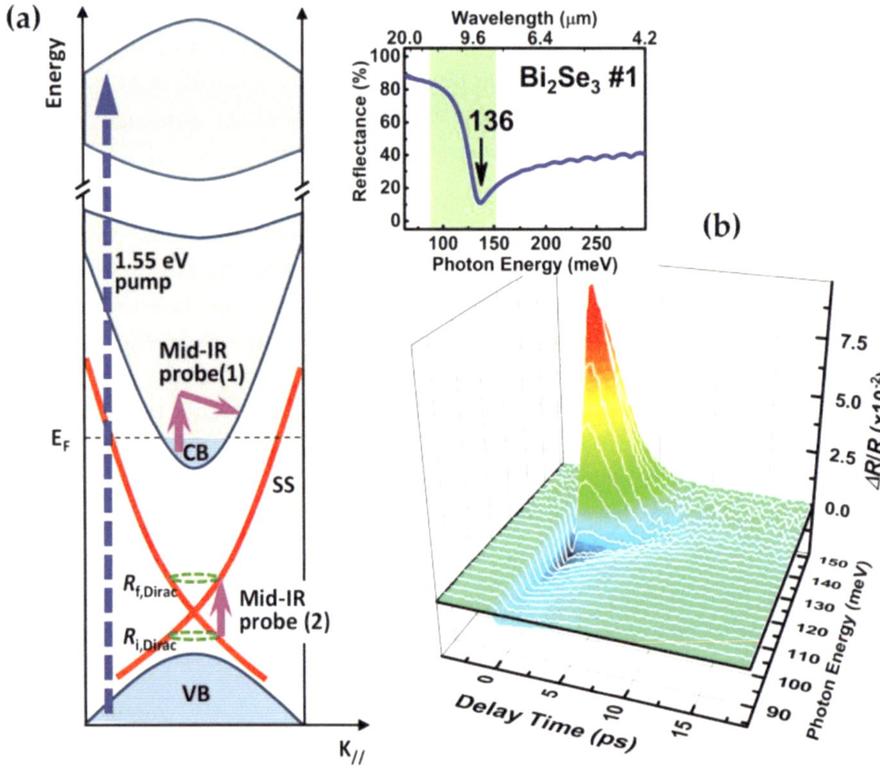

Figure 10. Schematic energy band structure and photon energy-dependent $\Delta R/R$ in a bulk state. (a) Schematic band structure of TIs based on the ARPES images in Figure 9(b) and the pump-probe processes. CB: Conduction band, VB: Valence band, SS: Surface state. $R_{i,Dirac}$ & $R_{f,Dirac}$: The circumferences of initial/final states in Dirac cone for probing. (b) With a fluence of 38 μJ/cm^2 for pumping, the $\Delta R/R$ of Bi_2Se_3 #1 at various photon energies (wavenumber) from 87 to 153 meV (700–1234 cm^{-1}). Inset: The Fourier transform infrared (FTIR) reflectance spectrum of Bi_2Se_3 #1. The gray area indicates the range of the mid-IR photon energy used in this study [26].

sample is studied and shown in **Figure 10(b)**. Clearly, $\Delta R/R$ gradually changes from positive to negative as decreasing the photon energy. At around 136 meV (1100 cm^{-1}), it appears that intermediate signals mixed with both positive and negative peaks, corresponding to deep in the Fourier transform infrared (FTIR) reflectance spectrum [the inset of **Figure 10(b)**]. The excited carriers after pumping suffer the so-called intervalley scattering, leading to the red shift of the reflectance spectra. Thus, the reflectivity increases as a function of time when probing photon energy is higher than the position of the 136 meV deep. In contrast, the reflectivity decreases as a function of time when probing photon energy is smaller than the 136 meV deep. Similar results were also observed in a typical semiconductor n-type GaAs [23].

As found in **Figure 9** and **Table 1**, the amplitude of positive peak in $\Delta R/R$ gradually decreases as bulk carrier concentrations reduce. Meanwhile, the negative peak of $\Delta R/R$ increases while reducing the bulk and surface carrier concentrations. Intriguingly, the negative peak increases substantially with an increasing ratio of the surface to total carrier concentration [$n_{surface}/(n_{surface} + n_{bulk}\cdot d)$ in **Table 1**], implying a close relation between the negative peak of $\Delta R/R$ and Dirac fermions. In addition, **Figure 11(a)** shows the $\Delta R/R$ signal as a function of the pumping fluences. The positive peak exhibits a stronger dependence on the pumping fluences than the negative peak does. For a pumping fluence of 3.3 μJ/cm², the maximum photo-induced carrier density Δn is around 2.54×10^{18} cm^{-3}. Indeed, if one absorbed photon generates one photo-induced carrier, the maximum photo-induced carrier density can be estimated by $\Delta n = (1-R) \times F/(E \times \delta)$, where $R = 0.55$ is the reflectance, $F = 3.3$ μJ/cm² is pumping fluence, $E = 2.48 \times 10^{-19}$ J (= 1.55 eV) is the pumping photon energy, $\delta = 23.5$ nm is the penetration depth. Consequently, the negative peak still subsists at the low pumping fluence of 3.3 μJ/cm, while the positive peak almost vanishes [see **Figure 11(b)**]. Namely, the process (1) associated with the positive peak can be suppressed and the process (2) associated with the negative peak can be preserved by reducing the pumping fluences. To quantitatively certify the relation between the negative peak and Dirac fermions, the amplitude of the negative peak dependence of probing photon energy is studied using low n samples #3 and #4 to avoid disturbance of the positive peak [**Figure 11(b)**]. According to Fermi's golden rule, the amplitude of the negative peak should be proportional to the transition probability ($T_{i \to f}$) between the

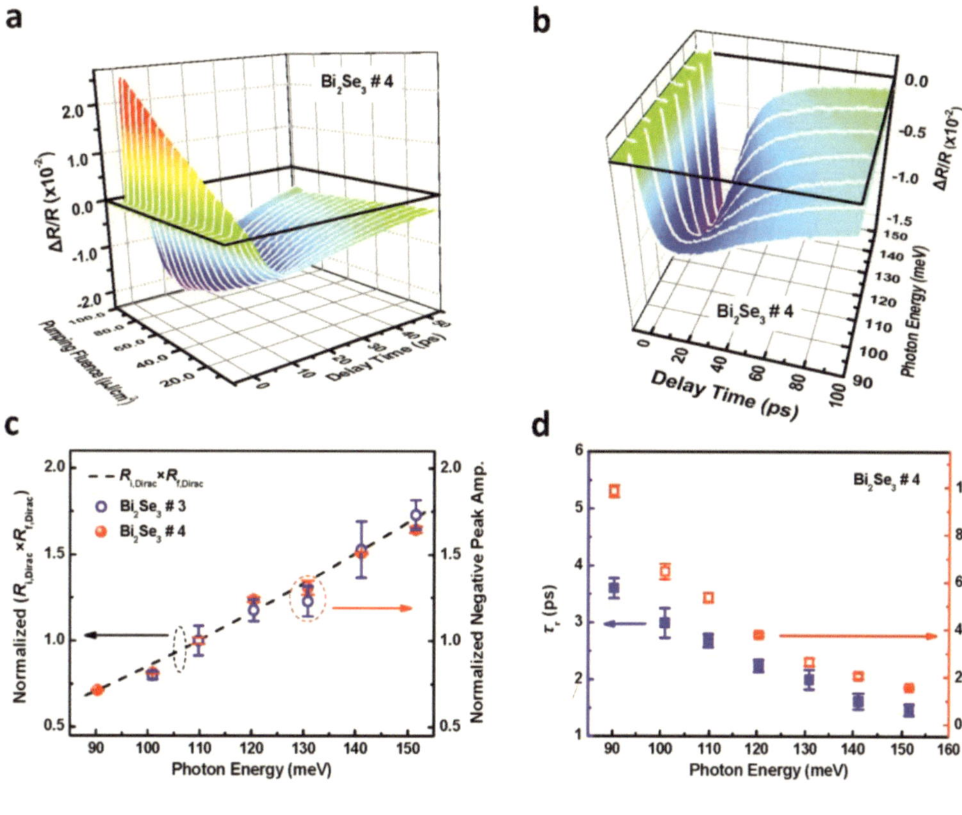

Figure 11. Pumping fluence and photon energy dependence of $\Delta R/R$ and its amplitude and rising (decay) time in the surface state. (a) With probing photon energy of 141 meV, the $\Delta R/R$ of Bi_2Se_3 #4 at various pumping fluences from 3.3 to 105 μJ/cm². (b) With pumping fluence of 3.3 μJ/cm², the $\Delta R/R$ of Bi_2Se_3 #4 at various photon energies from 90 to 152 meV. (c) The photon energy dependence of negative peak amplitude of $\Delta R/R$ in (b). The photon energy-dependent normalized absorption probability [dashed line, i.e. $R_{i,Dirac} \times R_{f,Dirac}$ in **Figure 10 (a)**] of the mid-IR probe beam in the Dirac cone. (d) The photon energy dependence of the rising time (τ_r) and decay time (τ_d) of $\Delta R/R$ in (b) [26].

initial and final density of states in the Dirac cone. Indeed, in **Figure 11(c)**, the $R_{i,Dirac} \times R_{f,Dirac}$ presents linear relation with probing photon energy, reflecting the proportional relation to the transition rate between the initial and final density of states for the mid-IR probe process (2) in the Dirac cone [**Figure 10(a)**]. This confirms that the negative peak of $\Delta R/R$ is predominantly attributed to the mid-IR probe process (2) in the Dirac cone. Consequently, the ultrafast dynamics of the Dirac fermions can be clearly disclosed by the negative peak of $\Delta R/R$.

As shown in **Figure 11(d)**, both rising time (τ_r) and decay time (τ_d) of the negative peak of $\Delta R/R$ strongly depend on the probing photon energy. The τ_r becomes longer when the probed regime is closer to the Dirac point (or smaller probing energy). The ultrafast relaxation picture for Dirac fermions in TIs can be established. The major process right after the 1.55 eV pumping is that the carriers in the bulk valence band (BVB) are excited to the bulk conduction band (BCB). The carrier recombination between the BCB and BVB can be ignored in this study because of the large timescale (typically >> 1 ns) for such a process. Consequently, the unoccupied states in BVB would mainly be refilled by carriers in the upper Dirac cone. Carriers in the Dirac cone can be easily transferred into the unoccupied states in BVB due to the overlapping between the Dirac cone and BVB [see **Figure 9(b)**], leading to the increase in the number of the unoccupied states near the Dirac point and thus enhancing the absorption channel for process (2) in the Dirac cone [**Figure 10(a)**]. Therefore, the reflectivity of the mid-IR probing light decreases within 1.47~3.60 ps, that is, the rising time of the negative peak in **Figure 11(b)** and **(d)**. Once the carriers in the Dirac cone relax into BVB, the excited carriers in the BCB are subsequently injected into the unoccupied states in the Dirac cone to diminish the absorption channel for the mid-IR process (2) [**Figure 10(a)**] and consequently lead to the increased mid-IR reflectivity within 14.8~87.2 ps. The timescale (τ_d) of this process is several tens of picoseconds, which is much longer than the τ_r of several picoseconds, because the carriers in BCB cannot directly transfer into the top of the Dirac cone without overlaps occurring between them and other auxiliaries, for example, phonons.

The relaxation of Dirac fermions has been demonstrated via phonon medium [38, 39]. The coupling strength (λ) between Dirac fermions and phonons varies at different positions of the Dirac cone, which can be revealed from the photon energy-dependent rising time. Based on the second moment of the Eliashberg function [40], the coupling strength (λ) is inversely proportional to the relaxation time (τ_e) of excited electrons:

$$\lambda\langle\omega^2\rangle \propto \frac{1}{\tau_e} \tag{2}$$

where ω is the phonon energy which couples with the electrons. For the estimate of $\langle\omega^2\rangle$, some vibrational modes are more efficiently coupled to Dirac fermions than the others. For Bi_2Se_3, the symmetric A^1_{1g} mode of ~8.9 meV is coherently excited by photoexcitation and efficiently coupled. Taking $\tau_e = \tau_r$ in **Figure 11(d)** and $T_e = 370$ K (obtained from [32] at the aforementioned low pumping fluence) to estimate the coefficient of $(\pi k_B T_e/3\hbar)$ in Eq. (2), photon energy dependence of the Dirac fermion-phonon coupling strength is $\lambda = 0.08$–0.19. The Dirac fermion-phonon coupling strength measured by the present OPMP becomes significantly smaller near the Dirac point (the point of $K_{//}$=0). As getting closer to the Dirac point, Dirac fermions will possess a weaker coupling with the phonons to reduce the scatterings with phonons. In addition, the effective mass of Dirac fermions in the surface state gradually decreases when approaching the Dirac point, which is consistent with the results in graphene [9].

5. Conclusion

We report the ultrafast dynamics of carriers and phonons in topological insulator Bi_2Se_3, $Cu_xBi_2Se_3$ (x = 0, 0.1, 0.125) single crystals. By time-resolved pump-probe spectroscopy, one damped fast oscillation was clearly observed in the transient reflectivity changes ($\Delta R/R$) for $Cu_xBi_2Se_3$, which is assigned to the coherent optical phonon (A^1_{1g}). The frequency of A^1_{1g} phonon decreases considerably with increasing Cu contents, suggesting the intercalation of Cu atoms between quintuple layers of Bi_2Se_3. The schematic illustration of the direct transitions and subsequent relaxation processes induced by optical excitation in Bi_2Se_3 single crystals is also reported here. The femtosecond snapshots of the relaxation processes were revealed by optical pump and mid-infrared probe spectroscopy. Especially, the Dirac fermion dynamics in the Dirac cone surface state near the Dirac point of Bi_2Se_3 was unambiguously revealed through the negative peak of $\Delta R/R$. The Dirac fermion-phonon coupling strength was found in range of 0.08–0.19 and the strength is reduced as getting closer to the Dirac point. These results are extremely crucial to the design of Dirac fermion devices and optoelectronics, especially in the mid-IR and THz ranges.

Acknowledgements

Financial support from Vietnam National Foundation for Science and Technology Development (NAFOSTED) under grant number 103.99–2015.17, the Ministry of Science and Technology of the Republic of China, Taiwan (Grant No. 103-2628-M-009-002-MY3, 103-2119-M-009-004-MY3, 106-2119-M-009-013-FS, 106-2628-M-009-003-MY3) and the Grant MOE ATU Program at NCTU are gratefully acknowledged.

Author details

Phuoc Huu Le[1]* and Chih-Wei Luo[2,3]*

*Address all correspondence to: lhuuphuoc@ctump.edu.vn and cwluo@mail.nctu.edu.tw

1 Faculty of Basic Sciences, Can Tho University of Medicine and Pharmacy, Can Tho, Vietnam

2 Department of Electrophysics, National Chiao Tung University, Hsinchu, Taiwan

3 Taiwan Consortium of Emergent Crystalline Materials, Ministry of Science and Technology, Taipei, Taiwan

References

[1] Fu L, Kane CL, Mele EJ. Topological insulators in three dimensions. Physical Review Letters. 2007;**98**:106803. DOI: 10.1103/PhysRevLett.98.106803

[2] Moore JE, Balents L. Topological invariants of time-reversal-invariant band structures. Physical Review B. 2007;**75**:121306(R). DOI: 10.1103/PhysRevB.75.121306

[3] Hsieh D, Qian D, Wray L, Xia Y, Hor YS, Cava RJ, Hasan MZ. A topological Dirac insulator in a quantum spin hall phase. Nature. 2008;**452**:970-974. DOI: 10.1038/nature06843

[4] Hsieh D, Xia Y, Qian D, Wray L, Meier F, Dil JH, Osterwalder J, Patthey L, Fedorov AV, Lin H, Bansil A, Grauer D, Hor YS, Cava RJ, Hasan MZ. Observation of time-reversal-protected single-Dirac-cone topological-insulator states in Bi_2Te_3 and Sb_2Te_3. Physical Review Letters. 2009;**103**:146401. DOI: 10.1103/PhysRevLett.103.146401

[5] Moore JE. The birth of topological insulators. Nature. 2010;**464**:194-198. DOI: 10.1038/nature08916

[6] Hasan MZ, Kane CL. Colloquium: Topological insulators. Reviews of Modern Physics. 2010;**82**:3045-3067. DOI: 10.1103/RevModPhys.82.3045

[7] Hasan MZ, Moore JE. Three-dimensional topological insulators. Annual Review of Condensed Matter Physics. 2011;**2**:55-78. DOI: 10.1146/annurev-conmatphys-062910-140432

[8] Qi X-L, Zhang S-C. Topological insulators and superconductors. Reviews of Modern Physics. 2011;**83**:1057-1110. DOI: 10.1103/RevModPhys.83.1057

[9] Novoselov KS, Geim AK, Morozov SV, Jiang D, Katsnelson MI, Grigorieva IV, Dubonos SV, Firsov AA. Two-dimensional gas of massless Dirac fermions in graphene. Nature. 2005;**438**:197-200. DOI: 10.1038/nature04233

[10] Coleman JN, Lotya M, O'Neill A, Bergin SD, King PJ, Khan U, Young K, Gaucher A, De S, Smith RJ, Shvets IV, Arora SK, Stanton G, Kim H-Y, Lee K, Kim GT, Duesberg GS, Hallam T, Boland JJ, Wang JJ, Donegan JF, Grunlan JC, Moriarty G, Shmeliov A, Nicholls RJ, Perkins JM, Grieveson EM, Theuwissen K, McComb DW, Nellist PD, Nicolosi V. Two-dimensional nanosheets produced by liquid exfoliation of layered materials. Science. 2011;**331**:568-571. DOI: 10.1126/science.1194975

[11] Xue Q-K. Nanoelectronics: A topological twist for transistors. Nature Nanotechnology. 2011;**6**:197-198. DOI: 10.1038/nnano.2011.47

[12] Le PH, Liao C-N, Luo CW, Lin J-Y, Leu J. Thermoelectric properties of nanostructured bismuth–telluride thin films grown using pulsed laser deposition. Journal of Alloy Compound. 2014;**615**:546-552. DOI: 10.1016/j.jallcom.2014.07.018

[13] Le PH, Liao C-N, Luo CW, Lin J-Y, Leu J. Thermoelectric properties of bismuth-selenide films with controlled morphology and texture grown using pulsed laser deposition. Applied Surface Science. 2013;**285P**:657-663. DOI: 10.1016/j.apsusc.2013.08.107

[14] Tuyen LTC, Le PH, Luo CW, Leu J. Thermoelectric properties of nanocrystalline Bi_3Se_2Te thin films grown using pulsed laser deposition. Journal of Alloys and Compounds. 2016;**673**:107-114. DOI: 10.1016/j.jallcom.2016.03.006

[15] Zhang H, Liu C-X, Qi X-L, Dai X, Fang Z, Zhang S-C. Topological insulators in Bi_2Se_3, Bi_2Te_3 and Sb_2Te_3 with a single Dirac cone on the surface. Nature Physics. 2009;**5**:438-442. DOI: 10.1038/NPHYS1270

[16] Le PH, Tzeng W-Y, Chen H-J, Luo CW, Lin J-Y, Leu J. Superconductivity in textured bi clusters/Bi_2Te_3 films. APL Mater. 2014;**2**:096105. DOI: 10.1063/1.4894779

[17] Le PH, Chiu S-P, Jian SR, Luo CW, Lin J-Y, Lin J-J, Wu KH, Gospodinov M. Nanomechanical, structural, and transport properties of Bi_3Se_2Te thin films. Journal of Alloys and Compounds. 2016;**679**:350-357. DOI: 10.1016/j.jallcom.2016.03.226

[18] Le PH, Liu P-T, Luo CW, Lin J-Y, Wu KH. Thickness-dependent magnetotransport properties and terahertz response of topological insulator Bi_2Te_3 thin films. Journal of Alloys and Compounds. 2017;**692**:972-979. DOI: 10.1016/j.jallcom.2016.09.109

[19] Miller GR, Li C-Y, Spencer CW. Properties of Bi_2Te_3- Bi_2Se_3 alloys. Journal of Applied Physics. 1963;**34**:1398. DOI: 10.1063/1.1729588

[20] Xia Y, Qian D, Hsieh D, Wray L, Pal A, Lin H, Bansil A, Grauer D, Hor YS, Cava RJ, Hasan MZ. Observation of a large-gap topological-insulator class with a single Dirac cone on the surface. Nature Physics. 2009;**5**:398-402. DOI: 10.1038/NPHYS1274

[21] Moss TS, Hawkins TDF, Burrell GJ. Use of plasma edge reflection measurements in the study of semiconductors. Journal of Physics C. 1968;**1**:1435-1446. DOI: 10.1088/0022-3719/1/5/333

[22] Choi HS, Ahn JS, Jung JH, Noh TW, Kim DH. Mid-infrared properties of a VO_2 film near the metal-insulator transition. Physical Review B. Condensed Matter. 1996;**54**:4621-4628. DOI: 10.1103/PhysRevB.54.4621

[23] Dantzig NAV, Planken PCM. Time-resolved far-infrared reflectance of n-type GaAs. Physical Review B. 1999;**59**:1586. DOI: 10.1103/PhysRevB.59.1586

[24] Nagai M, Shimano R, Kuwata-Gonokami M. Electron-hole droplet formation in direct-gap semiconductors observed by mid-infrared pump-probe spectroscopy. Physical Review Letters. 2001;**86**:5795-5798. DOI: 10.1103/PhysRevLett.86.5795

[25] Winnerl S, Orlita M, Plochocka P, Kossacki P, Potemski M, Winzer T, Malic E, Knorr A, Sprinkle M, Berger C, de Heer WA, Schneider H, Helm M. Carrier relaxation in epitaxial graphene photoexcited near the Dirac point. Physical Review Letters. 2011;**107**:237401. DOI: 10.1103/PhysRevLett.107.237401

[26] Luo CW, Wang HJ, Ku SA, Chen H-J, Yeh TT, Lin J-Y, Wu KH, Juang JY, Young BL, Kobayashi T, Cheng C-M, Chen C-H, Tsuei K-D, Sankar R, Chou FC, Kokh KA, Tereshchenko OE, Chulkov EV, Andreev YM, Gu GD. Snapshots of Dirac fermions near the Dirac point in topological insulators. Nano Letters. 2013;**13**:5797-5802. DOI: 10.1021/nl4021842

[27] Kaindl RA, Woerner M, Elsaesser T, Smith DC, Ryan JF, Farnan GA, McCurry MP, Walmsley DG. Ultrafast mid-infrared response of $YBa_2Cu_3O_7$. Science. 2000;**287**:470-473. DOI: 10.1126/science.287.5452.470

[28] Luo CW, Tseng PS, Chen H-J, Wu KH, Li LJ. Dirac fermion relaxation and energy loss rate near the Fermi surface in monolayer and multilayer graphene. Nanoscale. 2014;**6**:8575-8578. DOI: 10.1039/c4nr02205j

[29] Hajlaoui M, Papalazarou E, Mauchain J, Lantz G, Moisan N, Boschetto D, Jiang Z, Miotkowski I, Chen YP, Taleb-Ibrahimi A, Perfetti L, Marsi M. Ultrafast surface carrier dynamics in the topological insulator Bi_2Te_3. Nano Letters. 2012;**12**:3532-3536. DOI: 10.1021/nl301035x

[30] Crepaldi A, Ressel B, Cilento F, Zacchigna M, Grazioli C, Berger H, Bugnon P, Kern K, Grioni M, Parmigiani F. Ultrafast photodoping and effective Fermi-Dirac distribution of the Dirac particles in Bi_2Se_3. Physical Review B. 2012;**86**:205133. DOI: 10.1103/PhysRevB.86.205133

[31] Sobota JA, Yang S, Analytis JG, Chen YL, Fisher IR, Kirchmann PS, Shen Z-X. Ultrafast optical excitation of a persistent surface-state population in the topological insulator Bi_2Se_3. Physical Review Letters. 2012;**108**:117403. DOI: 10.1103/PhysRevLett.108.117403

[32] Wang YH, Hsieh D, Sie EJ, Steinberg H, Gardner DR, Lee YS, Jarillo-Herrero P, Gedik N. Measurement of intrinsic Dirac fermion cooling on the surface of the topological insulator Bi_2Se_3 using time-resolved and angle-resolved photoemission spectroscopy. Physical Review Letters. 2012;**109**:127401. DOI: 10.1103/PhysRevLett.109.127401

[33] Chen H-J, Wu KH, Luo CW, Uen TM, Juang JY, Lin J-Y, Kobayashi T, Yang H-D, Sankar R, Chou FC, Berger H, Liu JM. Phonon dynamics in $Cu_xBi_2Se_3$ (x = 0, 0.1, 0.125) and Bi_2Se_2 crystals studied using femtosecond spectroscopy. Applied Physics Letters. 2012;**101**: 121912. DOI: 10.1063/1.4754005

[34] Othonos A. Probing ultrafast carrier and phonon dynamics in semiconductors. Journal of Applied Physics. 1998;**83**:1789. DOI: 10.1063/1.367411

[35] Kumar N, Ruzicka BA, Butch NP, Syers P, Kirshenbaum K, Paglione J, Zhao H. Spatially resolved femtosecond pump-probe study of topological insulator Bi_2Se_3. Physical Review B. 2011;**83**:235306. DOI: 10.1103/PhysRevB.83.235306

[36] Zhang J, Peng Z, Soni A, Zhao Y, Xiong Y, Peng B, Wang J, Dresselhaus MS, Xiong Q. Raman spectroscopy of few-quintuple layer topological insulator Bi_2Se_3 nanoplatelets. Nano Letters. 2011;**11**:2407-2414. DOI: 10.1021/nl200773n

[37] Richter W, Köhler H, Becker CR. A Raman and far-infrared investigation of phonons in the rhombohedra1 V2-V13 compounds. Physica Status Solidi. 1977;**84**:619-628

[38] Zhu X, Santos L, Howard C, Sankar R, Chou FC, Chamon C, El-Batanouny M. Electron-phonon coupling on the surface of the topological insulator Bi_2Se_3 determined from surface-phonon dispersion measurements. Physical Review Letters. 2012;**108**:185501. DOI: 10.1103/PhysRevLett.108.185501

[39] Pan Z-H, Fedorov AV, Gardner D, Lee YS, Chu S, Valla T. Measurement of an exceptionally weak electron-phonon coupling on the surface of the topological insulator Bi_2Se_3 using angle-resolved photoemission spectroscopy. Physical Review Letters. 2012;**108**: 187001. DOI: 10.1103/PhysRevLett.108.187001

[40] Allen PB. Theory of thermal relaxation of electrons in metals. Physical Review Letters. 1987;**59**:1460. DOI: 10.1103/PhysRevLett.59.1460